IET TELECOMMUNICATIONS SERIES 75

Digital Television Fundamentals

Other volumes in this series:

Digital Television Fundamentals

Edited by
Stefan Mozar and Konstantin Glasman

The Institution of Engineering and Technology

Published by The Institution of Engineering and Technology, London, United Kingdom

The Institution of Engineering and Technology is registered as a Charity in England & Wales (no. 211014) and Scotland (no. SC038698).

The Institution of Engineering and Technology
Futures Place
Kings Way, Stevenage
Herts, SG1 2UA, United Kingdom

www.theiet.org

British Library Cataloguing in Publication Data
A catalogue record for this product is available from the British Library

ISBN 978-1-78561-250-3 (hardback)
ISBN 978-1-78561-251-0 (PDF)

Typeset in India by Exeter Premedia Services Private Limited

Contents

4   Digital TV system approach   **81**
Marcelo Knörich Zuffo, Roseli de Deus Lopes, and
Laisa Caroline Costa de Biase

Preface

The aim of this book is to cover the concepts of digital television from a practitioner's perspective. The target audience is anyone involved in television technology. The inspiration came from a 1950s series of books on television receiver design, which covered monochrome receivers. The series was published in the Philips Technical Library. It aimed at providing a reference to the then-current state of technology and concepts of how to design a TV receiver. It consisted of several books, each of which covered different stages of the receiver.

Since the 1950s, nobody has published a book on television receivers from a design perspective. Several books have been published on the theory of television signal processing. These books are limited to the signal processing aspects and do not cover the practical considerations of how to design a television receiver.

The reader may ask: Are the fundamental concepts and principles of digital television important for the design of television devices? 'There is nothing more practical than a good theory'. This statement is attributed to many famous scientists, including the outstanding physicists such as Gustav Robert Kirchhoff and Ludwig Eduard Boltzmann. They were apparently among the first to express this thought. A theory must explain existing facts. A 'good' theory should be able to predict the result of an experiment that has not yet been performed. The fundamental concepts of digital television presented in this book are 'good and essential' for engineers and scientists working on and advancing television technology. Theory not only describes the operation of modern systems but can also help in the design of new television technologies.

The first three chapters deal with the basic concepts that are important in understanding how television systems work. These basic theories provide important information for understanding how television has evolved and how human vision and colour theory contributed to the design of digital television systems and devices.

This book begins with a brief history of television. An understanding of the evolution of television systems helps us understand the constraints that shaped modern television systems. This chapter is written by David Burger, who was chair of the IEEE History Centre.

To understand television, it is important to have a firm understanding of human vision. Chapter 2 was provided by Professor Boon. She is an optometrist who has an active research interest in the effects of display technology on human vision. Television relies very much on making use of the limitations of human vision. The shortcomings of human vision are used to give the impression that moving images are seen. The limitation of human vision hides the fact that television is merely a sequence of still frames. These 'moving pictures' make us believe we see motion. The impact of human vision on display technology is covered in more detail than in any other television book. This is a very important chapter in understanding television concepts.

Professor Mizokami specializes in colour vision and colour science. She wrote Chapter 3. An understanding of optical colour theory is essential towards an understanding of why the chroma signal is processed the way it is and why the three primary colours (green, red, and blue) can produce any colour. It helps understand why hardware such as display technology or signal processing are done the way they are.

Chapter 4 provides an overview of television systems. It introduces the fundamental concepts of television systems and provides an overview of television concepts. It sets the stage for the following chapters that describe television technology.

Chapters 5, 6 and 7 describe the fundamental concepts of how digital technology has changed television signal processing. Chapter 5 provides an overview of digital signal processing concepts such as source and channel coding, Analogue to Digital and Digital to Analogue Conversion. It covers the effects of sampling on signal quality. It introduces audio and video frames. Chapter 6 describes error correction. This includes block and tree codes and transmission errors. Hamming and other codes (BCH, Low Density Parity Check (LDPC) and more) are covered in detail with easy-to-understand mathematics. Chapter 7 focuses on data compression. Huffman coding is covered followed by video compression basics. The compression theories are described with the aid of colour images. The chapter includes an overview of all commonly used video compression standards. Audio compression is also covered in this chapter.

Chapters 8 to 10 cover the application of video standards. Chapter 8 starts with a historic overview of MPEG development. This is followed by an overview of MPEG standards development. Then the importance of MPEG in the DTV systems (Advanced Television Systems Committee (ATSC) and Digital Video Broadcasting (DVB)) is covered.

DVB is covered in Chapter 9. This chapter provides an excellent overview of DVB Systems considerations in detail. This chapter is a valuable reference on DVB and its implementation.

Chapter 10 is the final chapter and it provides a sound overview of ATSC standard. This perspective is presented by a broadcast engineer and describes practical issues associated with ATSC broadcasting.

The editors and authors of this book trust that it will be a valuable reference for anybody involved in television technology.

Stefan Mozar, Sydney
Konstantin Glasman, St Petersburg

Acknowledgements

From the editors:

This book is dedicated to Swee Choo and Christine. Their support and patience have made this project possible.

The editors thank Professor Alexander Belozertsev, Department of Television, St Petersburg Electrotechnical University (LETI), for providing images of video signals that were used in this book.

Chapter 2:

The editor acknowledges Jacinta Wei Chin Chong for assistance with the illustrations.

Chapter 4:

The editors would like to acknowledge the invaluable contribution of Mr Mário Nagamura in the production of the figures used throughout the text. We would also like to thank the National Council for Scientific and Technological Development (CNPQ) Process 490088/2013-9 entitled, GLOBAL ITV: Interoperability of Interactive and Hybrid TV systems.

About the Editors

Stefan Mozar (Editor) Australia

Stefan Mozar is a Chartered Professional Engineer (C.P.Eng.), Life Fellow of the IEEE, and Fellow of Engineers Australia. He is based in Sydney, Australia. His expertise covers the fields of consumer and industrial electronics. He has spent most of his career in R&D labs but also taught for several Australian, Asian, and British Universities. He has worked on Television for Grundig, and NordMend in Germany, NEC in Australia, and Philips Video Labs in Singapore. He has taught Television and Video Systems for several institutions, including the RMIT in Melbourne, TAFE New South Wales, and the Singapore Polytechnic. He is active in the IEEE and has served as President of the IEEE Consumer Technology Society (2013–2014) and IEEE Product Safety Engineering Society (2020–2021). He is the founder of the International Symposium on Consumer Electronics (IEEE ISCE), the *IEEE Consumer Electronics Magazine*, and a past Associate Editor of the *IEEE Communications Magazine*, and the *IEEE Transactions on Consumer Electronics*. He obtained his engineering degrees from the University of New South Wales, Sydney, Australia, and Okayama University, Japan.

Konstantin Glasman (Editor) Russia

Konstantin Glasman is a Professor of the Television Department at St Petersburg Electrotechnical University (LETI), St Petersburg, Russian Federation; MIEEE and Member, BoG, IEEE Consumer Technology Society; Technical Papers Committee Member, IBC (2008–2022); Regional Governor EMEA, Central and South America region SMPTE (2021–2022).

See Bio below for Chapters 5 to 7.

Chapter Authors

David Burger (Chapter 1) Australia

David graduated from Ballarat University in 1979. His early career was in manufacturing with NEC and Ericsson, followed by project management with National Australia Bank, NDY, Macquarie Bank, PricewaterhouseCoopers, Ausgrid and University of Sydney.

David was the IEEE History Committee Chair and involved in Milestone commemorations in Tokyo and Huntsville AL.

David is a Fellow of Engineers Australia (2007) and Life Senior Member of IEEE (2021). David holds USA, New Zealand and Australia amateur radio licenses.

Mei Ying Boon (Chapter 2) Australia

Mei Ying Boon is an Associate Professor of Optometry and Head of Discipline of Vision Science and Optometry at the University of Canberra. She also holds Adjunct academic appointments at the Australian National University and the University of New South Wales. Dr Boon's expertise is in optometry and vision science with a research focus on vision throughout the lifespan, with interests in the development of colour vision, binocular vision, perception and functional vision in the presence of sight-threatening pathology. Dr Boon uses clinical and visual electrophysiology, visual psychophysics and qualitative and quantitative methods in her research.

Yoko Mizokami (Chapter 3) Japan

Yoko Mizokami is a Professor in the Department of Imaging Sciences, Graduate School of Engineering, Chiba University, Japan. She received a Ph.D. in Engineering in 2002 from Ritsumeikan University, Japan. From 2002 to 2006 she was a postdoctoral fellow at the University of Nevada, Reno, Department of Psychology. She moved to Chiba University in 2006.

Her research interests lie in colour vision, colour science and vision in natural environments. The current research topics cover colour constancy, colourfulness adaptation, the perception of image colour difference, the influence of lighting to object appearance, the appearance of skin colour and texture, and colour vision deficiency. She is a member of the International Colour Vision Society (ICVS), Vision Sciences Society, Optica, Vision Society of Japan, Color Science Association of Japan, Optical Society of Japan, and Illuminating Engineering Institute of Japan. She is currently a member of the Board of Directors of the International Colour Vision Society and an associate director and official member of CIE Division 1.

Marcelo Knörich Zuffo (Chapter 4) Brazil

Marcelo Knörich Zuffo (58), IEEE Member since 1989, is a Full Professor of the Electronics Systems Engineering Department of the Polytechnic School of University of São Paulo (USP) (EPUSP). He holds a degree in Electrical Engineering (1988), a Master's degree (1993) and a PhD (1997) in Electrical Engineering from EPUSP. He is currently the coordinator of Interactive Electronics Group at the Laboratory for Integrated Systems, where he leads research and developing projects on Consumer Electronics, including

virtual reality, digital image processing, Digital Health, multimedia hardware, Digital TV, embedded computing, Internet of Things, interactive and distributed systems. He coordinates a national facility for Public Key Infrastructure certification in Brazil. He is also a member of national and international scientific and professional organizations, such as SBC (Brazilian Computer Society), ACM SIGGRAPH (Special Interest Group in Computer Graphics) and Union of Engineers of the State of São Paulo and Brazil. He was responsible for the implantation of the first CAVE Digital (Cave Automatic Virtual Environment) of Latin America. He has been very involved in defining the Bazilian Digital TV System that has been adapted by nearly all countries in Latin America. He is currently Director of USP Innovation Center InovaUSP.

Roseli de Deus Lopes

Roseli de Deus Lopes is an Associate Professor in the Department of Electronic Systems Engineering, Polytechnic School of the University of São Paulo (EP-USP), and has been an IEEE Member since 1988; graduated, master's, doctorate and lecturer in Electrical Engineering from EP-USP. She is Vice-coordinator of the

Interdisciplinary Center for Interactive Technologies (CITI-USP), Research Support Center, created in 2011. She was Vice-Director (2006 to 2008) and Director (2008 to 2010) of Estação Ciência, Center for Scientific Diffusion, Technological and Cultural Institute of USP's Dean of Culture and University Extension. She has been a researcher at LSI Integrated Systems Laboratory since 1988, where she is the leader of the Interactive Electronic Media Research Group, which involves computer graphics, digital image processing, human computer interaction techniques and devices, virtual reality and augmented reality. She coordinates research projects in the area of Interactive Electronic Media, with an emphasis on applications aimed at Education and Health. She coordinates scientific dissemination projects and projects aimed at identifying and developing talents in Science and Engineering. She was responsible for initiating and establishing of FEBRACE (Brazilian Science and Engineering Fair). Since 2003, she has been the general coordinator of FEBRACE. She was part of the technical and pedagogical advisory working group of the UCA Program—One Computer per Student from MIT-MediaLab. She is currently the Vice Director of USP Institute for Advanced Studies IEA-USP.

Laisa Caroline Costa de Biasi

Laisa Costa Paula de Biase graduated in Electrical Engineering with emphasis in Telecommunications from the Polytechnic School of the University of São Paulo (2004). She obtained her master's degree in 2009 and her doctorate in 2015 from the same institution. She worked as a visiting researcher at the University of California at Berkeley in 2013 at Ubiquitous Swarm Lab. She is currently a research and development engineer at LSI-TEC, where she has served as a project leader. She is a collaborative researcher at CITI-USP (Interdisciplinary Center in Interactive Technologies at the University of São Paulo) and at the Ubiquitous Swarm Lab at UC Berkeley. She has experience in embedded systems, especially in the areas of Internet of Things, Accessibility and Digital Television. She participates in the SBTVD and IoT Forum. She is a member of IEEE and ACM, part of the Brazilian chapter of the consumer electronics society (CES). She has extensive experience in project leadership and team coordination for the development of research and development projects.

Konstantin Glasman (Chapter 5, 6, 7) Russia

Konstantin Glasman graduated from St. Petersburg Electro-technical University LETI with a degree in communication engineering. He received his Ph.D. degree in film and television technology from St Petersburg University of Film and Television. He held various research and teaching positions. He was Head of the University's Department of Broadcast Technology for 21 years. He is currently Professor and Deputy Head of the Education Department at St Petersburg Electrotechnical University LETI.

He was a TPC member of the IEEE ICCE, ISCE and ICCE-Berlin conferences for many years; a member of the Board of Governors of the IEEE Consumer Technology Society, the Chair of the Video/Multimedia Committee; a member of the IEEE.tv Advisory Committee; General Chair of the IEEE ISCE 2006 Symposium, a member of the Editorial Board and Columnist of the IEEE Consumer Electronics Magazine; scientific editor of MediaVision Magazine; an elected Board member, SMPTE (Society of Motion Picture and Television Engineers) Regional Governor: EMEA, Central & South America Region (2021–22) and IBC (International Broadcasting Convention) Technical Papers Committee Member (2008–22). He is the founder and general producer of the IEEE CESocTV and IEEE BTS-TV (Official YouTube Broadcast Channels of the IEEE CT Society and IEEE BT Society).

He has published over 100 papers in journals and conference proceedings. His main research interests include image processing, visual quality aspects of video coding, multimodal quality evaluation in digital television systems with video and audio compression, and multimedia resource allocation.

Michael Isnardi (Chapter 8) USA

Michael Isnardi is a graduate of the MIT Media Lab and is a Distinguished Computer Scientist at the Princeton, New Jersey, USA facility of SRI International.

His background is in image and video system design, analysis and simulation. He was a technical manager for DIRECTV's world's first real-time MPEG encoder and led the development of SRI's Emmy Award-winning MPEG Compliance Bitstreams. He developed SRI's DVD watermarking technology and Salience-Based Compression technologies. He was a technical contributor or lead on many commercial and U.S. government programmes with recent contributions to media forensics, hyper- dimensional computing AI, edge-friendly AI, advanced chemical detection and enhanced image preprocessing for improved OCR robustness. Dr Isnardi is an IEEE Fellow and has 41 US patents and 30 published papers.

Edward Feng Pan (Chapter 9) Canada

Edward Feng Pan received the B.Sc., M.Sc. and Ph.D. degrees in communication and electronic engineering all from Zhejiang University, China.

Since then, he has been teaching and researching in several universities in China, UK, Ireland, and Singapore. His research areas are digital image/video compression and quality evaluation. He has published 70 technical papers, over 20 IP disclosures, and many contributions to international standard organization. Dr Pan was the General Chairman of the 7th International Symposium on Consumer Electronics, Sydney, Australia, 3–5 December 2003. He was the Chapter Chairman of IEEE Consumer Electronics, Singapore, from 2002 to 2004.

Dr Pan is currently a senior member of the technical staff at Advanced Micro Devices, Inc., leading the team working on video encoding and signal processing algorithms. Before that he had been serving as a video architect and manager at ViXS Systems Inc. Prior to that he had been serving as a senior scientist at the Institute for Infocomm research in Singapore.

Guy Bouchard (Chapter 10) Canada

Guy Bouchard is 'directeur technique', transmission systems at Télé-Québec, He is responsible for digital delivery infrastructure at Télé-Québec. During his 33 years with the Canadian Broadcasting Corp, Guy has worked in analogue and digital television transmission and production systems, as well as satellite and terrestrial microwave communication systems. He holds a degree in telecommunications from the Université du Québec (ETS). He is a board member of the IEEE Broadcast Technology Society AdCom and was an active contributor to  the ATSC Technology & Standard group. Guy has served the broadcast industry since 1980 with a special interest in digital communications systems. He has written and delivered papers on DTV, Satellite and MPEG Transport Technology for the NAB, CCBE, IEEE, Canadian Digital Television (CDTV) and the Society of Motion Picture and Television Engineers (SMPTE). Well known for making technology understandable by non-engineers, Guy also delivered educational papers in Canada, Argentina, Uruguay and South Africa.

Glossary

Acronym	definition
ALP	Adaptation Layer protocol
AM	Amplitude Modulation
ATSC	Advanced television System Committee
AWGN	Additive White Gaussian Noise (thermal noise)
BCH	Bose–Chaudhuri–Hocquenghem (efficient FEC Scheme)
C/N	Carrier-to-Noise Ratio
DCT	Discrete Cosine Transform
EM	Electro-magnetic
EB/N0	Energy per Bit/Noise ratio (Modulation density to noise ratio)
FEC	Forward Error Correction
FFT	Fast Fourier Transform
FM	Frequency Modulation
GI	Guard Interval
LDM	LDM Layer Division Multiplex
LDPC	Low Density Parity Check (Efficient FEC scheme)
MER	Modulation Error Ratio
MODCOD	Modulation and FEC code Rate ex: QPSK ¾
PAPR	Peak-to-Average Power Ratio
PLP	Physical Layer Pipes
QEF	Quasi Error Free
QPSK	Quadrature Phase Shift Keying
RTSA	Real-Time Spectrum Analyzer
SSB	Single Side-band
TDM	Time division Multiplex
UDP	User datagram protocol
VHF	Very High Frequencies
UHF	Ultra-High Frequencies

Chapter 1

Historical development of television

David E. Burger[1]

Television! Teacher, mother... secret lover. Urge to kill... fading... Come, family, sit in the snow with daddy and let us all bask in TV's warm glowing warming glow.

Homer Simpson; 1994 Treehouse of Horrors

1.1 Early developments and exploration

The quest for conveying live images over distances coincided with the quest for text and voice communications over distance. The historical developments in wireless telegraphy and vacuum tubes were broadly running in parallel with those working on 'television'. Early image conveyance was heavily constrained by technological developments; indeed, the very early efforts were based purely on the mechanical doctrine of the industrial revolution.

Popular literature and the national pride of many Countries purport to be the 'first' in all manner of achievements; or being the 'underlying' basis for live image conveyances, i.e. television in its broadest term. Experience has taught that making any statements of being 'first' in the technology stakes, invites distracting rebuttals and arguments. This exploration of television history is intended to highlight timelines, and provide a salutary nod to the many inventors, engineers and scientists around the world who contributed to this great achievement.

1.2 The inventors and influencers

1.2.1 Henry Sutton, Ballarat, Australia

In 1885, Sutton's invention of the 'telephane' was the forerunner to television 3 years before John Logie Baird was born. Around 1871 at the age of 15, Henry first invented a method so that any big event in Melbourne could be seen in Ballarat by the medium of the telegraph. Henry was so sure of this that he wrote the particulars to Mr R.L.J. Ellery who was the government astronomer of Victoria so the invention could be in the hands of someone capable of stating his claim of being the first in this direction.

[1] Retired Consultant

Some years later in 1885, Mr R.L.J. Ellery was witness to the transmission of the images of the telephane. In 1885, Henry purportedly transmitted through the telephane the Melbourne Cup race to Ballarat, it was stated that it worked quite well. The telephane used a similar principle to the Nipkow spinning disk system.

In 1890, Henry went to England and France and demonstrated the telephane to the scientific community. Henry's paper on the telephane was published in England, France and America and Scientific America republished his paper again in 1910.

Henry did not patent the telephane but Baird did leverage Henry's principles to progress television developments some 43 years later. The telephane is considered to be Henry's magnum opus by some people.

In 1890, Henry's paper on the telephane system was published in the French science Journal La Lumiere Electrique and in 1890, Henry demonstrated his telephane system to the Royal Society of London. He then went to France and demonstrated his system there [1, 2].

1.2.2   Paul Gottleib Nipkow, Berlin, Germany

In 1883, Gottleib's concept of breaking down moving images into a mosaic pattern leveraged earlier work from the facsimile machine of Alexander Bain. A patent for Nipkow's invention was granted in Berlin on 15 January 1885 backdated to 6 January 1884 but lapsed soon afterwards. Nipkow's development underpinned most of the mechanical and hybrid mechanical/electronic systems [3].

1.2.3   Georges Rignoux and A. Fourier, Paris, France

The first recorded demonstration in France of the instantaneous transmission of images was by Georges Rignoux and A. Fourier in Paris in 1909. A matrix of 64 selenium cells, individually wired to a mechanical commutator, served as an electronic retina. In the receiver, a type of Kerr cell modulated the light and a series of variously angled mirrors attached to the edge of a rotating disc scanned the modulated beam onto the display screen. The 8×8 pixel proof-of-concept demonstration was sufficient to clearly transmit individual letters of the alphabet [4].

1.2.4   Lev Sergeyevich Termen / Leon Theremin, St. Petersburg, Russia

Theremin studied physics and astronomy at the university in St. Petersburg, and by 1920 he was heading the experimental electronic oscillation laboratory at the Institute of Physical Engineering in St. Petersburg. In 1925, Theremin began developing a mirror drum-based television, starting with 16 lines resolution, then 32 lines and eventually 64 lines using interlacing in 1926. In Theremin's thesis of 7 May 1926, he electrically transmitted and then projected near-simultaneous moving images on a 5 ft^2 screen at a demonstration at the Leningrad Polytechnic Institute [5].

1.2.5   John Logie Baird, London, UK

Baird was educated at Larchfield Academy, now part of Lomond School in Helensburgh; the Glasgow and West of Scotland Technical College; and the University of Glasgow.

On 25 March 1925, Baird gave the first public demonstration of televised silhouette images in motion, at Selfridge's department store in London. On 26 January 1926, Baird repeated the transmission for members of the Royal Institution and a reporter from *The Times* in his laboratory at 22 Frith Street in the Soho district of London [6].

Baird's system was an improved version of the Nipkow spinning disk system and leveraged developments in signal processing by Arthur Korn. Baird is credited with developing the first system to record images [7].

In 1926, Baird invented the 'noctovisor', which used infrared light to make objects visible in the dark or through the fog.

Baird demonstrated the world's first colour transmission on 3 July 1928, using scanning discs at the transmitting and receiving ends with filtered primary colour light on three spirals of apertures [8].

By 1927 Baird successfully transmitted moving images over 705 km of telephone wires between London and Glasgow and formed the 'Baird Television Development Company'. The following year he transmitted the first transatlantic and first shore-to-ship moving images. As his technology improved, resolution rose from 30 lines of definition to 240 lines. In late 1927, his 'televisor' receiver went on sale to the public.

Initially viewing Baird as a competitor, the BBC was resistant to working with him. Starting in 1929, they collaborated on experimental broadcasts. But in 1935, the BBC set up a side-by-side competition between Baird's system and Marconi's all-electric system. By 1937 they had dropped Baird's system altogether [9].

In 1939, Baird showed a system known later as a hybrid colour projection using a cathode ray tube in front of which revolved a disc fitted with colour filters, a method adopted at that time by the Columbia Broadcasting System (CBS) and Radio Corporation of America (RCA).

Around 1940, Baird commenced work on a fully electronic system he called the 'telechrome'. Early telechrome devices used two electron guns aimed at either side of a phosphor plate. The phosphor was patterned with cyan and magenta phosphors, offering a limited-colour image. He called it a 'stereoscopic' at the time. In 1941, Baird patented and demonstrated this system of a monochrome 3-dimensional (3D) television at a definition of 500 lines.

On 16 August 1944, Baird gave the world's first demonstration of a practical fully electronic colour television display. His 600-line colour system used triple interlacing, using six scans to build up each picture.

1.2.6   Arthur Korn, Breslau, Germany/Hoboken, New Jersey, USA

With a background in physics, mathematics and a keen interest in photography, Korn was well grounded in the basics of images. At age 69, Korn was forced to the USA in 1939 due to World War II conflicts.

On the 18th October 1906, Korn managed to transmit a photograph of Crown Prince William over a distance of 1800 km [10].

At a 1913 conference in Vienna, Korn demonstrated the first successful visual telegraphic transmission of photographs, known as telephotography or in German the Bildetelegraph. In 1923, he successfully transmitted an image of Pope Pius XI across the atlantic ocean, from Rome to Bar Harbor, Maine [11].

Korn invented and built the first successful signal-conditioning circuits for image transmission. The circuits overcame the image-tearing time lag effect that is part of selenium photocell light detection. Korn's compensation circuit allowed him to send photographs by cable or wireless, noting his circuits operated without the benefit of electronic amplification [12].

1.2.7   Charles Francis Jenkins, Washington, DC, USA

In 1913, Charles published an article on 'Motion Pictures by Wireless', but it was not until December 1923 that he transmitted moving silhouette images. On 13 June 1925, Charles publicly demonstrated synchronised transmission of silhouette pictures. In 1925, Jenkins used the Nipkow disk system with a lensed disk scanner with a 48-line resolution, transmitting a silhouette image of a toy windmill in motion, over a distance of 5 miles from a naval radio station in Maryland to his laboratory in Washington, DC.

1.2.8   Philo T. Farnsworth, Utah, USA

Circa 1921, Philo T. Farnsworth drew something on the blackboard at Brigham Young High School that would change the world. It was an 'Image Dissector' comprising a lens and glass tube with an electron beam. Years later, his chemistry teacher redrew that sketch and helped Farnsworth win a long courtroom battle against the RCA [9].

Farnsworth first proposed his TV system to his chemistry class in 1922. In 1927, at the age of 21, he produced his first electronic television transmission. He took a glass slide, smoked it with carbon and scratched a single line on it. This was placed in a carbon arc projector and shone onto the photocathode of the first camera tube [13].

Farnsworth raised money from friends and family to build his TV system. Meanwhile, another inventor, Vladimir Zworykin, developed an electronic TV for his employer, RCA. Farnsworth aged 30, filed Patent #1,773,980 for his camera tube, entitled 'television system', on 7 January 1927 and was granted the patent on 25 August 1930, after a long battle with RCA. RCA's long-time leader, the hard-nosed David Sarnoff, fought Farnsworth's patent claims in court for years. In 1939, Farnsworth won [9].

Zworykin applied for a patent for his television camera, the 'iconoscope', in 1923; but his patent was not granted until 1938 because of Farnsworth's earlier patent [13, 14].

A second Farnsworth patent was granted; Patent #1,773,981 for the cathode ray tube (CRT) providing the display tube in the receiver, now known as television screen.

In 1939, with World War II looming, the US government stopped television development. Television manufacturing did not start again until after the war. By then Farnsworth's patents had expired and his company subsequently collapsed.

Time magazine called Farnsworth one of the best mathematicians and inventors of the 20th century. Philo junior described his father as having a 'romance with the electron'. That romance resulted in over 150 patents for television, amplifiers, CRTs, vacuum tubes, fusion reactors, satellites and more.

1.2.9 David Sarnoff, Belarus and New York, USA

In 1923, Sarnoff, as RCA's general manager, perceived the potential of television, which the contributions of several inventors were now making technically feasible. His meeting in 1929 with Westinghouse engineer Vladimir Zworykin convinced him that home television was possible, and Sarnoff persuaded Westinghouse to back Zworykin's work. In 1930 Westinghouse's television research and Zworykin were transferred to RCA. By 1939 Sarnoff was able to give a successful demonstration of the new medium at the New York World's Fair.

Sarnoff was instrumental in protecting the RCA Patents, which underpinned television broadcasting, launching many protracted legal battles, but particularly battles with both Zworykin and Edwin Armstrong.

RCA became a market leader in the television industry, but balked in the new field of colour television. In 1950, the Federal Communications Commission (FCC) approved the colour television standard developed by the rival Columbia Broadcasting System (CBS). Existing black-and-white sets, including those of RCA were incompatible with the new colour program format. Sarnoff committed RCA to developing a set that would be compatible with both black-and-white and colour formats but that experienced major delays. Sarnoff initiated a crash program to develop the compatible system, and in 1953 RCA's system was adopted as the standard for colour television by the FCC. Sarnoff became chairman of the RCA board in 1949 [15].

1.2.10 Edwin Armstrong, New York, USA

While Armstrong's contribution to television is tangential, they are pivotal in the delivery and reception of television signals. These included the developments with the superheterodyne receivers and the use of wideband frequency modulation (FM) for the delivery of the high quality 'audio' component of historic television signals. Indeed, the eventual utilisation of the 300 MHz to 800 MHz Ultra High Frequency (UHF) frequency band for television transmission was enabled by the superheterodyne developments of Armstrong. Armstrong was embroiled in a patent battle over the discovery of FM with Radio Corporation of America (RCA), eventually losing the case. These battles wore down Armstrong, with his demeanour ultimately poisoned by the bitter legal processes, selfishly resulting in his suicide in 1954.

1.2.11 Elmer W. Engstrom, Minneapolis, USA

Circa 1930, Dr Engstrom assumed supervision of RCA's intensified program of television research and development. He brought to this program the concept of television as a complete system. This program led to the development of a practical commercial black and white television system by 1939. In the post-war years, as the head of RCA Laboratories, Dr Engstrom directed a similar systems program, which led to the achievement of the compatible all-electronic colour television system.

In 1943 Dr Engstrom became Director of Research of RCA Laboratories and in 1945 he was elected Vice President in Charge of Research. After several executive assignments, he was appointed Senior Executive Vice President of RCA in 1955. Six years later, he became President of RCA and on 1 January 1966, he assumed the position of Chief Executive Officer of the RCA and Chairman of the Executive Committee of the RCA Board.

1.2.12   Edward W. Herold, NJ, USA

After graduating in 1924, Herold worked for several years as a technical assistant at the Bell Telephone Laboratories, on picture transmission; at the same time, he became familiar with the television experiments under Ives. In 1927, he joined an electron tube company, E.T. Cunningham, Inc., later that year, he enrolled as a physics major at the University of Virginia. After graduating in 1930 with his B.Sc. and a Phi Beta Kappa key, Herold joined RCA as a development engineer at Harrison, NJ. He became known in the 1930 for the development of a number of receiving tubes and for his analyses of frequency converters and signal-to-noise problems.

In 1949, Herold directed an RCA corporate-wide 'crash' program to develop a colour television picture tube, a project which, up to that time had appeared to be close to impossible. In 6 months, several approaches had shown success, among them the shadow-mask tube of H.B. Law. The latter tube was first publicly demonstrated in March 1950 and made possible the adoption of the compatible CRT based colour systems. The shadow-mask tube that was developed was the basis for all CRT television until the CRT's demise in the early 2000s.

1.2.13   Vladimir Zworykin, St. Petersburg and Washington

The involvement of Zworykin began with the emergence of electronic image television systems. In 1924, he created the iconoscope, the first practical, and all-electronic television camera tube. In 1929 he invented a part of the receiver called the kinescope, a cathode-ray picture tube. Zworykin's work significantly moved television away from the old mechanical systems.

Zworykin was born in Russia on 30 July 1889. In 1912, Zworykin was in France working with Paul Langevin, a Nobel Prize winner in theoretical physics. As a young engineering student at the Institute of Technology in St. Petersburg, he worked for physicist Boris Rosing who was trying to send pictures through the air. The system Zworykin and Rosing developed was a Nipkow transmitter and utilising a CRT receiver.

In 1919, following the Russian revolution, Zworykin moved to the United States. He worked at Westinghouse Electric Corporation in Pittsburgh. In 1929, when Zworykin did not get the support or encouragement he needed to build electronic televisions, he moved to RCA. With the strong support of RCA's head David Sarnoff, another Russian immigrant, Zworykin continued developing electronic television. His all-electronic television system was introduced to the public at the 1939 New York World's Fair.

Zworykin's and RCA's work were directly challenged by the patent claims of Philo T. Farnsworth who worked on electronic television. Farnsworth won a major patent suit against RCA in 1939 and Zworykin's designs for electronic television would not, therefore, be considered the first. Despite the legal battle, Zworykin's contributions to television were significant. While some called him the father of television, Zworykin stressed that television was the creation of hundreds of inventors and researchers. Zworykin's contributions were recognized by the AIEE who awarded Zworykin its Edison Medal in 1952 'for outstanding contributions to concept and design of electronic components and systems'. In 1950 he was awarded the prestigious Medal of Honor by the Institute of Radio Engineers (IRE) 'for his outstanding contributions to the concept and development of electronic apparatus basic to modern television, and his scientific achievements that led to fundamental advances in the application of electronics to communications, to industry and national security' [9, 14].

1.2.14   M. Georges Valensi, France

Valensi patented a method of transmitting color images in 1938 utilising separate luminance and chrominance signals that were compatible with both color and black and white television receivers. The luminance and chrominance image decomposition was subsequently used by the PAL, NTSC and SECAM television standards.

1.3   Television timeline overview

1.3.1   1880s and 1890s

The earliest published proposal for colour television was one by Maurice Le Blanc in 1880, including the first mentions in television-related literature of line and frame scanning, although he gave no practical details [16]. Polish inventor Jan Szczepanik patented a colour television system in 1897, using a selenium photoelectric cell at the transmitter and an electromagnet controlling an oscillating mirror and a moving prism at the receiver. His design contained no means of analysing the spectrum of colours at the transmitting end and therefore remained as a theoretical development [17].

1.3.2   1900s

In 1899 Russian military engineer and scientist K.D.Perskiy had submitted the paper "The Modern condition of a question on electrovision on distance (televising)" to the First All-Russia Electrotechnical congress in Saint-Petersburg [18]. Konstantin Perskyi is credited with coining the word 'television' in a paper read to the International Electricity Congress at the International World Fair in Paris on

25 August 1900. Perskyi's paper reviewed the existing electromechanical technologies in detail, mentioning the work of Nipkow and others [3, 19, 20].

1.3.3   1920s

The first demonstration of television in Germany occurred at the Berlin radio show in 1928, based on equipment from the Baird Television Company. Nipkow is credited with attending the demonstration and seeing his early concepts becoming a reality.

In 1928, General Electric launched their experimental television station W2XB, broadcasting from the GE plant in Schenectady, New York. The station was popularly known as 'WGY Television', named after their radio station WGY. The station eventually converted to an all-electronic system in the 1930s and in 1942, the FCC license call-sign was updated to WRGB. The station is still operating today.

In Australia 1929, Melbourne radio stations 3DB and 3UZ began demonstrating television using the radiovision system by Gilbert Miles and Donald McDonald. This took place at the Menzies Hotel on 30 September 1929 [21, 22].

1.3.4   1930s

Germany's first public television channel, commenced transmitting in Berlin in 1935, was named Fernsehsender Paul Nipkow, as a nod to the German inventor Paul Nipkow.

In 1931, television broadcasts commenced in France.

In 1933, television broadcasts progressed quickly in Europe ahead of the USA. Zworykin traveled at that time quite a deal from RCA because of connections with the laboratory but in 1934 Zworykin went to Russia and succeeded to sell them the idea that they were realizing the importance of television for propaganda. The government was very interested in getting the equipment. Zworykin returned and reported to Sarnoff about their interest, wanting to pay cash for the equipment. After long discussions with the state department, Zworykin returned to Russia to re-negotiate the deal. In 1937, RCA sent their equipment and it was all installed in Moscow. A hundred receivers and television transmitters were operational around Moscow. Zworykin noted the televised content was 'trash'.

In 1936, RCA created the National Broadcasting Corporation (NBC), the first American TV network. NBC held its first broadcast at the 1939 World's Fair and the public was intrigued. World War II put a temporary damper on television development, re-emerging in the early 1950s.

In 1936, Kálmán Tihanyi described the principle of plasma illuminated display, the first key component of a flat panel display system [23].

In 1939, Telefunken demonstrated television to the public at the Exhibition Fair in Rio de Janeiro, Brazil [24].

The vacuum tubes developments by Eitel-McCullough (EIMAC) in San Francisco were growing in popularity and dependability. The EIMAC founder, William Eitel and Edwin Armstrong were close and regularly met at Columbia University to discuss problems and fixes to vacuum tubes to support both radio and television transmitters at that time.

1.3.5   1940s

During this period, all of the earlier work on mechanical based television systems was obsoleted, superseded by fully electronic systems. The mechanical basis of ras-terizing and scanning the 'coding logic' of an image, for want of a better term, continued into the realm of vacuum tube electronics, were faster, more accurate, and more practical systems could be implemented.

Major developments in television were undertaken by the USA and the UK using electronic standards that were incompatible, but did not differ too greatly; however, the commercial socio-political models adopted were so vastly different. The UK adopted a television receiver license tax to fund their BBC broadcaster, while the USA adopted a commercial private ownership model, funded by advertising making television freely accessible.

Australia followed the UK socio-political funding model in 1948, then reverting to the USA model in December 1949.

Zworykin participated in the first television research group in Camden, comprising Arthur Vance, Harley Iams and Sanford Essig [14]. The group expanded with Leverenz joining the group in 1944.

Circa 1945, EIMAC engineers had evolved new vacuum tube types, including production of the new beam tetrode (4X150A) which performed well in the VHF region for both radio and television. At the same time, a new Salt Lake City plant was opened to make CRT television picture tubes, with subcontracts to RCA as well [25].

In 1948, the IRE Audio Professional Group was formed on 2 June 1948, and the Broadcast Engineers Group was the second one to be approved on 7 July 1948. The name was changed in August 1949, to the IRE Professional Group on Broadcast Transmission Systems. The name was later changed to the IRE Professional Group on Broadcasting. Today it is known as the IEEE Broadcast Technology Society, with a 1982 offshoot known as the IEEE Consumer Electronics Society.

Launched in 1948, Technology and Engineering Awards (Emmy) honor development and innovation in broadcast technology and recognize companies, organizations and individuals for breakthroughs in technology that have a significant effect on television engineering [26].

Oddly, there were no new television broadcast deployments cited for any Country in the 1940s, probably due to World War II conflicts.

1.3.6   1950s

The official Australian launch of black and white television occurred on 16 September 1956 in Sydney, operated by the channel Nine Network station TCN-9. This was a pre-cursor to the Olympic Games held in Melbourne, Australia.

The USA introduced colour television in 1950, followed by the 'National Television Standards Committee' (NTSC) standard deployment in 1953.

NHK Japan commences black and white television transmission on 1 February 1953 [27].

In 1953, Dr. Norikazu Sawazaki developed a prototype helical scan video tape recorder that led to the development of domestic VCRs in the 1970s [28].

In 1956, the French SECAM colour system 'Sequential Colour with Memory' was developed. It was one of three major colour television standards widely used, the others being the European PAL 'Phase Alternate Line' and North American NTSC formats. The SECAM development was led by Henri de France working at Compagnie Française de Télévision. Ownership moved when bought by Thomson, and then later by Technicolor. The first SECAM colour broadcast was made in France in 1967, and was obsoleted around 2002.

In 1958, a new EIMAC plant was built in San Carlos, California, for production of new tetrode tubes, plus larger tubes for broadcast and TV service. At the same time, television klystron transmitters were developed for high power television broadcasting [25].

In 1958, Toshiba produces their first 14 inch black and white television for sale [27].

1.3.7   1960s

NHK Japan commenced advertising limited colour television programming on 10 September 1960, in the Tokyo Shimbun. A copy of this newspaper guide is on display at the Toshiba Museum [27].

1.3.8   1970s to 1980s

In 1978, James P. Mitchell described, prototyped and demonstrated one of the earliest monochromatic flat panel LED television displays by an invitation from the Iowa Academy of Science at the University of Northern Iowa. The operational prototype was displayed at the Eastern Iowa SEF on March 18 [27]. The colour version of this display later became known as the jumbotron.

In 1989, as a result of the energy-efficient performance of the klystrode inductive output tube (IOT), the EIMAC Division of Varian Associates was awarded an Emmy Award for technological achievement by the Academy of Television Arts and Sciences. The klystrode IOT used in television transmitters at WCES in Augusta, GA was the first of the high-power output devices to be on-air for UHF television broadcasting [25].

1.3.9   1990s

Digital terrestrial television was launched in the United Kingdom on 15 November 1998, just after the digital satellite television launch on 1 October 1998.

The HD Video MPEG-4 v2 became an International Standard in the first months of 1999 after commencing in July 1993 [29, 30].

Key developments in the 1990s were consolidations of the network ownership's of broadcasting businesses. The proliferation of cable television programming in the USA, described by Encyclopedia Britannica as 'the loss of shared experiences', where viewers up until then had a very restricted choice in viewing content.

The FCC Telecommunications Act, 1996 deregulated television station licence periods and permitted larger network ownership penetration, and introduced regulations on childrens programming and restricting violent and adult content.

On 1 June 1995, the IEEE Sendai Section dedicated a milestone titled 'Directive Short Wave Antenna, 1924, Miyagi, Japan'. In these laboratories, beginning in 1924, Professor Hidetsugu Yagi and his assistant, Shintaro Uda, designed and constructed a sensitive and highly-directional antenna using closely-coupled parasitic elements. The antenna, which is effective in the higher-frequency ranges, has been important for radar, television, and amateur radio.

1.3.10 2000s and onwards

Television and video in this period slowly became a commodity, merging crossover technologies with digital systems traditionally used in computing, internet and data management. Major inroads bringing television and video services onto the internet and mobile telephony with most laptop computers and mobile phones having cameras circa 2007. Telecommunication carriers have introduced video calling, and made available subscription based television streaming services.

Digital television broadcasting quickly supplanted old analogue systems, driven largely by making advances in display technology, high definition and physical size of consumer television receivers. The days of analogue television receivers suffering noise interference, ghosting and multiple images during times of enhanced radio propagation are now forgotten. The remainder of this book describes these technologies in detail.

A summary of historic television 'milestone' recognition's are being described which focuses on key achievements.

- On 29 November 2001, the IEEE History Committee dedicated a milestone titled 'Monochrome-Compatible Electronic Color Television, 1946–1953'. The plaque resides in Princeton, NJ, USA [31, 32].
- On 1 July 2002, the IEEE History Committee dedicated a milestone titled 'First Transatlantic Television Signal via Satellite, 1962'. There are two plaques, residing at Goonhilly Downs, Cornwall, England and Andover, Maine, USA. The signals were received in Pleumeur-Bodou, France using the TELSTAR satellite [32, 33].
- On 30 September 2006 – IEEE Princeton/Central Jersey Section dedicated a Milestone 'Liquid Crystal Display, 1968 Princeton, NJ, U.S.A.' Between 1964 and 1968, at the RCA David Sarnoff Research Center in Princeton, NJ, a team of engineers and scientists led by George H. Heilmeier with Louis A. Zanoni and Lucian A. Barton, devised a method for electronic control of light reflected from liquid crystals and demonstrated the first liquid crystal display. Their work launched a global industry that now produces millions of LCDs annually for

watches, calculators, flat-panel displays in televisions, computers and instruments [32].

- On 12 November 2009 – the IEEE History Committee dedicated a milestone titled 'Development of Electronic Television, 1924–1941', Hamamatsu, Japan. Professor Kenjiro Takayanagi started his research program in television at Hamamatsu Technical College, now Shizuoka University in 1924. He transmitted an image of the Japanese character イ on a cathode-ray tube on 25 December 1926 and broadcast video over an electronic television system in 1935. His work, patents, articles and teaching helped lay the foundation for the rise of Japanese television and related industries to global leadership [32].
- On 23 November 2009 – the IEEE History Committee dedicated a milestone titled 'First Transpacific Reception of a Television (TV) Signal via Satellite, 1963', Takahagi City, Japan. The Ibaraki earth station received the first transpacific transmission of a television Signal from Mojave earth station in California, USA, using the Relay 1 communications satellite. The Ibaraki earth station used a 20 m Cassegrain antenna, the first use of this type of antenna for commercial telecommunications. This event helped open a new era of intercontinental live TV programming relayed via satellite [32].
- On 27 September 2010 – the IEEE History Committee dedicated a milestone titled 'TIROS-1 Television Infrared Observation Satellite, 1960', Princeton, NJ, USA. TIROS-1 was a 'Television Infrared Observation Satellite'. On 1 April 1960, the National Aeronautical and Space Administration launched TIROS-1, the world's first meteorological satellite, to capture and transmit video images of the Earth's weather patterns. RCA staff at Defense Electronics Products, the David Sarnoff Research Centre and Astro-Electronics Division designed and constructed the satellite and ground station systems. TIROS-1 pioneered meteorological and environmental satellite television [32].
- On 6 November 2010 – the IEEE History Committee dedicated a milestone titled 'First Television Broadcast in Western Canada, 1953, North Vancouver, BC, Canada'. On 16 December 1953, the first television broadcast in Western Canada was transmitted by the Canadian Broadcasting Corporation's CBUT Channel 2. The engineering experience gained here was instrumental in the subsequent establishment of the more than one thousand public and private television broadcasting sites that serve Western Canada today [32].
- On 18 November 2011 – the IEEE History Committee dedicated a milestone titled 'First Direct Broadcast Satellite Service, 1984'. NHK began research on the main unit of a broadcasting satellite and a home receiver in 1966 and started the world's first direct satellite broadcasting in 1984. This enabled TV broadcasts to be received at homes throughout Japan, even in the mountainous regions and remote islands and established a foundation for the satellite broadcasting services currently used by people around the world [32, 34].
- On 10 June 2014 – IEEE Kansai Section dedicated a Milestone 'Sharp 14-inch Thin-Film-Transistor Liquid-Crystal Display (TFT-LCD) for TV, 1988 Nara, Japan'. Sharp demonstrated a fourteen-inch TFT-LCD for TV in 1988 when the display size of the mass-produced TFT-LCD was three inches. The high display

quality in Cathode Ray Tube size convinced other electronic companies to join the infant TFT-LCD industry aimed at emerging full-color portable PCs. Two decades later, TFT-LCDs replaced CRTs, making the vision of RCA's LCD group in the 1960s a reality [27].

- On 18 December 2014 – the IEEE North Jersey Section dedicated a Milestone for Bell Telephone Laboratories, Inc., 1925–1983 Murray Hill, NJ, USA, Wireless and Satellite Communications, Transatlantic Transmission of a Television Signal via Satellite, 1962 Telstar – first active communications satellite (1962) [27].

- On 21 September 2015 – IEEE New Hampshire Section dedicated a Milestone 'Interactive Video Games, 1966 Nashua, NH, USA'. The 'Brown Box' console, developed at Sanders Associates – later BAE Systems – between 1966 and 1968, was the first interactive video game system to use an ordinary home television set. This groundbreaking device and the production-engineered version Magnavox Odyssey game system (1972) spawned the commercialization of interactive console video games [27].

- On 11 May 2016 – the IEEE History Committee dedicated a milestone titled 'The High Definition Television System, 1964–1989'. NHK began basic research on a high-quality television system in 1964 and conducted research and development based on a wide range of psychophysical experiments, including investigating the relationship between the angle of view and the sense of reality, to system development. In 1989, we started the world's first Hi-Vision broadcasting trials, which formed the basis of the broadcasting standard of 1,125 scanning lines with a 16:9 image aspect ratio. With the system of 1,125 scanning lines adopted as a unified worldwide studio standard in 2000, Hi-Vision has become widespread throughout the world [32, 34].

- On 11 May 2016 – the IEEE History Committee dedicated a milestone titled 'Emergency Warning Code Signal Broadcasting System, 1985'. NHK implemented emergency warning broadcasting, which automatically turns on televisions or radios and provides information to the public in the case of a large-scale earthquake or tsunami. Emergency warning broadcasting has been integrated into the technical standards of international satellite and terrestrial broadcasting during digital TV standardisation. It is still in use today, supporting disaster broadcasting [32, 34].

- On 10 June 2016 – the IEEE History Committee dedicated a milestone titled 'Ampex Videotape Recorder – 1956', Palo Alto, CA, USA. In 1956, Ampex Corporation of Redwood City, California, introduced the first practical videotape recorder for television stations and networks to produce and time-shift broadcasts, replacing impractical 'kinescope' movie film previously used to record TV. The Emmy-award-winning Ampex 'VTR' analogue-video standard ruled broadcasting and video production worldwide for twenty years [32].

- On 26 January 2017 – the actual anniversary, the IEEE History Committee dedicated a milestone titled 'First Public Demonstration of Television, 1926' London, UK. Members of the Royal Institution of Great Britain witnessed the world's first public demonstration of live television on 26 January 1926 in this building at 22 Frith Street, London. Inventor and entrepreneur John Logie Baird used the first floor as a workshop during 1924–1926, for various experimental activities, including the development of his television system. The BBC adopted Baird's system for its first television broadcast service in 1930 [32].

- On 27 July 2017 – IEEE Sendai Section dedicated 'The Discovery of the Principle of Self-Complementarity in Antennas and the Mushiake Relationship, 1948 Tohoku University'. In 1948, Prof. Yasuto Mushiake of Tohoku University discovered antennas with self-complementary geometries. These were frequency independent, presenting a constant impedance, and often a constant radiation pattern over wide frequency ranges. This principle is the basis for many very-wide-bandwidth antenna designs, with applications that include television reception, wireless broadband, radio astronomy, and cellular telephony [27].
- On 5 April 2018 – IEEE United Kingdom and Ireland Section dedicated a Milestone 'Amorphous Silicon Thin Film Field-Effect Transistor Switches for Liquid Crystal Displays, 1979 Dundee, Scotland, UK'. A research team in the Physics department of Dundee University, Scotland demonstrated in 1979 that amorphous silicon field-effect transistors were able to switch liquid crystal arrays. Other semiconductor thin film materials had been found to be unsuitable for deposition on large area substrates. The invention laid the foundation for the commercial development of flat panel television displays [27].
- On 11 October 2019 – IEEE New South Wales Section dedicated a Milestone 'Parkes Radiotelescope, 1969 Parkes, Australia'. Parkes radiotelescope and Honeysuckle Creek stations in Australia received voice and video signals from the Apollo 11 moonwalk, which were redistributed to millions of viewers. Parkes' televised images were superior to other ground stations, and NASA used them for much of the broadcast. One of the first to use the newly developed corrugated feed horn, Parkes became the model for the NASA Deep Space Network large aperture antennas [27].

Oddly, the Solomon Islands located in the Pacific Ocean only implemented Free To Air television in 2008. Up until then, individual viewers originally utilised VCR tapes, then DVDs augmented with the Solomon Islands Broadcasting Commion (SIBC) Amplitude Modulated (AM) and High Frequency (HF) radio sound broadcasting. More recently streaming video over the Internet became available to their 6 major islands and then in 2010 SATSOL deployed a digital satellite television subscription service.

1.4   Spectrum management pressures

1.4.1   Early days

Developments in early black and white television transmission bandwidths were based on the basic spectrum of an analogue amplitude modulated (AM) signal; modulated by a raster scan which generated about 8 MHz of bandwidth. Black and white television developments coincided in a timely manner with the emergence of single sideband (SSB) development in high frequency (HF) radio voice transmission, spectrum engineers determined that one sideband of the television signal could be largely suppressed, and the AM carrier reduced in amplitude without degrading the television channel. Given practicalities of both televsion transmitters and receiver design, parts of the unwanted sideband were still transmitted, known as the vestigial sideband. With that in mind, and the quest to assign as many televsion channels as possible in the very high frequency (VHF) spectrum, i.e. 48 – 224 MHz, nominal channel bandwidths of 5 MHz to 5.5 MHz were widely adopted in many Countries.

The 5 MHz channel bandwidth included part of the AM vestigial sideband and an FM sound carrier offset by approximatey 4.6 MHz.

Individual Countries adopted slightly different channel bandwidths, vestigial sideband suppression and sound carrier offset standards. In today's environment that may sound peculiar, but the television bandwidth development internal to specific countries and markets were very insular, with little opportunity for globalised markets.

1.4.2   Pressure of colour transmission

By the end of 1948; however, technical and regulatory pressures forced the Bell Labs' researchers to start finding ways to compress the bandwidth for colour down to the nominal 5.5/6 MHz used for monochrome in the VHF section of the electromagnetic spectrum. Until then, they and the FCC had counted on broadcasting colour in the UHF section, using up to 14.5 MHz of spectrum per channel. Princeton staff under Dr George Brown discovered the difficulties in using the UHF band for television in terms of greater propagation losses propagation and the need for greater transmitter RF power outputs, while the growing demand for channels indicated that broadcasters would use 6 MHz allotments in UHF as a direct extension of VHF bandwidth allocations.

Careful selection of the colour information coding techniques leveraged developments in information coding theory supporting spread spectrum communications occurring around that time. This meant that interleaving colour information between the monochrome signal spectrum 'spurs' resulted in no overall additional channel bandwidth being consumed.

While most analog television transmitters have since disappeared, now in favour of digital transmissions. It is interesting to note the old video carrier frequencies of the vestigial sideband signals were made unique to each transmitter. These were often offset from nominal channel frequencies in a 2 KHz range. This permitted regulators to immediately identify the source of a possible interfering television transmitter. Indeed, many television 'channel 0' transmitters operating around 48 MHz were used by short wave listeners to monitor and identify international radio propagation.

1.5   3D television and images

Conventional television principles are based on the application of dimensional analysis to obtain time dependent 2 dimensional (2D) images from a single time dependent radio signal. 3D television pushes the envelope where time dependent 3D images are synthesized using a single time dependent radio signal. The synthesised extra dimension solution can be coded in a number of ways, alternate images synchronised with shutter glasses, alternate images using 90° polarisation switching or fast LCD blanking in 3D goggles.

While not television as such, one of the earliest documented renditions of 3D images was the haunted house effects of John Pepper dating from 1862, colloquially known as 'Peppers Ghost'.

3D television was conceived in the 1930s, and from the very beginning, 3D televisions were a dubious and under-supported technology that was mostly dismissed as a gimmick. Despite much-touted 3D Olympics, cable TV channel promotions and passionate endorsements from award-winning filmmakers, actual consumer endorsement and acceptance never eventuated.

Much of the 3D movie content available has been specifically filmed using twin cameras feeding complex encoders; however, some films have been converted to 3D using partial image shifting editing. Oddly, there have been no 3D films produced in black and white to date.

Indeed, the 2009 3D movie titled 'Avatar', which was arguably the best 3D film produced at the time, caused many viewers to enter a state of euphoria immediately after the movie ended.

On 16 April 2010, Samsung released their first 3D television system utilising time synchronised alternating images. These required glasses where the alternate lens was synchronously blanked with the composite 3D image. Oddly, at the same time; Samsung issued a warning to users that it may 'trigger epileptic fits or cause ailments ranging from altered vision and dizziness to nausea, cramps, convulsions and involuntary movements such as eye or muscle twitching' [35].

A later clarification by Samsung highlighted people's perception of 3D is largely dependent on the distance between a person's eyes, and the immense reality a 3D television can convey means that some people will suffer disorientation. Indeed, a small percentage of people are unable to perceive 3D images at all. It is certainly clear that 3D television has a profound effect on the human brains perception of reality.

In 2015, 3D was still present in most HD TVs, but advertisements had conspicuously stopped mentioning it. In 2016, the 3D feature has been quietly killed off by most major television and DVD/Blu-ray manufacturers. After years of trying to convince consumers that 3D was the next big thing, it's finally been consigned to the dustbin of history – for the moment.

The key issues to 3D television failure were defined broadly as:

1. Lack of content, lack of skilled movie producers, with the acknowledged exceptions of the movies 'Gravity' and 'Avatar';
2. The technology was considered gimmicky and quirky;
3. Eye strain and headaches, and given that some peoples inability to even perceive 3D; and

4. Considered a hassle, with limited viewing angles, lack of brightness and often weird glasses [36].

The IEEE reported in 2014 that manufacturers were now re-focussing development into UHD TV, and indeed the NHK Science and Technology Research Laboratories, Setagaya-ku, Tokyo, recognised UHD resolution and beyond as the burgeoning developments in 2011.

It's unlikely that the dream of 3D TV will ever die. Indeed, there are already people talking about how 8K video is what's needed to make 3D television work, so expect inevitable resurrection attempts to come in the future [37].

1.6 Slow scan television

A variant of television transmission, known as 'slow scan' was developed in the 1960's by amateur radio operators and operators of the early earth-orbiting spacecraft. While only a niche group of amateur radio operators use slow scan television, it is one of the peculiar variants of communication protocols available to enthusiast radio operators.

Slow scan television (SSTV) development leveraged the traditional fast scan television methods and those of Facsimile transmission, but with massive reductions to the scan rates and image resolution, it became possible to transmit images over a narrow voice channel with the 3 kHz bandwidth typically allowed by regulators on HF Single Side Band radios.

A typical monochrome image frame could be transmitted in 36 seconds with the 240-line resolution, while colour images required 72–188 seconds. Slow scan colour later leveraged the development of solid-state memory and the availability of surplus hard drives to store the 3 primary colour composite images in the mid-1970s. While the image rendition period is too slow to show motion, it is a hybrid development using analogue television transmission protocols, while generating a sequence of facsimile-like images.

The FCC reviewed the transmit bandwidths and legalised the use of SSTV for advanced-level amateur radio operators in 1968, with other telecommunications regulators around the world following suite.

Historic SSTV transmission protocols included AVT, Martin, HQ, Robot, Scottie, WinPixPro, JV Fax and AX480 with later PC applications developed included MMSSTV, RDFT/SCAMP and Multiscan.

Early slow scan receivers utilised the long persistence phosphor CRT screens found on storage oscilloscopes manufactured in the 1960s and 1970s, the conventional vidicon transmission tubes were actually re-purposed from surplus fast scan television cameras. In more recent years, personal computer and Android applications provide the protocol conversion and image freeze and storage, with the ability to use PAL and NTSC analogue cameras and more recently digital Internet Protocol 'IP' cameras. Slow Scan television images are extremely susceptible to interference and noise corruption of the image, despite many attempts at protocol enhancements [38].

Variants of SSTV implementations were used in the image transmissions from the 1960 VOSTOK spacecraft known as Seliger-Tral-D TV system, configured to produce 10 frames per second. It was developed by OKB MEI, A.F. Bogomolov [39].

More recently there is an online library of SSTV images captured from the International Space Station on 18 February 2017 [40].

1.7   Further reading

While any book is a snapshot in time, development and recognition in the television industry continues unabated. Given the scope of this chapter, some key people and achievements may have been inadvertently overlooked, not intentional. A great resource is the technical Emmy's website [26].

References

[1]   Branch M.L. *Biography of Henry Sutton*. Ballarat Heritage Services; 2017.
[2]   Withers W.B. *The history of Ballarat*. Ballarat: Ballarat Heritage Services; 1887.
[3]   Perskyi C. 'Television by means of electricity' in *International universal exhibition*. Paris: Bureau International des Expositions (BIE); 1900.
[4]   Varigny H.d. *La vision à distance*. Paris: L'Illustration; 1909. p. 451.
[5]   Glinsky A. *Theremin: ether music and espionage*. Urbana, IL: University of Illinois Press; 2000.
[6]   BBC Director. *Historic figures: John Logie Baird (1888–1946)*. [Film].
[7]   Nature. 'Current topics and events'. *Nature*. 1925, vol. 115, pp. 505–06.
[8]   *Baird. USA patent US1925554* [US Patent Office]. 1928.
[9]   IEEE History Staff. 2016. Available from http://ethw.org/John_Logie_Baird [Accessed 22 Aug 2016].
[10] Welle D. *Director, first photoelectric fax transmission, 17.10.1906* [film]. Germany: Deutsche Welle; 1906.
[11] Available from https://teslaresearch.jimdofree.com/invention-of-radio/
[12] Baker T.T. 'Wireless pictures and television' in *Wireless pictures and television*. London: Constable & Company; 1926. p. 28.
[13] Available from http://www.byhigh.org/history/farnsworth/philot1924.html
[14] IEEE History Center [online], 2017. Available from http://ethw.org/Oral-History:Vladimir_Zworykin [Accessed 22 Jan 2017].
[15] *The editors of encyclopædia britannica, David Sarnoff*, Televiaion Pioneer. London: Encyclopædia Britannica; 2016.
[16] Blanc M.L. *Etude sur la transmission électrique des impressions lumineuses*. Vol. 11. Paris: La Lumière; 1880. pp. 477–81.
[17] Burns R.W. *Television: an international history of the formative years*. London: IET; 1998 Jan 1. Available from https://digital-library.theiet.org/content/books/ht/pbht022e

[18] Perskiy K.D. 'The modern condition of a question on electrovision on dis-
      tance (televising), works of the first all-russia electrotechnical congress'. *SPb.*
      1901, vol. II, pp. 346–62.

[19] HOSPITALIER M.E. 'General rapporteur'. *General Rapporteur.* 1901.

[20] Staff N.Y.T. 'Sunday magazine'. New York, NY: New York Times. 1907.

[21] Bielby P. *Australian TV – the first 25 years.* Melbourne, Victoria: Nelson in
      association with Cinema Papers; 1981. p. 173.

[22] Luck P. *50 years of Australian television.* Sydney, NSW: New Holland
      Publishers; 2006. p. 15.

[23] IEEE Region 2 Trivia. Available from http://ewh.ieee.org/r2/johnstown/
      downloads/20090217_IEEE_JST_Trivia_Answers.pdf

[24] Fazano C.A. *Fazano.pro.br* [The Electron Age website]. 2022

[25] Burger D.E. *Eimac* [online]. 2013. Available from http://ethw.org/Eimac

[26] Available from https://theemmys.tv/tech/

[27] Mitchell J.P. 1978. 'Light emitting diode television screen'. University of
      Northern Iowa.

[28] 'SMPTE journal: publication of the society of motion picture and television
      engineers'. *SMPTE Journal.* 1984, vol. 93(12).

[29] Chen P.T. 'Coding, systems, and networking presentation'. *Multimedia
      Communications.* 2002.

[30] Pagarkar M.H., Shah C. 'MPEG-4'. *SE-IT VESIT Conference Proceedings*;
      2002.

[31] *IEEE milestone 1* [online]. Available from http://ethw.org/Milestones:
      Monochrome-Compatible_Electronic_Color_Television,_1946-1953

[32] *List of IEEE milestones* [online]. Available from http://ethw.org/Milestones:
      List_of_IEEE_Milestones

[33] *IEEE milestone 2* [online]. Available from http://ethw.org/Milestones:First_
      Transatlantic_Television_Signal_via_Satellite,_1962

[34] *IEEE milestones awarded for NHK's technical achievements* [online].
      Available from https://www.nhk.or.jp/strl/open2016/tenji/f4_e.html

[35] Journalist Staff. 'Samsung warns of dangers of 3D television' in *The tele-
      graph (UK)*; 2010 Apr 16. p. 1.

[36] Jager C. 'Why 3D TV failed (and why we don't care)'. *Lifehacker.* 2016, p. 1.

[37] Cass S. '3-D TV is officially dead (for now) and this is why it failed'. *IEEE
      Spectrum.* 2014, p. 1.

[38] ARRL. *ARRL handbook.* Newington, CT: ARRL; 2009. p. 9.

[39] *Unknown VOSTOK* [online]. Available from http://www.svengrahn.pp.se/
      radioind/mirradio/mirradio.htm

[40] Club I.F. *ISS fan club* [online]. 2017 Feb 18. Available from https://www.
      issfanclub.com/image

Chapter 2

Fundamentals of human vision

Mei Ying Boon[1]

2.1 Introduction

When we open our eyes, we immediately start seeing. How do we see? How is that we are also able to immerse ourselves in other worlds through the medium of television? This chapter aims to cover the basic anatomy of the eyes and visual system and how they work to produce visual perception. This chapter will then discuss how visual system capability and image production technologies may interact.

The human visual system allows us to see the world around us in colour during the day time, in grey scale at night time, in the harsh sunlight of a summer day and in the muted tones of a half-moon. We can see depth, moving objects, near objects and far objects, recognise faces and read words, numbers and symbols. We can also have similar visual experiences while viewing a device such as a television from a fixed distance while indoors.

We are still learning about how the visual system works. The earliest artists understood that an outline could indicate form (figure-background) [1] and then later artists found that perspective could indicate depth [2], that adjacent colours can result in different hue perceptions [3] and that successive still images can provide the illusion of moving images [4]. This chapter provides a brief review of how the visual system works covering the basic ocular anatomy and physiology that underlies the abilities of our visual system. The relevance of each aspect of the visual system and how it interacts with television images during television viewing will then be described.

2.2 Anatomy of the eye

The eyes are the key human sensory interface with the world, which allows the detection of objects in the environment that may be a great distance away. This differs from the other human senses such as smell, taste and hearing, which have shorter detection ranges. The visual system is stimulated by visible light, comprising photons of electromagnetic energy from the visible range of the electromagnetic spectrum (400–760 nm). The eyes focus this light onto the retinal photoreceptors, which then capture this energy, and process these signals, before

[1]Optometry, University of Canberra, Australia

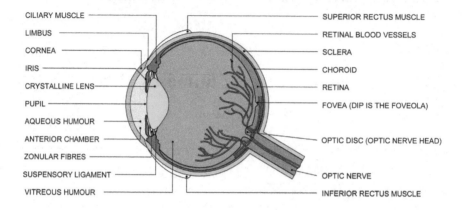

CILIARY MUSCLE
LIMBUS
CORNEA
IRIS
CRYSTALLINE LENS
PUPIL
AQUEOUS HUMOUR
ANTERIOR CHAMBER
ZONULAR FIBRES
SUSPENSORY LIGAMENT
VITREOUS HUMOUR

SUPERIOR RECTUS MUSCLE
RETINAL BLOOD VESSELS
SCLERA
CHOROID
RETINA
FOVEA (DIP IS THE FOVEOLA)
OPTIC DISC (OPTIC NERVE HEAD)
OPTIC NERVE
INFERIOR RECTUS MUSCLE

Figure 2.1   Cross-sectional diagram of a human eye

sending this information onwards to the lateral geniculate nucleus (LGN) and there-after the visual brain for further processing.

The electromagnetic spectrum that interacts with the eyes, termed optical radia-tion, is not limited to visible light but also includes ultraviolet (100–400 nm) and infrared radiations (760 to above 10 000 nm) [5]. Our eyes are vulnerable to hazard-ous doses from both visible light and invisible optical radiations. High-energy vis-ible light elicits pupil constriction, eye movement and eye shutting as reflex actions that act to rapidly reduce the dose and hazard [6]. The eyes and visual system do not have such strong mechanisms to reduce the dose of invisible radiations, but at lim-ited doses, the eye has the capacity to recover from damage by the optical radiation through a process of homeostasis. In common with the rest of the human body, the eye has components that are designed to maintain the healthy function of the eye. The eye also has other components that are dedicated to the focussing of light infor-mation (optical components) and the capture and processing of this light information (sensory neural components).

Figure 2.1 shows a cross section of the human eye showing the main structures involved with maintaining health, focusing of light and capturing light.

2.2.1   Key components of the eye responsible for maintaining health of the eye

The eyes are separated from the external world by the cornea and the sclera, which together form the structural envelope that protects the internal structures of the eye. The cornea and sclera meet at the limbus, which has cells that contribute to the nour-ishment and wound healing of the cornea. The strength of these structures is mainly due to the arrangement and properties of collagen fibrils. In the case of the cornea, the fibrils are structured as regular layers that allow light to be transmitted so that the cornea is essentially transparent. In the case of the sclera, the collagen is laid down more irregularly resulting in increased tensile strength compared with the cornea at the expense of transparency [7].

The ciliary body is a ring of tissue 5–6 mm wide. The posterior portion contains the ciliary processes that produce aqueous humour. Aqueous humour is a transparent fluid that provides nourishment to the eye and removes wastes from those structures in the eye with which it comes into contact. Aqueous humour circulates through the eye before exiting by the outflow pathways of the trabecular meshwork, which is located behind the limbus.

The blood circulatory system within the eye provides oxygen and nourishment and also removes wastes from the eye. It is only present in those ocular structures for which transparency is less important. This includes the iris, ciliary body, peripheral retina and choroid. The other transparent structures rely on diffusion of oxygen and nutrition from these structures or from the circulation of the aqueous humour.

The vitreous is the largest structure in the eye. It is a transparent viscoelastic gel that comprises collagen fibres, which are very fine in diameter, and water. It is mostly spherical in shape with an indentation anteriorly to make room for the crystalline lens.

The retinal pigmented epithelium forms a layer of cells that are located between the retinal photoreceptors and the choroid. They provide structural and metabolic support to the activities of the photoreceptors, clearing away by-products from the biochemical reactions that involve the capture of photons.

2.2.2   Key optical imaging components of the eye

The healthy human adult eye is approximately 23.8 mm long from its anterior to posterior poles [8], with some variation in length accounting for refractive errors (myopic eyes are longer than usual and hyperopic eyes are shorter than usual). At the posterior pole is a region of the retina called the foveola, which is structured to resolve fine spatial variations in spectral energy across the retina. Thus, for distantly viewed objects to be focussed on the retina, the eye must refract light to focus onto the retina.

The cornea is the first refracting surface of the eye and has the form of a meniscus lens, with an anterior convex surface and a posterior concave surface. The anterior surface of the cornea is covered by a thin tear film layer. The cornea contains toricities and other asphericities that result in some astigmatism (cylindrical power) and higher order aberrations, respectively [9]; these distort the incoming light in predictable ways to produce images that are clearer in some meridians and not others. The cornea transmits all visible light and infrared in the range of 760–3 000 nm. The cornea absorbs ultraviolet radiation below 295 nm but transmits ultraviolet in the 295–380 nm range.

The light then passes through the anterior chamber that is filled with circulating aqueous humour. The light that falls on the iris is obstructed from reaching the retina, whereas the light that falls within the pupil, the hole formed at the centre of the iris, is transmitted into the posterior chamber, the region between the iris/pupil plane and the retina. The iris controls incoming light levels falling on the retina to maintain sufficient dynamic range within the retinal photoreceptors to capture spatial light information. In darker environments, the iris dilates the pupil to allow sufficient light in to stimulate the photoreceptors. In bright environments, the iris constricts the pupil to limit incoming light, including straylight [10], so that the photoreceptors are not overwhelmed, and to improve focus [11] by removing paraxial light rays that are more likely to be aberrated.

The next refracting structure within the eye is the crystalline lens. In children and young adult eyes, the crystalline lens can accommodate, a process by which the lens becomes more positive in power to allow light to be imaged clearly from close objects. During accommodation, the anterior portion of the ciliary body (see Figure 2.1) contracts, which releases tension on the lens zonules, thereby allowing the crystalline lens to take on a more rounded shape. The net effect of this action is that the crystalline lens increases in equivalent positive power to allow the incoming light rays from near objects to be focussed onto the retina. Thus, accommodation of the crystalline lens facilitates clear vision across a range of viewing distances.

The crystalline lens continues to grow throughout life, laying down new collagen fibrils around an embryonic nucleus [12]. Along with this increase in crystalline lens thickness with increasing age, its overall refractive index and power decrease (from 24.5 to 23 D on average) [8] and lens stiffness increases [13]. A consequence of these age-related changes is that the ability to accommodate gradually decreases, leading to the phenomenon called presbyopia. The main symptom of presbyopia is an inability to read at near without additional reading glasses to compensate for the deficit in positive power by the ageing crystalline lens.

Another consequence is that the crystalline lens may change in transmission characteristics with increasing age. In the young adult eye, total transmission by the crystalline lens is typically 95% and work by Artigas *et al.* showed that may decrease by 20% with each decade following [14]. Further, the transmission of short wavelengths (blue light for people with normal colour vision) was found to decrease at a more rapid rate with age than the other wavelengths. Therefore, older people tend to demonstrate blue-yellow colour vision deficiencies due to the relative change in spectral transmission properties of the crystalline lens with age. Ultraviolet light is nearly completely filtered at most ages, except for wavelengths at around 320 nm where there is approximately 8% transmittance, which appears to decrease with increasing age [14].

Light that has been focussed by the crystalline lens is then transmitted through the vitreous onto the retina. The focussed light is projected onto the retina, as a movie may be projected onto a cinema screen with a few differences. In contrast with the cinema, the image that falls on the retina is inverted, and the retina is not flat or passive. Instead, the retina is curved as it lines the inside of the approximately spherical eye, and the retina actively captures and processes the pattern of the incident light.

2.2.3   *Key sensory neural components of the eye*

The retina is located between the vitreous and the choroid, and its main sensory cells are the cone and rod photoreceptors (see Figure 2.2).

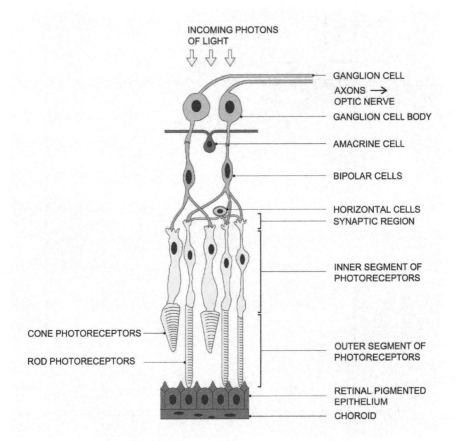

INCOMING PHOTONS
OF LIGHT

GANGLION CELL
AXONS →
OPTIC NERVE

GANGLION CELL BODY

AMACRINE CELL

BIPOLAR CELLS

HORIZONTAL CELLS
SYNAPTIC REGION

INNER SEGMENT OF
PHOTORECEPTORS

CONE PHOTORECEPTORS

ROD PHOTORECEPTORS

OUTER SEGMENT OF
PHOTORECEPTORS

RETINAL PIGMENTED
EPITHELIUM
CHOROID

Figure 2.2   Rod and cone photoreceptors within the retina

The normal healthy eye has three classes of cone photoreceptors, long (L-), medium (M-) and short (S-) wavelength sensitive, and one class of rod photoreceptor. They each contain different photopigments that are erythrolabe, chlorolabe, cyanolabe and rhodopsin, respectively. The slight differences in the photopigment molecules of the photoreceptor classes mean that they interact slightly differently with incoming electromagnetic energy. Short-wavelength sensitive cones have their peak absorption at wavelengths of about 440 nm, M-cones at 530 nm, L-cones at 560 nm [15] and rod photoreceptors at 495 nm [16]. The responsiveness of each cone class to light energy of different wavelengths is described by the spectral sensitivity curve (see Figure 2.3).

It can be seen that the spectral sensitivity functions for the three cone classes overlap, which is significant as it means that for many wavelengths of light, the retina will produce a response in more than one cone class. The principle of univariance applies to each photoreceptor class such that each can only transmit information

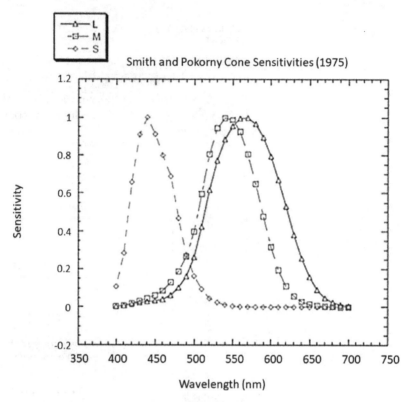

Figure 2.3   Spectral sensitivity functions of cone photoreceptors

about the quanta of photons it absorbs, and each photoreceptor on its own cannot transmit information about the wavelength of the absorbed light energy [17].

Cone photoreceptors are active under bright conditions (termed photopic conditions, typically > 3 cd/m²), whereas rod photoreceptors are active under dim conditions (termed scotopic conditions, typically < 0.01 cd/m²). Both cone and rod photoreceptors are active under intermediate light levels (mesopic conditions, e.g. full moon or early twilight). On absorption of photons, chromophores are activated and undergo photo-isomeric change. It takes time for the photo-excited pigments to recover back into their inactivated forms, to be able to be re-excited by photons. This delay partially explains why when moving initially from a bright environment to a dark environment, nothing is visible until the cone photoreceptors recover function (reaching their maximal sensitivity at about 7 minutes) and then rod photoreceptors recover function so further increase sensitivity (in fact, account for all improvements after about 10 minutes in the dark) to resume signalling the presence of spatial variations in light in the low-light conditions. Maximum rod photoreceptor sensitivity is achieved following 40 minutes in the dark. Neural processing to adjust the sensitivity of the ability to detect spatial variation in light levels explains the remaining improvements in function when moving from lighter to darker environments (dark adaptation) and when moving from darker to lighter environments (light adaptation) [18].

If there is only one photoreceptor class present or active in the retina, as occurs at night time with only the rod photoreceptors active, only brightness comparisons can be made and a monochromatic perception will result. The presence of two cone class types allows the comparison of relative cone excitation values by later neural circuitry within the visual system. If there are three cone types, there is the potential to see colours that are described by three relative cone excitation values. Some people are born with fewer than three cone classes, which can result in colour vision that is different from normal. Therefore, the same light can be perceived to be different in appearance depending on whether one cone class, two cone classes, three cone classes, rods only, rods and combinations of cones are collecting the light information. Under mesopic conditions, rods are thought to desaturate the sensation of colour [17].

The photoreceptors signal their level of excitation to bipolar cells within the retina, which in turn transmit signals to the retinal ganglion cells. The retina also contains horizontal and amacrine inhibitory cells that link adjacent cells and suppress excitation. These cells allow the retina to detect and code information about the spatial variations in quantal catch levels and characteristics as contrast. Ganglion cells are organised to collect information from the combined activity of the bipolar and horizontal cells over a certain area or receptive field. One way in which information is processed is in a centre–surround configuration. In such a configuration, the information is collated by the ganglion cell comparing the information from a central region with information from its immediate surround. It is only when the quality of the excitation of the central region is different from the surround, i.e. contrast is present within the centre/ surround receptive field, that the ganglion cell becomes excited. Excitation is indicated by increased electrophysiological activity. If the excitation of the centre and surround is identical, indicating uniformity of light quality within the centre/surround receptive field, then the ganglion cells remain at baseline levels of activity (see Figure 2.4).

Different ganglion cell types make different kinds of contrast comparisons and over different receptive field sizes and regions of the retina. For example, the retinal ganglion cells may send information about contrast between light and dark (luminance contrast), L- vs M-cone excitation (red-green opponent colour contrast) and S- vs (L- + M-) cone excitation (blue-yellow opponent colour contrast) across the retina to the brain through their axons. The axons of the ganglion cells make up the optic nerve when they exit the eye (Figure 2.1). The brain then builds a picture of the retinal image based on these different kinds of contrast samplers. One consequence of this opponent organisation is that objects cannot be light-and-dark, bluish-and-yellowish or reddish-and-greenish at the same time for people with normal trichromatic colour perception as they comprise opposite pairings. In contrast, people who are born with deficient colour vision due to the absence of one or more photopigment types will make typical colour confusions, e.g. perceiving objects that people with trichromatic vision would describe as red or green as being the same in colour [19, 20].

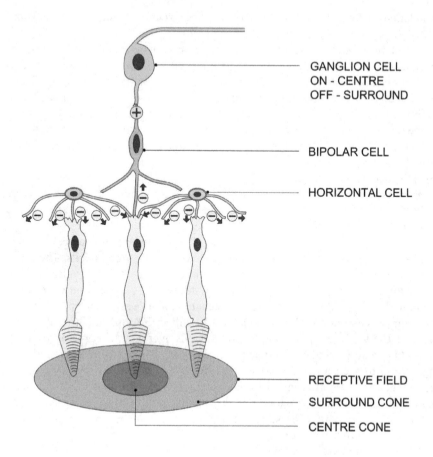

GANGLION CELL
ON - CENTRE
OFF - SURROUND

BIPOLAR CELL

HORIZONTAL CELL

RECEPTIVE FIELD

SURROUND CONE

CENTRE CONE

Figure 2.4   If the centre and surround have the same illumination, then the on-centre off-surround ganglion cell is not excited

The photoreceptors vary in their distribution and density across the retina, meaning that the building blocks for perception also differ across the retina (Figure 2.5). The cone photoreceptors are most densely packed at the foveola, and the neural circuitry is organised in a way that allows for the detection of contrast between single photoreceptors. This means that at the foveola, the limit for spatial acuity is the width of a single cone photoreceptor. To detect a dark stimulus between two lighter stimuli (e.g. a black dot on a white page), a ganglion cell that samples from the foveola must compare information from three adjacent cone photoreceptors, and it must have light falling on the two outer photoreceptors and no light falling on the middle photoreceptor. This kind of sampling is performed by the parvocellular ganglion cells and requires high light levels to ensure that sufficient light reaches the individual photoreceptors. This explains the high spatial acuity of the foveola. With increasing eccentricity from the foveola, the ganglion cells sample over higher numbers of photoreceptors so the best achievable spatial acuity

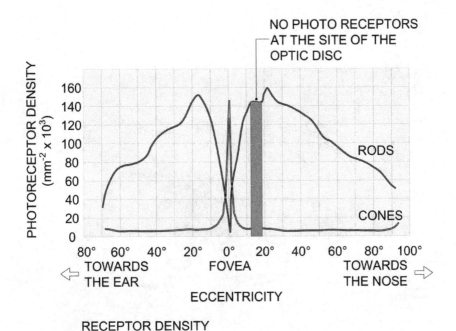

RECEPTOR DENSITY

Figure 2.5   Photoreceptor density varies across the retina. Note that the cone density (green line on the plot) is maximum, whereas there are no rod photoreceptors (red line in the diagram), at the fovea.

decreases. At 3° eccentricity, the best acuity is typically halved. The average human adult can resolve a stimulus of 1 minute of arc at the foveola but only 2 minutes of arc at 3° eccentricity.

The rod photoreceptors are absent from the foveola and are more numerous in the non-central retina, particularly the peripheral retina. Many rod photoreceptors may converge to a single ganglion cell with a large receptive field. This means that the quantal catch zone is very large. This kind of sampling is performed by magnocellular ganglion cells and does not require high light levels to work as the many rod photoreceptors that feed into the ganglion cell cause the magnocellular ganglion cells to be more efficient at capturing a threshold number of light photons than parvocellular ganglion cells. The trade-off is that spatial acuity is poorer in human peripheral vision during dim lighting conditions.

2.2.4   Post-receptoral visual processing

Different ganglion cell types make different kinds of contrast comparisons and over different receptive field sizes and regions. Therefore, the retinal ganglion cell population sends information about contrast between light and dark (luminance contrast, parvocellular and magnocellular systems), L- vs M-cone excitation (red-green

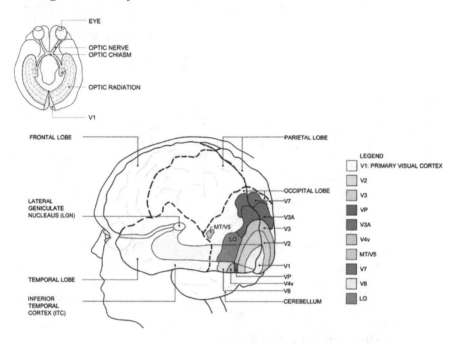

Figure 2.6   Visual pathway

colour contrast, parvocellular system [21]) and S- vs L- + M-cone excitation (blue-yellow colour contrast, koniocellular system [22]) across the retina to the brain through their axons about the same regions of the retina. These axons leave the eye at the optic nerve head organised retinotopically (in a way that maintains the spatial relationships of the sources of their inputs on the retina) and transmit information to the LGN and the brain for further processing. The neurons within the brain then build pictures of the retinal images based on the signals from these different kinds of retinotopically processed contrast samplers as they vary over time. At the optic nerve head, the exit point of the ganglion cell axons from the eye, there are no photoreceptors, so there is a small region of blindness in each eye located about 12° temporal (towards the ears) from fixation within the visual field (Figure 2.5).

Ganglion cells first send information to the magnocellular and parvocellular layers and koniocellular blobs at the LGN. The LGN sends information to V1, located within the occipital cortex of the brain (Figure 2.6).

V1 contains information about the spatial origin of the signals and further processes orientations and directions. Because humans have two eyes that are separated in space, the two eyes will have slightly different views of the same external world and the differences in the projected location in the visual field are termed disparity (measured as an angle). The retina, combined with the anatomical spatial configurations of the eyes within the head and the dimensions of the face, allows vision in a field that extends approximately 60° upwards, 75° downwards, 60° medially (towards the nose) and 100° laterally (towards the ears) [23] for each eye when it

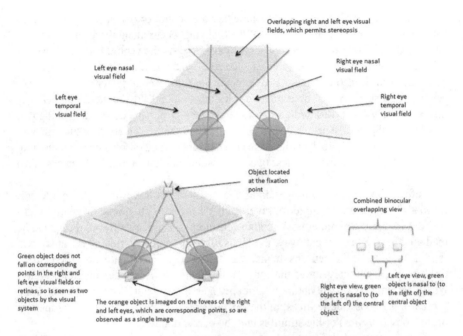

Figure 2.7   *Nasal, temporal and overlapping visual fields. Overlapping visual fields serve the perception of stereopsis and depth through disparity.*

is stationary. The extraocular muscles allow the eyes to rotate towards each other (converge) to look at near objects and to rotate away from each other (diverge) to look at more distant objects (Figure 2.7).

For the central and medial visual fields, there is an overlap of the visual fields where the disparity in the views from each eye allows for an additional layer of processing that utilises this binocular information to allow the monocular visual information from the right and left eyes to be merged to form an expanded binocular visual field with a central region of three-dimensional (3D) depth perception. Objects that fall in both the right and left eye visual fields and on corresponding points (i.e. the same location) on the retinotopically organised visual systems are projected to binocular cells in the brain, which underlie stereo perception, and seen as single objects. Hence, when the two eyes are fixated on the same object, the object and objects slightly in front and behind will be fused by the visual system into a single image. However, objects greatly in front of or behind the plane of the object will fall on non-corresponding points on the right and left eye visual fields (and retinas), so will be observed as existing in two locations in the visual field simultaneously, which is called physiological diplopia [24]. The lower panel in Figure 2.6 demonstrates what happens when a person fixates on a pink object. The green object in front will appear in the nasal field of both the right and left eyes and will therefore appear twice in the visual field.

Information at V1 is then sent along either the ventral or dorsal (occipito-parietal) pathways. The connections and areas of the brain that occur along these pathways suggest broadly different functions for the pathways [25]. As the ventral pathway traverses V1, V2, V4 and the inferior temporal (IT) cortex, it is thought to process object quality, including what is stable in the visual scene or those perceptual features that are characteristic and stable of that object that enables recognition [25]. As the dorsal pathway traverses V1, V2 and the middle temporal and posterior parietal cortex, it is thought to process visually guided action, spatial orientation and navigation and working memory and those aspects that are dynamic or unstable about the relationships between the multiple objects [25]. Historically, these two broadly different functions have been described as the 'what' and 'where' pathways [25].

Information travels rapidly, taking about 100 ms to travel from the eye to V1. Similar timelines are required to travel from V1 through the dorsal or ventral pathways. A mystery still remains: Although conscious visual perception is thought to be mediated by the ventral pathway, the locus of visual perception is as yet unknown. Is it the activity of the neurons themselves or a soul or spirit that makes the perception? Fechner, who developed the concept of psychophysics, understood that even if we do not understand the nature of perception as it is a quality of the spirit or mind (*psycho*), it is possible to measure the relationship between the physical properties of the body (*physics*) or the stimulus and the perceptual response [26].

Coordination of timing of events is important for our visual perception. When an object moves in the environment, or when our eyes move across a view, the image of objects shifts on the retina in accordance with their motion in the external world. We need to be able to sample the visual scene frequently enough so that we do not miss noting shifts in locations of moving objects as a function of time. One measure of the perceptual temporal resolution ability of the visual system is the critical fusion frequency rate, which is presented in cycles per second or Hz. The critical fusion rate is the rate at which a stimulus presented on and off in a cyclic manner first ceases to appear intermittently but appears as a steady stimulus. This rate varies according to light levels and the location on the retina that is stimulated, as the photoreceptor types, density and connections and their subsequent processing by the visual system differ across the retina. In general, this is lower for scotopic levels of vision, about 12 Hz, for vision mediated only by rod photoreceptors and higher for photopic levels. The critical fusion frequency increases for increasing levels of retinal illuminance and for increasing spatial frequency of the components, so estimates have ranged from 45 [27] to 90 [28] Hz for photopic levels. This characteristic is particularly important when watching pictures on a television screen, as the television images are consecutive frames made up of pixels of light that are updated at a temporal frequency called the refresh rate, for which the pixels are changed in light levels and colour over time. Where parts of the picture are not supposed to change over time, in the real world, we would expect to see a steady image free of flicker; therefore, the images presented on televisions should be presented at a refresh rate that provides flicker-free viewing to maintain realism.

Our visual system is also limited in its ability to perceive fast motion, which is thought to be limited by the rate at which ganglion cells can transmit information about the brief and rapidly oscillating light stimulation, slow motion (about 20° per second), and to resolve changes in space and location [29]. For example, it is only when we view slow motion footage that we can see the detailed flapping of the wings of a bird rather than just observing a smeared image of the wings in motion.

2.3 Using vision to watch television

Colour television displays should produce images that facilitate comfortable trichromatic and detailed vision, adequate variation in luminance output for viewing in bright and dim environments, with smooth representation of motion and stability of the image. For colour vision, the television screen must emit photopic light levels sufficient to stimulate the cone photoreceptor system and suppress the rod photoreceptor systems to maintain maximum saturation of the colour images.

2.3.1 Perceiving detail and the visual extent

Television viewers must be able to perceive sufficient detail (contrasts over small distances) for the images to appear continuous, as they do in the real world. As explained earlier, the adult with average visual acuity is able to resolve objects of 1 minute of arc in visual angle, so will be able to perceive individual pixels if they are at least 1 minute of arc in visual angle in size. As most objects on the screen are likely to be larger than one pixel in size, and each pixel is made up of subpixels, this means that most objects depicted on the screen should be composed of multiple pixels, and we would not be able to perceive the subpixels within those pixels. However, it is possible that for some images, we would be aware of the individual pixels, although not the subpixels. If pixels are observable, either the viewing distance must be increased, to reduce the visual angle of each pixel, or the pixel size must be reduced and pixel density correspondingly increased.

There is a relationship between the original object or object motion that is being sampled and the sampling rate of the recording and playback devices in order to accurately capture the spatial and temporal frequency and morphology of the object and its motion accurately. To accurately measure or depict a frequency, the original frequency must be sampled at a rate at least two times that frequency: called the Nyquist frequency. Otherwise, an 'aliased' form will be recorded. However, this does not cover all situations and when sampling the natural world, as there are variable spatial and temporal frequencies within one scene and it is not possible to know in advance the highest spatial frequency that we will be recording [30]. This is because the spatial frequency of objects will vary with distance from the recording device. For example, recordings of people wearing stripy clothes recorded at multiple viewing distances and levels of zoom may appear aliased for some combinations. There is more control when considering computer-generated images (CGI). Therefore, while it may not be possible to be able to avoid aliasing for all situations from the recording point of view and from the playback viewpoint, if the pixel size is sufficiently small to capture the morphology of the recorded images, and the smallest details presented are at least smaller than 1 minute of arc, for the majority of the population, this should result in sufficient detail for *naturalistic* viewing. This even holds for microscopic objects; generally, we enlarge microscopic objects to allow us to see what is usually invisible with our natural visual system, e.g. views of pores on the skin or the biological structures within single cell bodies. Images of those objects are magnified through optics to be visible with our human visual systems. Most people prefer images presented on television screens to be naturalistic and realistic. This preference for a naturalistic and realistic depiction of the visual world starts at a young age [31–33]. However, naturalistic viewing is subject to preferences in image presentation, which appear to be related to the visual memory of colours [34]. For example, some people may prefer the blue of the sky to appear one way, and other people another way, based on their memory of how skies should look.

Field of view is still important for the television viewer. Not only must there be a sufficient density of pixels, but this must also be related to the overall size of the television image as the television images will be presented to fill the size of the television screen. The human visual field is limited to a vertical extent of 135° vertically and 200° horizontally; however, much of that visual field has poor resolution acuity and may be reduced further due to the facial landscape blocking aspects of the visual field. For example, a prominent forehead and large nose can reduce the superior (upward) and nasal (towards the nose) monocular visual fields respectively. The central visual field, being the central 30°, is typically what we would use for viewing objects of interest within the larger visual environment. Within this central field, the best acuity is located at the point of fixation. Therefore, we might consider that when watching a television, the height should fall within the central 30° of the visual field in order to be viewed with our central vision, which will also minimise the need to make shifts in fixation while watching television. However, some viewers may like to look around within the visual scene, to feel fully immersed, so one recommendation is that immersion may require the television to at least subtend a 40° viewing angle. Some viewers prefer even larger television screens, so it appears that personal preference as to level of immersion [35] and viewing comfort [36] are key.

2.3.2   Perceiving colours and light and dark

The image of a television is made up of pixels, which are themselves made up of subpixels that usually emit light that is typically red, green or blue light. Some televisions have additional subpixels which emit other colours, such as cyan or white. How the image is composed spectrally is less important than its effect on the cone photoreceptors. This is because the cone photoreceptors do not detect wavelength, only quantal catch, so it is possible for lights reflected from objects to be spectrally different in composition yet produce the same relative excitation of the cone photoreceptor classes and appear to have the same hue. To achieve trichromatic vision from television screens, a minimum of three stimuli types are required to differentially affect the three cone photoreceptor types and to activate the colour opponent systems, i.e. red vs green and blue vs yellow (red + green).

These stimuli may be produced through additive or subtractive colour mixing. In additive colour mixing, light sources of different wavelengths are mixed together to produce stimuli suitable for trichromatic vision. The use of spectrally red, green and blue light wavelengths is sufficient to produce red, green, blue and yellow stimuli as yellow may be produced by the additive mixture of spectrally red and green wavelengths. In subtractive colour mixing, pigments or dyes are used to partially subtract wavelengths from incident white light (which is composed of all wavelengths) so that the reflected wavelengths remaining produce the stimuli for trichromatic vision. In the subtractive colour mixture, the best primary colours are cyan, magenta and yellow. Cyan is a bluish-green, and magenta is a purplish-red. These primaries are useful because post-receptoral colour processing is tuned to slightly different colours (i.e. magenta and cyan) [37–39] compared to cone photo-receptor sensitivities.

Achromatic stimuli (greys, white and black) may also be produced using additive and subtractive colour mixture of the primary colours. However, the best achromatic vision is served by having the greatest range (i.e. the darkest blacks and the lightest whites). The greater the difference in luminance between the black and the white in the display, the greater the achromatic contrast is presented to the visual systems by the display. Recent work has indicated that the visual system does not necessarily produce luminance contrast opponency from the same L- and M-cones that produce colour contrast opponency, but that different individual L- and M-cones may be recruited for a parallel parvocellular system [21].

The colour gamut of present screen display technology is limited compared with the natural gamut of our visual system; however, this does not appear to impact on the enjoyment of televised images. The way images are presented on television currently is in part due to an economical approach towards the design of displays, utilising the minimal configuration of subpixels to correspond with the minimal number of stimulus types, and due to historical capabilities of recording and playback devices. Now that technology has improved, that same economy may be holding back the potential of screen display. Newer television displays are able to improve on the original image data from the recorded image signal, but there is also a commercial need to be able to display images that were recorded in more popular and/or older formats. Improvements on the image data can include post-processing and hardware initiatives such as increasing pixel density, improving colour purity of pixels and interpolating between encoded pixel values in the recorded image. Because in the real world, our visual system must make judgements about colours under different prevailing light conditions [40, 41], this skill may have the added benefit of facilitating our acceptance of colours used in televised images [34].

One consequence of the human visual system seeing a much higher dynamic range of luminances than can be presented on most VDT displays is that recording devices, such as cameras, will need to process the raw light pattern information and compress the range of luminances to within the screen's capabilities. This step is called 'gamma encoding', which facilitates playback on a 'gamma corrected' device (e.g. Reference 42). It allows luminance information of the visual world to be encoded by one device and decoded by the display device, so that the relationship between luminance levels of the external world and the displayed visual scene is approximately maintained. The number of levels of luminance within the presented display range can impact the accuracy of the luminance relationships presented, which affects the realism of the images presented. One reason why this further manipulation of light levels by video recording and playback by devices such as televisions may be accepted by our visual systems is that our visual systems are already designed with some ability to adapt to the prevailing range of light levels. As described earlier, our visual system carries out neural and optical manipulation of incoming light information so that we are able to maximise our ability to detect contrast as evidenced by the processes of light and dark adaptation and pupil size changes with light-level changes, respectively. In theory, the processes of camera image capture, image processing followed by the television display is assisting the visual system in this process.

2.3.3   Temporal aspects of viewing

Consider the scenarios of self-motion or the making of large amplitude rapid eye movements (called saccades); we might expect to perceive blurred images due to the images being smeared across the retina with motion. We might also expect that our vision should black out (or at least dramatically reduce in luminance) every time we blink due to the obstruction of incoming light by the eyelids; however, the human visual system does some pre-processing of the retinal signal for us so that we do not notice these disruptions. Our visual system suppresses (i.e. reduces the strength of the perception) the incoming information just prior to (about 75 ms) and for the duration of the blink or the saccade [43, 44]. Prior to making the saccade, the activity of V1 in the visual cortex is greatly reduced so that incoming retinal signals from the retina during the saccade are suppressed [44]. Eye closure is also associated with altered brain activity, including increased alpha waves and reduced primary visual cortex V1 activity [45].

Our visual system also pre-processes other aspects of vision. For example, during suppression of the retinal signal during saccades, it has been found that we are unable to accurately judge object motion. However, when there is missing information, our visual system has the ability to extrapolate the most likely motion [46], thereby maintaining the stability of our visual percept. Post-saccade, the visual system maintains visual stability by comparing the pre-saccadic and post-saccadic properties of the target object near the landing saccade position. If they are found to be similar, visual stability is maintained. If they are found to be dissimilar in terms of expected motion or timing or location, as might be the case for CGI, the difference is noted and the pre- and post-saccadic targets are treated as separate and different objects [47].

In fact, perception of complex CGI presentations, such as for 3D television viewing, where slightly different non-natural images are presented to the right and left eyes with in-built disparity, appears to be different from the perception of naturalistic images, both in the real world and on television screens. For naturalistic images, where successive frames in video images are similar with very small changes in luminance between frames, flicker is not generally perceived for frame rates such as 60 Hz. Naturalistic images presented on newer high-resolution wide-screen television screen display technologies now permit very high spatial frequency stimuli to be presented to the peripheral visual field, and critical flicker fusion frequency is reportedly higher, approximately 90 Hz [28]. When considering non-natural images (such as in CGI), the presentation of different images to each eye, micro-saccades [48] may become more important such that flicker artefacts are still perceptible at 500 Hz refresh screen rates [49]. Further, the disparity between the right and left images will induce changes in convergence and divergence in the viewer in order to maintain single binocular vision, and if the images are not as expected as in the natural world, this may bring symptoms of discomfort [24]. This shows that it is important to test the setup for novel screen display technologies and dimensions as not all aspects of the visual system are fully understood [49, 50].

2.3.4   Attention influences the percept

Another aspect of television viewing is attention. It is the art of television content providers to produce interesting images for viewing. When viewing television images, not all aspects of the scene are equally perceived; most viewers prefer to fixate and attend to the centre of the television screen unless something more salient is happening elsewhere on the screen. Saliency describes the attention bias that objects in a visual scene may possess [51], which tend to be living objects, faces (eyes, nose, mouth), cars and text, towards the centre of the screen and objects that may be interacting (i.e. action-oriented). When attention is directed towards one aspect, perception of other objects in the visual system may be greatly reduced [52]. Earlier in the chapter, we also noted that there is a progressive drop-off in visual acuity (ability to see fine spatial detail) with increasing eccentricity from fixation, yet most of the time we do not perceive this to be true (unless tested under highly controlled laboratory conditions). Within the natural world, many of us perceive the peripheral world to be as vivid and present as our central vision [53]. Might this be due to a similar mechanism that allows us to fill in perception from the physiological blind spot? A recent theory is that filling in the details, when there is insufficient information, may be explained by predictive coding of the visual system for natural systems [54]. That is, the visual system does not only generate percepts from incoming retinal signals, but it also uses prior knowledge of typical characteristics of the behaviour of the natural world to generate a percept when there is missing information. This is thought to be possible as the visual system has evolved to perceive the natural world and both feedback and feedforward connections have been found between and within the retina and visual system [54]. Others have suggested a different concept of 'visual inertia' or 'visual momentum' where there is continuity with the known details according to the laws of physics [46].

2.3.5   Visual problems and television viewing

2.3.5.1   Presbyopia

For eye care professionals, the most common complaints are eye strain when viewing visual display terminals (VDTs) such as on personal computers or smartphones, rather than from television [55]. Therefore, most of the work is on the visual comfort of viewing VDTs rather than televisions; however, some of what we know from VDT viewing is applicable to television viewing. The main difference between VDT and television viewing is the distance at which the screens are viewed. Televisions are typically viewed at longer viewing distances (e.g. 1.5–6 m, mean 2.9 m [56]) than personal computer (e.g. 50–81 cm, mean 65 cm [57]) or smartphone displays (on average 32–36 cm [58]). Generally, preferred viewing distances are shorter for smaller viewing details, lower luminance and smaller screen sizes [55–57].

For computer users, discomfort has been attributed to the nature of the task, i.e. sustained concentration on work-related matters, sustained accommodation and convergence, problematic workstation ergonomics and the use of multiple screens or documents. All of the above are less important for watching television because television viewing is usually for entertainment rather than work. There is usually less need to adjust accommodation between several working distances while watching television, and the working distances are longer than for computer viewing, requiring less accommodation. Accommodation required is equivalent to 1/viewing distance in metres. Therefore, at an average viewing distance of 2.9 m, the eyes must accommodate by about 0.3 dioptres but at an average computer working distance of 60 cm, the eyes must exert about 6x more effort to accommodate by about 1.7 dioptres [55]. Further, the long viewing distances for television viewing mean that the ageing condition of presbyopia is less of a problem for viewing television than computers for viewers over 45 years, as most people will still have sufficient residual accommodation to view the television with their customary distance optical correction. At the closer working distances of computers and smartphones, viewers will need external lenses to help them see at closer distances, to make up for the lack in accommodating ability of the crystalline lens [55]. Such lenses may be multifocal (lenses that have two or more focal lengths within the lens) or progressive lenses (which have a progression of power to allow clear focus from distance to near viewing distances). Both multifocal and progressive lens designs result in trade-offs in the field of view for the convenience of having multiple focal lengths.

2.3.5.2   Low vision

Viewing distance and visual angle may also differ if a person's visual acuity (ability to see the smallest spatial details) is below average. A study comparing a group of people with relatively normal vision and a group of people with visual impairment (people who on average have visual acuity at least three times poorer than normal) found that they used a viewing distance and television size combination that produced a median (range) viewing angle of 38° (~12 to ~127) and 18° (~6 to ~44), respectively [59]. Even with these adaptations, people with low vision experienced difficulties in watching television, specifically difficulty in seeing details and frequently missing important information [59]. Their main adaptations were to watch larger television screens and use closer working distances. An alternative method, which was less frequently used by the sampled group, is to use binocular telescopic devices that typically optically make the television screen appear about twice as large or half the distance away (e.g. [60]).

People with low vision may also have other problems within the overall visual field, such as local areas for which vision is missing (termed absolute scotomas) or local areas where vision is present, but contrast detection or ability to see detail may be reduced (termed relative scotomas). Researchers have explored other ways of enhancing images to assist people with low vision to appreciate the details [61]. They found that the kinds of enhancements that may be appreciated depended on personal preference (e.g. whether they preferred sharper images or smoother images). The level of enhancement preferred also differed depending on the content of the image; significantly, faces were preferred with less high-pass filtering of the image or adjustment of luminance and saturation than other objects [61]. These preferences are similar to observed preferences by children in book illustrations for realistic and recognisable objects followed by realism [33]. In adults, as televised images are often linked with a strong narrative, realism [34] is likely to be appreciated above abstraction, even though there are differences in personal preferences as to the level of realism liked in images [62].

There are some people who have a visual impairment that is so great that optical aids or video enhancement will not help. For those individuals who are blind, descriptive video is helpful. Descriptive video describes the insertion of narration, into the natural pauses in dialogue, of the actions and appearances of objects and characters in the video image and the scene changes [63]. These narrations can assist to satisfy the need for people with profound visual impairment to not miss out on 'important information' [59] and promote engagement with the televised material. Descriptive video utilises a sight-substitution method (using hearing) rather than enhancing the visual experience. Similarly, descriptive closed captioning for sounds is also useful for people with hearing impairments. For people with dual sensory loss (e.g. hearing and vision), both watching and hearing the television are difficult tasks and may require new solutions [64].

2.3.5.3   Glare sensitivity

Glare describes an interference with visual perception due to reflected light or nearby light sources. These can impact the television viewing experience by reducing comfort (termed discomfort glare) and reducing contrast of the image perceived (termed disability glare). The experience of glare tends to worsen with increasing age, due to the onset of age-related changes in the crystalline lens such as opacification and cataract [65, 66]. The placement of the television away from light sources, including windows, or reflected light or the use of anti-reflective coatings on screen technology may assist viewers with glare sensitivity.

2.3.5.4   The special case of 3D television viewing

The viewing of two-dimensional (2D) television images is relatively passive. We are asked to view an image. If we are not sitting too close, we do not need to make large eye or head movements. We are not required to change our accommodation and convergence as the television is located at a single plane or close to a single plane. However, the viewing of 3D images is a much more dynamic experience. Up to now, it has been explained that even when we may be physically passive, or visual system is actively processing incoming information. The visual information presented on 2D television screens is relatively realistic in the demands it makes on our visual systems. In contrast, 3D television viewing challenges us to converge for near objects, diverge for distance objects, look where the action is (it may be travelling across the screen, travelling towards or away from us or in the more peripheral corners of the screen) and do this without changing our accommodation. This presents a problem for us as viewers. The purpose of our visual system in one sense is to tell us about the world around us, so that we can respond appropriately. Physiologically, if we see something is close by or if we think something is closeby, our visual system starts accommodating and converging. If we converge, this is a stimulus to our visual system to also accommodate. If we accommodate, this is a stimulus to our visual system to converge. If we see that we are moving closer to an object, we prepare our body for that movement. What happens then, when we are presented with visual cues that tell us we are closer to or farther away from objects (convergence and divergence cues respectively) but we have not changed relative location? There is a conflict of information. When we have to learn a new way of doing things, learn to suppress the learned relationship between accommodation and convergence, between our visual system and our sense of where we are, this can result in a disruption of our natural sense of planning.

Despite these challenges, people like to feel transported into different worlds, immersed in the television world, a process which is facilitated through the viewing of which 3D television images. Unfortunately, 3D viewing is usually less comfortable than 2D viewing. One survey found that the probability of experiencing discomfort such as headaches and eye strain when viewing 3D images is about 10% compared to 1.5% when viewing 2D images [67]. Other symptoms found to be more frequently experienced during 3D than 2D viewing may be blurred or doubled vision, dizziness, disorientation and nausea [36, 67]. This section will describe some of the factors that may contribute to the greater discomfort experience with the 3D viewing experience as compared to the 2D viewing experience.

When presenting images intended for 3D perception on television, the developers of the television shows have to make some assumptions as to how to do this. The key parameter with respect to stereopsis is how much disparity to present to the right and left eyes and record to produce the minimum of aberrations [68]. Presenting a large range means the viewer has to perform high levels of convergence and divergence during viewing. Presenting too little range may beg the question of why bother with 3D presentation? The question of how much disparity to present is that it should vary according to the individual as individuals have developed their visual system to match the placement of their two eyes within their own body. In fact, stereopsis only starts to develop after eye opening following birth for this reason. The developers of 3D television programs must also consider how to present different images to the right and left eyes. Incomplete separation of the right and left eye images can lead to the perception of both images simultaneously by both the right and left eyes, producing 'ghost' images [68, 69]. If presenters decide on using a temporal separation, incorrect timing of alternate right and left eye presentations can elicit 'retinal rivalry' phenomenon, where the visual system alternates between the right and left eye views as the fusing of the images is too disturbing [68]. It can be seen that for 3D television, the characteristics of the viewer become much more crucial to ensure a clear and comfortable experience by the viewer.

Disparity is related to the location of the same object on the right and left eye visual fields and whether they fall on corresponding points, near-corresponding points or non-corresponding points. Ideally, zero disparity should correspond with the object image falling on the foveas of the right and left eyes of the viewer for the primary position of gaze (i.e. looking ahead with the visual axes of both eyes parallel to each other). However, mismatches between intended and actual achieved disparity are possible for individuals depending on the technology. Technologies that are intended for personal use, such as Nintendo 3DS or 3D virtual reality head-mounted displays, allow some adjustment of the location of the right and left eye images to achieve the desired percept. However, where there are multiple people viewing the same screen, the potential for mismatch is greater.

For example, the inter-pupillary distance (the distance between the viewing axes of the right and left eyes) is one factor that can contribute to the comfort or discomfort of the viewer with respect to convergence [68]. Inter-pupillary distance varies between males and females and children and adults so if a person's inter-pupillary distance does not match with the intended, then that person's eyes will need to converge or diverge to a greater extent than intended.

Another factor is location of the viewer in relation to the screen [70]. Persons who view from a more peripheral location will experience an oculomotor imbalance in the demand for convergence and divergence compared to a person who views from a more central location [70]. A third factor is the strength of the relationship between accommodation and convergence. Most people converge when they accommodate, as convergence is necessary to keep the images of nearer objects on the fovea. However, in the case of 3D television, the accommodative state should not change when convergence changes as both near and far objects are always at the plane of the television screen [70]. Different people have different accommodation–convergence relationships and so may experience this disruption of the accommodation–convergence relationship in different ways. Interestingly, this is one situation where presbyopia may be an advantage as an inability to accommodate may reduce this conflict, which may result in fewer complaints of this symptom [36]. It has been suggested that some customisation for viewing location and displays that incorporate multiple distances (rather than being on a flat plane) might be able to overcome some of these problems, but what works for one viewer, will not work for another [71] so are likely to be impractical for universal application.

Apart from the stimulus to converge, discomfort can also arise when viewing 3D television if the viewer has a propensity for motion sickness [36, 67]. Motion sickness is thought to arise in sensitive individuals when there is a mismatch between the visual and vestibular percepts. The vestibular system provides information about motion, balance and posture based on the relative motion of fluids in the vestibular apparatus. The vestibular apparatus comprises canals at different orientations. If the vestibular apparatus indicates that the body is not moving, but the visual system indicates that the body is moving, then this conflict can result in motion sickness and nausea. Although this can also occur for 2D viewing, it may be reduced by increasing the viewing distance to reduce the level of immersion in the visual world [36]. Motion sickness and nausea are less easy to overcome with 3D viewing. One promising suggestion that appears to assist when viewing virtual worlds (for which viewpoint must shift with head motion) is to add a virtual nose as a stable point of reference to the display [72]; however, this would not work well with the narrative demands of 3D television viewing, which is not a first-person game environment.

2.4   Conclusions

It is hoped that this chapter has described to the reader that our perception of television images is not passive. Instead, our visual system is dynamically processing incoming light information to maximise the extraction of meaningful information that is of interest to us and actively using this information to plan our next actions. Context plays an important role in how we perceive and react to visual information. We select what is of most interest to us, and then a percept is constructed that is consistent with all the available and present information. In the special case of 3D television, the visual system's dynamic processing may result in the stimulation of responses that are not actually required (e.g. convergence may stimulate accommodation and the perception of visual motion may stimulate movement), resulting in conflict in the way this information is integrated, which may result in discomfort. For 2D and 3D television viewing, realism in visual appearance, as well as accommodation and postural cues is desirable for viewing pleasure.

Acknowledgements

Jacinta Wei Chin Chong for assistance with illustrations

References

[1]   Biederman I., Kim J.G. '17,000 years of depicting the junction of two smooth shapes'. *Perception*. 2008, vol. 37(1), pp. 161–64.
[2]   Pepperell R., Haertel M. 'Do artists use linear perspective to depict visual space?'. *Perception*. 2014, vol. 43(5), pp. 395–416.
[3]   Valberg A., Lange-Malecki B. '"Colour constancy" in mondrian patterns: a partial cancellation of physical chromaticity shifts by simultaneous contrast'. *Vision Research*. 1990, vol. 30(3), pp. 371–80.
[4]   Galifret Y. 'Visual persistence and cinema?'. *Comptes Rendus Biologies*. 2006, vol. 329(5–6), pp. 369–85.
[5]   Youssef P.N., Sheibani N., Albert D.M. 'Retinal light toxicity'. *Eye (London, England)*. 2011, vol. 25(1), pp. 1–14.

[6]   Lin Y., Fotios S., Wei M., Liu Y., Guo W., Sun Y. 'Eye movement and pupil size constriction under discomfort glare'. *Investigative Ophthalmology & Visual Science*. 2015, vol. 56(3), pp. 1649–56.

[7]   Forrester J.V., Dick A., McMenamin P., Roberts F., Pearlman E. 'Anatomy of the eye and orbit' in *The eye, basic sciences in practice*. Edinburgh UK: WB Saunders; 1996.

[8]   Atchison D.A., Markwell E.L., Kasthurirangan S., Pope J.M., Smith G., Swann P.G. 'Age-related changes in optical and biometric characteristics of emmetropic eyes'. *Journal of Vision*. 2008, vol. 8(4), 29.

[9]   Navarro R. 'The optical design of the human eye: a critical review'. *Journal of Optometry*. 2009, vol. 2(1), pp. 3–18.

[10] Franssen L., Tabernero J., Coppens J.E., van den Berg T.J.T.P. 'Pupil size and retinal straylight in the normal eye'. *Investigative Ophthalmology & Visual Science*. 2007, vol. 48(5), pp. 2375–82.

[11] Mathôt S., Van der Stigchel S. 'New light on the mind's eye: the pupillary light response as active vision'. *Current Directions in Psychological Science*. 2015, vol. 24(5), pp. 374–78.

[12] Dubbelman M., Van der Heijde G.L., Weeber H.A., Vrensen G.F.J.M. 'Changes in the internal structure of the human crystalline lens with age and accommodation'. *Vision Research*. 2003, vol. 43(22), pp. 2363–75.

[13] Heys K.R., Friedrich M.G., Truscott R.J.W. 'Presbyopia and heat: changes associated with aging of the human lens suggest a functional role for the small heat shock protein, alpha-crystallin, in maintaining lens flexibility'. *Aging Cell*. 2007, vol. 6(6), pp. 807–15.

[14] Artigas J.M., Felipe A., Navea A., Fandiño A., Artigas C. 'Spectral transmission of the human crystalline lens in adult and elderly persons: color and total transmission of visible light'. *Investigative Ophthalmology & Visual Science*. 2012, vol. 53(7), pp. 4076–84.

[15] Smith V.C., Pokorny J. 'Spectral sensitivity of the foveal cone photopigments between 400 and 500 nm'. *Vision Research*. 1975, vol. 15(2), pp. 161–71.

[16] Kraft T.W., Schneeweis D.M., Schnapf J.L. 'Visual transduction in human rod photoreceptors'. *The Journal of Physiology*. 1993, vol. 464, pp. 747–65.

[17] Rushton W.A.H. 'Pigments and signals in colour vision'. *The Journal of Physiology*. 1972, vol. 220(3), p. 1–31.

[18] Kalloniatis M., Luu C. 'Light and dark adaptation' in *Webvision: the organization of the retina and visual system*. University of Utah Health Sciences Center; 2007.

[19] Bento-Torres N.V.O., Rodrigues A.R., Côrtes M.I.T., Bonci D.M. de O., Ventura D.F., Silveira L.C. de L. 'Psychophysical evaluation of congenital colour vision deficiency: discrimination between protans and deutans using mollon-reffin's ellipses and the farnsworth-munsell 100-hue test'. *PloS One*. 2016, vol. 11(4), e0152214.

[20] Davidoff C., Neitz M., Neitz J. 'Genetic testing as a new standard for clinical diagnosis of color vision deficiencies'. *Translational Vision Science & Technology*. 2016, vol. 5(5), p. 2.

[21] Sabesan R., Schmidt B.P., Tuten W.S., Roorda A. 'The elementary representation of spatial and color vision in the human retina'. *Science Advances*. 2016, vol. 2(9), e1600797.

[22] Chatterjee S., Callaway E.M. 'Parallel colour-opponent pathways to primary visual cortex'. *Nature*. 2003, vol. 426(6967), pp. 668–71.

[23] Spector R.H. *Visual fields, in clinical methods: the history, physical and laboratory examinations*. Boston, MA: Butterworths; 1990. pp. 565–72.

[24] Howarth P.A. 'Potential hazards of viewing 3-D stereoscopic television, cinema and computer games: a review'. *Ophthalmic & Physiological Optics*. 2011, vol. 31(2), pp. 111–22.

[25] Kravitz D.J., Saleem K.S., Baker C.I., Ungerleider L.G., Mishkin M. 'The ventral visual pathway: an expanded neural framework for the processing of object quality'. *Trends in Cognitive Sciences*. 2013, vol. 17(1), pp. 26–49.

[26] Fechner G.T. *Elemente der psychophysik*. Amsterdam: E.J. Bonset; 1860.

[27] Wilson A.J., Kohfeld D.L. 'Critical fusion frequency as a function of stimulus intensity: and wavelength composition'. *Perception & Psychophysics*. 1973, vol. 13(1), pp. 1–4.

[28] Emoto M., Sugawara M. 'Critical fusion frequency for bright and wide field-of-view image display'. *Journal of Display Technology*. 1973, vol. 8(7), pp. 424–29.

[29] Lappin J.S., Tadin D., Nyquist J.B., Corn A.L. 'Spatial and temporal limits of motion perception across variations in speed, eccentricity, and low vision'. *Journal of Vision*. 2009, vol. 9(1), 30.

[30] Wescott T. *Applied control theory for embedded systems*. Elsevier; 2006.

[31] Boon M., Dain S.J, *et al.* 'The development of colour vision' in *Learning from Picturebooks: Perspectives from Child Development and Literacy Studies*. London: Routledge; 2015. pp. 71–95.

[32] Kimura A., Wada Y., Yang J., *et al.* 'Infants' recognition of objects using canonical color'. *Journal of Experimental Child Psychology*. 2010, vol. 105(3), pp. 256–63.

[33] Rudisill M.R. 'Children's preferences for color versus other qualities in illustrations'. *The Elementary School Journal*. 1952, vol. 52(8), pp. 444–51.

[34] Boust C *et al.* 'Does an expert use memory colors to adjust images?'. *Color and Imaging Conference, 12th Color and Imaging Conference Final Program and Proceedings*; Springfield, USA: Society for Imaging Science and Technology, 2004. pp. 347–53.

[35] Morrison G. *How big a TV should I buy?* [online]. 2016 Dec 21. Available from https://www.cnet.com/how-to/how-big-a-tv-should-i-buy/

[36] Yang S., Schlieski T., Selmins B., *et al.* 'Stereoscopic viewing and reported perceived immersion and symptoms'. *Optometry and Vision Science*. 2012, vol. 89(7), pp. 1068–80.

[37] Boon M.Y., Suttle C.M., Dain S.J. 'Transient VEP and psychophysical chromatic contrast thresholds in children and adults'. *Vision Research*. 2007, vol. 47(16), pp. 2124–33.

[38] Conway B.R., Moeller S., Tsao D.Y. 'Specialized color modules in macaque extrastriate cortex'. *Neuron.* 2007, vol. 56(3), pp. 560–73.

[39] Rabin J., Switkes E., Crognale M., Schneck M.E., Adams A.J. 'Visual evoked potentials in three-dimensional color space: correlates of spatio-chromatic processing'. *Vision Research.* 1994, vol. 34(20), pp. 2657–71.

[40] Logvinenko A.D., Funt B., Mirzaei H., Tokunaga R. 'Rethinking colour constancy'. *PloS One.* 2015, vol. 10(9), e0135029.

[41] Witzel C., van Alphen C., Godau C., O'Regan J.K. 'Uncertainty of sensory signal explains variation of color constancy'. *Journal of Vision.* 2016, vol. 16(15), 8.

[42] Bodduluri L., Boon M.Y., Dain S.J. 'Evaluation of tablet computers for visual function assessment'. *Behavior Research Methods.* 2017, vol. 49(2), pp. 548–58.

[43] Ridder W.H., Tomlinson A. 'A comparison of saccadic and blink suppression in normal observers'. *Vision Research.* 1997, vol. 37(22), pp. 3171–79.

[44] Vallines I., Greenlee M.W. 'Saccadic suppression of retinotopically localized blood oxygen level-dependent responses in human primary visual area V1'. *Journal of Neuroscience.* 2006, vol. 26(22), pp. 5965–69.

[45] Boon M.Y., Chan K.Y., Chiang J., Milston R., Suttle C. 'EEG alpha rhythms and transient chromatic and achromatic pattern visual evoked potentials in children and adults'. *Documenta Ophthalmologica. Advances in Ophthalmology.* 2011, vol. 122(2), pp. 99–113.

[46] Ramachandran V.S., Anstis S.M. 'Extrapolation of motion path in human visual perception'. *Vision Research.* 1983, vol. 23(1), pp. 83–85.

[47] Tas A.C., Moore C.M., Hollingworth A. 'The representation of the saccade target object depends on visual stability'. *Visual Cognition.* 2014, vol. 22(8), pp. 1042–46.

[48] Beer A.L., Heckel A.H., Greenlee M.W. 'A motion illusion reveals mechanisms of perceptual stabilization'. *PloS ONE.* 2008, vol. 3(7), e2741.

[49] Davis J., Hsieh Y.-H., Lee H.-C. 'Humans perceive flicker artifacts at 500 Hz'. *Scientific Reports.* 2015, vol. 5, 7861.

[50] Wong B.P.H., Woods R.L., Peli E. 'Stereoacuity at distance and near'. *Optometry and Vision Science.* 2002, vol. 79(12), pp. 771–78.

[51] Ramanathan S., Katti H., Sebe N., Kankanhalli M., Chua T.-S. 'An eye fixation database for saliency detection in images' in Daniilidis K., Maragos P., Paragios N. (eds.). *ECCV 2010, part IV, lecture notes in computer science.* Vol. 6314. Berlin, Heidelberg: Springer; 2010. pp. 30–43.

[52] Simons D.J., Chabris C.F. 'Gorillas in our midst: sustained inattentional blindness for dynamic events'. *Perception.* 1999, vol. 28(9), pp. 1059–74.

[53] Solovey G., Graney G.G., Lau H. 'A decisional account of subjective inflation of visual perception at the periphery'. *Attention, Perception & Psychophysics.* 2015, vol. 77(1), pp. 258–71.

[54] Raman R., Sarkar S. 'Predictive coding: a possible explanation of filling-in at the blind spot'. *PloS ONE.* 2016, vol. 11(3), e0151194.

[55]   Thomson W.D. 'Eye problems and visual display terminals–the facts and the fallacies'. *Ophthalmic & Physiological Optics*. 1998, vol. 18(2), pp. 111–19.

[56]   Lee D.-S. 'Preferred viewing distance of liquid crystal high-definition television'. *Applied Ergonomics*. 2012, vol. 43(1), pp. 151–56.

[57]   Jaschtnski-kruza W. 'On the preferred viewing distances to screen and document at VDU workplaces'. *Ergonomics*. 1990, vol. 33(8), pp. 1055–63.

[58]   Bababekova Y., Rosenfield M., Hue J.E., Huang R.R. 'Font size and viewing distance of handheld smart phones'. *Optometry and Vision Science*. 2011, vol. 88(7), pp. 795–97.

[59]   Woods R.L., Satgunam P. 'Television, computer and portable display device use by people with central vision impairment'. *Ophthalmic & Physiological Optics*. 2011, vol. 31(3), pp. 258–74.

[60]   *Products telescopes galilean detail 1624-11 maxtv* [online]. 2016 Dec 30. Available   from   http://www.eschenbach.com/904f1353-8a6c-48fc-9137-d09776df514d/products-telescopes-galilean-detail.htm

[61]   Satgunam P.N., Woods R.L., Bronstad P.M., Peli E. 'Factors affecting enhanced video quality preferences'. *IEEE Transactions on Image Processing*. 2013, vol. 22(12), pp. 5146–57.

[62]   Kettlewell N., Lipscomb S. 'Neuropsychological correlates for realism-abstraction, a dimension of aesthetics'. *Perceptual and Motor Skills*. 1992, vol. 75(3 Pt 1), pp. 1023–26.

[63]   *BBC. audio description on TV* [online]. 2016 Dec 30. Available from http://www.bbc.co.uk/aboutthebbc/insidethebbc/howwework/policiesandguidelines/audiodescription.html

[64]   Brennan M., Bally S.J. 'Psychosocial adaptations to dual sensory loss in middle and late adulthood'. *Trends in Amplification*. 2007, vol. 11(4), pp. 281–300.

[65]   Elliott D.B., Bullimore M.A. 'Assessing the reliability, discriminative ability, and validity of disability glare tests'. *Investigative Ophthalmology & Visual Science*. 1993, vol. 34(1), pp. 108–19.

[66]   Haegerstrom-Portnoy G. 'The Glenn A. Fry award lecture 2003: vision in elders–summary of findings of the SKI study'. *Optometry and Vision Science*. 2005, vol. 82(2), pp. 87–93.

[67]   Read J.C.A., Bohr I. 'User experience while viewing stereoscopic 3D television'. *Ergonomics*. 2014, vol. 57(8), pp. 1140–53.

[68]   Zeri F., Livi S. 'Visual discomfort while watching stereoscopic three-dimensional movies at the cinema'. *Ophthalmic & Physiological Optics*. 2015, vol. 35(3), pp. 271–82.

[69]   Seuntiëns P.J.H., Meesters L.M.J., IJsselsteijn W.A. 'Perceptual attributes of crosstalk in 3D images'. *Displays*. 2005, vol. 26(4–5), pp. 177–83.

[70]   Aznar-Casanova J.A., Romeo A., Gómez A.T., Enrile P.M. 'Visual fatigue while watching 3D stimuli from different positions'. *Journal of Optometry*. 2017, vol. 10(3), pp. 149–60.

[71] Terzic K., Hansard M. 'Methods for reducing visual discomfort in stereo-scopic 3D: A review'. *Signal Processing: Image Communication*. 2016, vol. 47, pp. 402–16.

[72] Whittinghill D.M., Ziegler B., Case T., Moore B. *System and method for reducing simulator sickness*. 2016.

Chapter 3

Colour vision

Yoko Mizokami[1]

3.1 Human colour vision mechanism

As Newton said, the light rays themselves are not coloured. Colour sensation is created in our brain as the result of complex visual information processing by our visual system. Therefore, it is important to understand the characteristics and mechanism of colour vision to understand colour appearance. It is also a critical matter of how to describe colour in an objective way. In this chapter, we first describe the basic colour vision theories, mechanisms and the variety of colour appearances influenced by viewing conditions. Then the basic colourimetry including colour appearance models is introduced.

Our colour perception is initiated by the stimulation of L (long-wavelength sensitive), M (middle-wavelength sensitive) and S (short-wavelength sensitive) cones in the retina. Change in the ratio of L, M and S responses evokes a difference in colour perception. Examples of L, M and S cone sensitivities [1] as well as the photopic luminance efficiency $V(\lambda)$ and scotopic luminance efficiency $V'(\lambda)$ corresponding to rod sensitivity are shown in Figure 3.1.

We begin with historically important models in colour vision that became the basis of various colour notation systems.

3.2 Colour vision model

3.2.1 Trichromatic theory of colour vision

The trichromatic theory was proposed in the nineteenth century by Thomas Young and Hermann von Helmholtz. It is based on the observation that it is possible to match all colours in the visible spectrum using an appropriate mixture of three primary colours. The choice of primary colours was not critically important as long as mixing two of them does not produce the third. It is hypothesised that there were three classes of photoreceptors for enabling colour vision based on this fact. Now, we know these photoreceptors correspond to L, M and S cones in the retina. This theory is paramount for colour science since many colour spaces and appearance models developed later are based on data taken from colour matching experiments that adopted this theory.

[1]Graduate School of Engineering, Chiba University, Japan

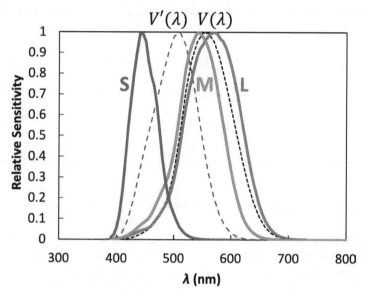

Figure 3.1   Spectral sensitivity of rod and cones, and photopic and scotopic spectral luminous efficiencies

3.2.2   Opponent processing theory of colour vision

Ewald Hering proposed the opponent-colour theory based on some observations of colour appearance that could not be explained by the trichromatic theory [2]. There are certain pairs of colours one never sees together at the same place and at the same time (e.g. colours such as reddish yellow and yellowish green exist, but reddish-green or yellowish-blue does not exist). The colour of afterimages also shows a distinct pattern of opponent relationships between red and green, and blue and yellow (e.g. afterimage of red is now green and vice versa, and likewise, for blue and yellow). Hering hypothesised that two major classes of processing are involved as follows: (1) spectrally opponent processes, e.g. red vs. green and yellow vs. blue and (2) spectrally non-opponent processes, e.g. black vs. white.

Hurvich and Jameson [3] quantitatively showed the opponent characteristics by applying the hue-cancellation method in psychophysical experiments, as shown in Figure 3.2. The discovery of electrophysiological responses showing opponent processing in the 1950s also supported the theory.

Thus, the opponent-colour theory was also a major theory. Opponent-colour and trichromatic theories competed for a long time. However, in fact, both theories helped to explain how our colour vision system works. The trichromatic theory operates at the receptor level, and the opponent processes theory applies at the subsequent neural level of colour vision processing. This two-stage model became common afterwards.

However, it should be noted that the physiological opponent process in the post-receptoral neural stage does not directly correspond to Hering's opponent theory. Opponent colour channels, L-M and S-(L+M), are the subtraction of L and M cone signals and that of S and (L+M) cone signals, respectively. Therefore, they are not the same as

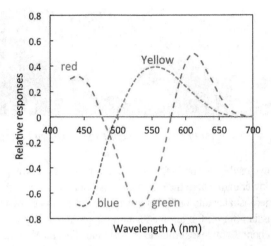

Figure 3.2   *Theoretical opponent characteristics of red-green and yellow-blue responses for CIE average observer (based on figure 6 in Hurvich and Jameson [3])*

unique hues. The unique hue perception would be associated with higher-level processes at the cortical level.

3.2.3   Colour vision processing in the neural stage.

In the neural stage after passing through the retina, the luminance and the colour information are transmitted separately and transferred to the brain, as shown in Figure 3.3.

Together they contribute to the efficient coding of visual information. However, there are many differences in the characteristics of luminance and chromatic channels. For example, their spatial and temporal resolutions are quite different. The Campbell-Robson contrast sensitivity test pattern in Figure 3.4 demonstrates how the spatial sensitivity of luminance and chromatic components is different. As shown in Figure 3.5, the contrast sensitivity function (CSF) of the luminance channel has a band-pass characteristic with a

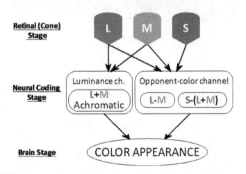

Figure 3.3   Colour vision processing

Figure 3.4   The Campbell-Robson contrast sensitivity test pattern

peak around 3–5 cycles/degree (cpd). Chromatic channels have a low-pass shape with a critical resolution lower than that of the luminance channel. This means that the luminance channel has a higher visual acuity or higher spatial resolution. As shown in Figure 3.6, an image with a Gaussian blur in the luminance component looks blurred, but images with blur in chromatic channels appear similar to the original. This means that an image can be represented without degradation of its appearance if a high-resolution luminance component and lower resolution chromatic component are used. Therefore, it would be possible to code or transfer information efficiently, controlling the resolution of those channels. As shown in Figure 3.7, the temporal characteristics of both channels show a similar trend. The temporal sensitivity of the luminance channel has a band-pass shape with a peak

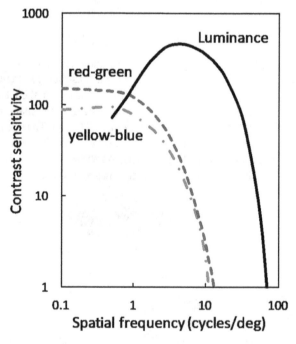

*Figure 3.5   Spatial CSF of luminance channel, and chromatic red-green and
yellow-blue channels (based on figure 7 in Reference 4 and figure 2 in
Reference 5)*

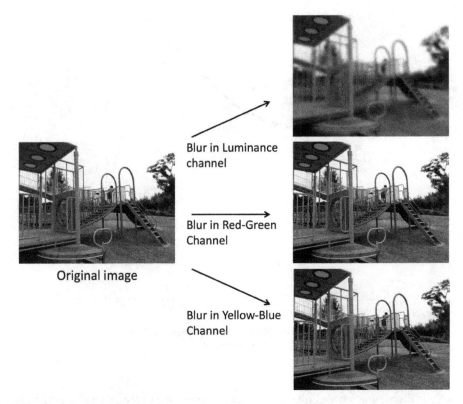

Figure 3.6   Appearance of an image with blur in luminance channel and chromatic channels

of around 20 Hz, whereas the chromatic channel has a low-pass shape. In addition, the critical fusion frequency is lower in chromatic channels than in the luminance channel. Note that the spatial and temporal CSF changes depending on the viewing condition. For example, both spatial and temporal CSFs acquire low-pass shape when the luminance level decreases. The spatial CSF acquires a low-pass shape for stimuli flickering with a high frequency.

The chromatic information is transferred from the retina to the visual cortex via a lateral geniculate nucleus in these opponent channels. There, the signals are separated into multiple channels to cater for higher visual mechanisms resulting in our sense of colour appearance.

3.3   Various colour perception

3.3.1   Categorical colour perception

We have high sensitivity to perceive small colour differences. We can also perceive a group of colours in unison, called categorical colour perception. It is one of the important aspects of colour perception. In our daily life, we usually do not need fine

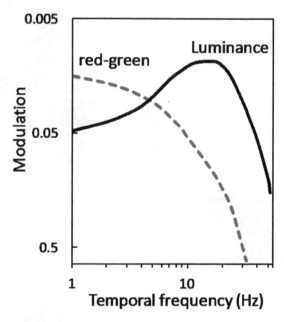

Figure 3.7   Temporal CSF of luminance and chromatic channels (Figure 7.4 in Reference 6, original: Reference 7)

colour discrimination and often use colour categories for recognising objects and for communication purposes such as colour codes, signs and traffic signals. Berlin and Kay [8] investigated colour names in different languages and countries over the world and found 11 basic colour terms common in many languages: White, Black, Red, Green, Yellow, Blue, Brown, Purple, Orange, Pink and Grey. These categories are often used for research on categorical colour perception.

3.3.2   Mode of colour appearance

The mode of colour appearance, a concept originally proposed by Katz [9], is manifested due to variations in a medium that emits, transmits or reflects light. For example, a surface that appears to reflect light is described to be in the object colour mode, surface colour mode or related colour mode. A surface that appears to emit light is described to be in the light-source colour mode, aperture colour mode or unrelated colour mode. Different modes can cause differences in colour appearance. For example, the same lights that appear brown when reflected from a given object surface may appear orange when emitted from a light source. Black does not exist in the light source colour mode. The mode of colour appearance does not always correspond to the actual mode of media. A TV display emits light, but it would appear to be in the object colour mode in a bright environment and in the light-source colour mode in a dark environment. A paper usually appears to be in the object colour mode, but it could appear to be in the light-source colour mode when a spotlight illuminates it, focusing only on the paper.

3.3.3 Chromatic adaptation

We can perceive colour based on the ratio of the LMS cone response. However, the ratio does not necessarily correspond to one particular colour, and the appearance changes depending on the viewing conditions.

One important mechanism in our colour vision is chromatic adaptation. While we can adapt to various light levels, we also change our sensitivity in adapting to a colour property of an environment. For example, one would see a golden (Kinkakuji) temple surrounded by green trees when fixating on the red spot in the right picture after fixating on the red spot in the left picture around 20–30 seconds in Figure 3.8. This phenomenon of opponent colours by chromatic adaptation is called colour (or chromatic) afterimage. Traditionally, adaptation is often explained by the gain control of the LMS response. However, the contributions of lower and higher visual mechanisms including cortical level are also important, as predicted by various models [10].

There are various models for the visual mechanism of chromatic adaptation and methods to achieve adaptation in imaging. One of the most basic models is the von Kries model or the chromatic adaptation transform. The working principle is that to keep the same appearance when the illumination colour changes, the LMS cones need to maintain the same responses. Typically, this is possible with a diagonal scaling of the axes in the LMS cone space. The LMS cone response changes when the illumination changes, but the changed cone responses L, M and S are normalised by L_0, M_0 and S_0, which correspond to the cone response for the colour of the illuminant, resulting in the compensated L, M and S cone responses L', M' and S' after perfect adaptation:

$$\begin{pmatrix} L' \\ M' \\ S' \end{pmatrix} = \begin{pmatrix} \frac{1}{L_0} & 0 & 0 \\ 0 & \frac{1}{M_0} & 0 \\ 0 & 0 & \frac{1}{S_0} \end{pmatrix} \begin{pmatrix} L \\ M \\ S \end{pmatrix} \tag{3.1}$$

In the case of image processing, we simply need to divide out the estimated illuminant and apply the desired illuminant separately for each channel. The choice of the transformation matrix to be used is the subject of research. It can be optimised

Figure 3.8   Colour afterimage (colour adaptation)

utilising a variety of criteria, such as mean colour, brightest area, or statistical distribution of an image.

3.3.4   Influence of context

Colour perception is also influenced by surroundings. Simultaneous lightness and colour contrast and colour assimilation are well-known factors affecting colour perception. However, there are phenomena that cannot be explained by simultaneous contrast such as the White effect as shown in Figure 3.9. The upper grey rectangles look darker than the lower ones even though they are bordered by more black than white. This is often explained by the influence of 'belongingness' which means the perceived lightness of an area is influenced by the part of the surroundings to which the area appears to belong.

Colourfulness perception can be influenced by surroundings, too. Colourfulness appears reduced when surrounded by saturated colours, called the Gamut expansion effect [11], as shown in Figure 3.10. Rectangles on a grey background look more colourful than the same rectangles surrounded by saturated colours. As shown, any attributes of colour, lightness, hue and saturation could be influenced by the context.

Many other effects influence the colour and brightness appearance. These include the 'Hunt effect' wherein colourfulness perception increases with luminance, the 'Stevens effect' wherein contrast perception increases with luminance, the 'Helmholtz–Kohlrausch effect' wherein more saturated stimuli appear brighter than less saturated stimuli at the same luminance and so on. It is important to consider that colour appearance changes depending on temporal and simultaneous effects.

Figure 3.9   White effect

Figure 3.10   Example of Gamut expansion effect

3.3.5   Colour constancy

Colour constancy is the ability to perceive the colours of objects, invariant to the colour of illumination. In our daily life, we usually see a stable colour of an object even if the colour of illumination substantially changes. For example, a white paper looks white whether it is under daylight or a reddish incandescent lamp. Multiple mechanisms from a low to a high level of visual system were thought to be involved in achieving capabilities such as chromatic adaptation, a clue of illumination in a scene (e.g. specular reflection) and spatial recognition. Colour and other perceptual constancies such as size and shape would be the result of our object-based recognition and suggest that our visual system is not primarily interested in the retinal image itself but in the properties of the objects and scenes out there. The degree of colour constancy depends on situations such as viewing conditions and stimuli [12]. It is also shown that the colour constancy usually works pretty well in the normal environment, but it becomes weaker for images and photographs [13].

3.4   Individual differences of colour vision

3.4.1   Colour deficiency

Colour deficiency or colour blindness is a symptom that makes it difficult to differentiate certain colour combinations. The L, M and S cones all work together allowing a human to see the whole spectrum of colours. People with normal colour vision have all three types of cones and pathways working correctly, but colour deficiency occurs when one or more of the cone types are faulty. For example, if the L cone is faulty, a person would not be able to discriminate the combination of greenish and reddish colours, as well as the combination of red and black because of lower sensitivity to longer wavelengths as shown in Figure 3.11. Most people with colour

Figure 3.11   *Simulation of the discriminability of colour deficiency. (a) Normal colour vision (b) Protan (c) Deutan (d) Tritan.*

deficiency have L or M cone impairments; therefore, they cannot distinguish certain shades of red and green.

People with dichromatism have only two types of functional cones. The total functional absence of one of the L, M or S cones causes protanopia, deuteranopia or tritanopia, respectively. People with anomalous trichromatism will be colour blind to some extent. In this condition, all of the three cones are used to perceive light colours, but one type of cone perceives light slightly out of alignment (shift of sensitivity of L, M or S cones corresponds to protanomaly, deuteranomaly or tritanomaly, respectively). The effects of colour vision deficiency can be mild, moderate or severe depending on the defect. Colour deficiency is a genetic (hereditary) condition mostly. About 4–8% (percentages vary depending on ethnicity or method of calculation) of the male population has colour deficiency.

Figure 3.11 is a simulation of the discriminability of colour efficiency. Note that it is not a simulation of 'colour appearance'. People with colour deficiency do not see the world without colour or two types of colours, but they would see some colourful world and perceive various categories of colours. It should also be noted that there is a large individual difference in the severity of the deficiency.

It would be important to design products and environments by selecting colours distinguishable for people with colour deficiency from the view of 'colour' universal design.

3.4.2   Ageing of the eye

Colour vision is also influenced by age. Some functions decline with age [14]. Our ability to discriminate small colour differences declines as we age. Gradual yellowing of the lenses decreases the light in the short wavelength range of the spectrum, while the cones on our retinas slowly lose sensitivity.

Although the eye physically deteriorates, our brain recalibrates or compensates our vision (at least some of these physical frailties) to maintain the quality of colour throughout ageing [15–17]. In other words, although the colour signal being sent from the eye changes significantly with age, the perception of colour remains almost constant regardless of how old a person is. This suggests that somewhere between the retina and the brain, the visual system must recalibrate itself. It could be possible that our brain might use some external standard in the environment such as the sky or sunlight as a reference.

Age would not affect all aspects of the visual system equally. While 18-year-olds and 75-year-olds were found to be equally good at picking pure red or green among others, older people were found to be less able to distinguish between subtly different colours, particularly in the bluish range. Constant perception of colours would be helpful for elderly people to communicate colours effectively when describing objects. However, it would be important for designing products and environments to select colours distinguishable for elderly people considering the reduced discrimination ability.

3.5 Colourimetry

It is important to describe the colour numerically for communication, industry and product management. There are many colour notations available to describe colour systematically or numerically. Known examples include the Munsell colour system and natural colour system (NCS); both are based on colour appearance.

The Munsell colour system is a colour space that specifies colours based on three colour attributes: hue, value (lightness) and chroma (colour purity). These three colour attributes are separated into perceptually uniform and independent dimensions. Thus, colours can be systematically illustrated in three-dimensional space.

NCS is a colour space and a logical colour system that builds on how humans see colour. It is based on six elementary colours, which are perceived by humans as 'pure'. The four chromatic elementary colours are yellow (Y), red (R), blue (B) and green (G), and the two non-chromatic elementary colours are white (W) and black (S). Other colours can be described by the combination of the elementary colours. A three-dimensional NCS colour space is constructed by hue described by the ratio of two chromatic elementary colours, white (W) to black (S) and chromaticness (C) within each hue.

The Munsell colour system and NCS are useful for colour communications because it is easier to deal with colour intuitively and to label colour in systematic order based on colour appearance in these colour systems. However, they are not convenient to express the colour quantitatively and mathematically. On the other hand, the colourimetric system of the International Commission on Illumination (CIE, Commission Internationale de l'Eclairage) is based on the data obtained by colour matching experiments by using colour mixture, and they have the benefit that colour can be described using metrics, which makes a variety of calculations possible. Therefore, it is used widely, including in industrial fields. CIE colour spaces start from the CIE 1931 rgb and CIE 1931 xyz standard colourimetric systems and have been developed to CIE 1976 $L*a*b*$ (CIELAB) and CIE 1976 $L*u*v*$ (CIELUV) colour spaces to improve the uniformity of colour space. It has been further developed with colour appearance models such as CIECAM02 considering a surrounding environment, colour difference formulas such as CIEDE2000 and other more advanced models. Here we explain CIE 1931 xyz, CIELUV and CIELAB colour spaces, which are the most basic and common colour spaces.

3.5.1   *CIE colour spaces (such as CIEXYZ and CIELAB)*

The CIE colourimetric system is based on the trichromacy of colour vision in which any colour can be matched with a mixture of three primary colours R, G and B. In the colour matching experiment, a circular split screen (a bipartite field) 2° in diameter is used, which corresponds to the angular size of the human fovea. Each monochromatic light on one side of the field was matched with an adjustable colour on the other side, which was a mixture of three primary colours, R, G and B, and the ratio of R, G and B primary colours was obtained to match the monochromatic light. Based on these experiments, colour matching functions $\bar{r}(\lambda)$, $\bar{g}(\lambda)$, $\bar{b}(\lambda)$ of the CIE standard observer were obtained and its modified version, the colour matching functions $\bar{x}(\lambda)$, $\bar{y}(\lambda)$, $\bar{z}(\lambda)$ were defined as shown in Figure 3.12. Here, the shape of $\bar{y}(\lambda)$ is equated with the spectral sensitivity function $V(\lambda)$ so that $\bar{y}(\lambda)$ exclusively represents luminance information for simplicity. The colour matching functions do not exactly correspond to the cone sensitivities, but this is one way to represent the sensitivity of three sensors for colour vision: $\bar{x}(\lambda)$, $\bar{y}(\lambda)$, $\bar{z}(\lambda)$ that have peak sensitivities at long, middle and short wavelengths, respectively. The response ratio of these sensors represents the difference in colour.

Tristimulus values X, Y, Z that indicate certain colours are obtained by integrating the spectral power distribution of light coming into the eyes and the colour matching functions. The spectral power distribution of light coming into the eyes is that of a light

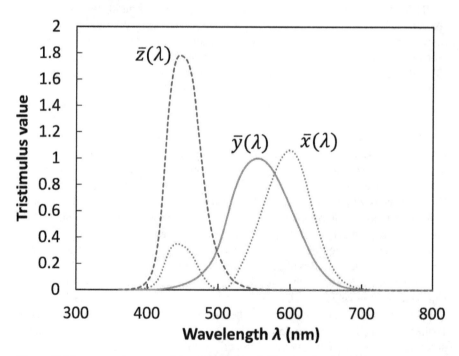

Figure 3.12   xyz *colour matching functions of CIE standard colourimetric observer*

source itself in the case of a self-illuminant colour such as a self-emitting display, but it consists of the spectral power distribution of an illuminant and spectral reflectance of an object in the case of surface colour with reflected light. The calculation of X, Y and Z for a self-illuminant colour and surface colour is described as follows. The standard calculation method is defined as summation at 1-nm intervals over the wavelength range from 360 to 830 nm. Alternative abridged methods are defined for larger intervals (up to 5 nm) and shorter ranges (down to 380–780 nm).

- Self-illuminant colour

$$X = K \int_{\lambda} L_e(\lambda) \bar{x}(\lambda) \, d\lambda = K \sum_n L_e(\lambda_n) \bar{x}(\lambda_n) \, \Delta\lambda \tag{3.2}$$

$$Y = K \int_{\lambda} L_e(\lambda) \bar{y}(\lambda) \, d\lambda = K \sum_n L_e(\lambda_n) \bar{y}(\lambda_n) \, \Delta\lambda \tag{3.3}$$

$$Z = K \int_{\lambda} L_e(\lambda) \bar{z}(\lambda) \, d\lambda = K \sum_n L_e(\lambda_n) \bar{z}(\lambda_n) \, \Delta\lambda \tag{3.4}$$

where L_e indicates spectral radiance (W/sr•m²•nm) and coefficient K corresponds to the maximum spectral luminous efficiency of radiation for photopic vision K_m,

$$K = K_m = 683 \text{ lm/W} \tag{3.5}$$

The constant, K_m, relates the photometric quantities with radiometric quantities and is called the maximum spectral luminous efficacy of radiation for photopic vision. The value of K_m is defined as 683 lm/W at 555 nm.

Luminance L (cd/m²) is defined by the following formula:

$$L \text{ (cd/m}^2\text{)} = K_m \int_{\lambda} L_e(\lambda) V(\lambda) \, d\lambda = K_m \sum_{\lambda} L_e(\lambda_n) V(\lambda_n) \, \Delta\lambda = Y \tag{3.6}$$

- Surface colour

$$X = K \int_{\lambda} S(\lambda) \rho(\lambda) \bar{x}(\lambda) \, d\lambda = K \sum_n S(\lambda_n) \rho(\lambda_n) \bar{x}(\lambda_n) \, \Delta\lambda \tag{3.7}$$

$$Y = K \int_{\lambda} S(\lambda) \rho(\lambda) \bar{y}(\lambda) \, d\lambda = K \sum_n S(\lambda_n) \rho(\lambda_n) \bar{y}(\lambda_n) \, \Delta\lambda \tag{3.8}$$

$$Z = K \int_{\lambda} S(\lambda) \rho(\lambda) \bar{z}(\lambda) \, d\lambda = K \sum_n S(\lambda_n) \rho(\lambda_n) \bar{z}(\lambda_n) \, \Delta\lambda \tag{3.9}$$

where $\rho(\lambda)$ indicates spectral reflectance and $S(\lambda)$ is the spectral intensity of illuminant (relative value of spectral radiance). When we define perfect reflecting diffuser $\rho(\lambda) = 1$ corresponds to $Y = 100$, coefficient K is indicated as follows:

$$K = \frac{100}{\int_{\lambda} S(\lambda) \bar{y}(\lambda) \, d\lambda} = \frac{100}{\sum_n S(\lambda_n) \bar{y}(\lambda_n) \, \Delta\lambda} \tag{3.10}$$

In the case of object colour, the spectral reflectance of an object $\rho(\lambda)$ is multiplied in addition to the spectral energy of illuminant $S(\lambda)$ and a colour matching function. The luminance of object colour is often described as the luminance factor since it is convenient to treat it as a relative value to white. The luminance factor corresponds to the reflectance of an object and it is defined as the Y value of an ideal white (perfect reflecting diffuser) whose maximum value is 100. Thus, coefficient K is calculated by dividing 100 by the luminance of the perfect reflecting diffuser.

Based on this calculation, a CIE 1931 XYZ colour space, which consists of tristimulus values XYZ, is constructed, as shown in Figure 3.13.

In CIE 1931 XYZ colour space, each colour is expressed as a vector with an XYZ component (arrow in the figure), where the Y component corresponds to luminance, and the direction of the vector indicates information of the hue and saturation. The intersection of the vector and the unit plane (x, y, z) represents the hue and saturation of a certain colour. Thus, the CIE 1931 xy chromaticity diagram was created by extracting and distorting the unit plane, as shown in Figure 3.14. A colour is described by the chromaticity coordinates (x, y) on the diagram. Any actual colour is in the gamut of the spectral locus. Now, we can describe any colour quantitatively using a combination of the luminance Y and chromaticity coordinates (x, y). Here, chromaticity coordinates of equal energy white are defined as $(x, y) = (0.333, 0.333)$. The curved line on the chromaticity diagram represents monochromatic light and is called the spectrum locus.

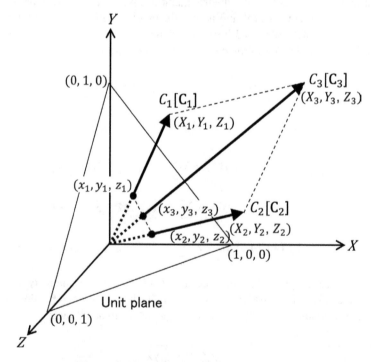

Figure 3.13   CIE 1931 XYZ colour space

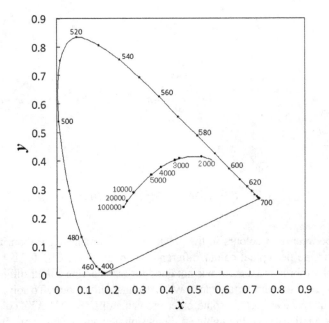

Figure 3.14   CIE1931 xy chromaticity diagram

The equation that transforms the tristimulus value *XYZ* in the CIE 1931 colour space to the *xy* chromaticity coordinates is as follows.

- Tristimulus value *XYZ* to *xy* chromaticity coordinates,

$$x = \frac{X}{X+Y+Z} \tag{3.11}$$

$$y = \frac{Y}{X+Y+Z} \tag{3.12}$$

$$\left(z = \frac{Z}{X+Y+Z} \right) \tag{3.13}$$

$$x+y+z = 1 \tag{3.14}$$

$$X = \frac{x}{y}L \tag{3.15}$$

$$Y = L \tag{3.16}$$

$$Z = \frac{z}{y}L = \frac{1-x-y}{y}L \tag{3.17}$$

The CIE 1931 *xy* standard colourimetric system realised that numerical representation of colour is very useful for the reproduction of colour. However, it has a disadvantage. The perceived equal colour difference at a different position in the diagram does not correspond to the same distance. For example, the colour difference (geometrical

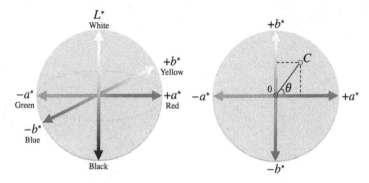

Figure 3.15   CIE 1976 L*a*b* (CIELAB) colour space

distance) between two colours in the green region is much longer than in the blue region when the perceptual colour difference is equal. Thus, CIE 1931 *XYZ* colour space is not uniform and therefore is not good for representing colour difference. The colour difference is an important factor when dealing with colours. To solve distortion of the colour space as well as some other problems of CIE 1931 *XYZ* colour space, such as the nonlinearly of the lightness dimension and adaptation, new chromaticity diagrams and colour spaces have been developed. Here, we introduce the CIELAB colour space and CIELUV colour space, which are commonly used.

3.5.1.1   CIELAB

The CIE 1976 *L*a*b** (CIELAB) colour space is shown in Figure 3.15. CIELAB consists of an *L** axis representing lightness, *a** axis representing the red-green component and *b** axis representing the yellow-blue component. The centre of the colour space represents the achromatic colour. *L**, *a** and *b** are values relative to the colour of the white surface under the same illumination.

XYZ represents the colour of light entering the eyes; the values change depending on the illumination colour even when looking at the same object surface, but the *L*a*b** values are constant regardless of the illumination colour since *L*a*b** is normalised by white (X_0, Y_0, Z_0) under the same illumination. In other words, CIELAB can be considered as a colour space of an object surface or colour space of perfect adaptation.

Equations of transform from *XYZ* to *L*a*b** are as follows:

$$L^* = \begin{cases} 116 \left(\frac{Y}{Y_0}\right)^{1/3} - 16 & \left(\frac{Y}{Y_0} > 0.008856\right) \\ 903.3 \left(\frac{Y}{Y_0}\right) & \left(\frac{Y}{Y_0} \le 0.008856\right) \end{cases} \tag{3.18}$$

$$a^* = 500 \left[\left(\frac{X}{X_0}\right)^{1/3} - \left(\frac{Y}{Y_0}\right)^{1/3} \right] \tag{3.19}$$

$$b^* = 200 \left[\left(\frac{Y}{Y_0}\right)^{1/3} - \left(\frac{Z}{Z_0}\right)^{1/3} \right] \tag{3.20}$$

L^* indicates lightness (metric lightness) and L^* of white is 100. a^* indicates the red-green opponent colour component, and b^* indicates the yellow-blue opponent colour component. When the values of a^* and b^* are both zero, it is an achromatic colour. In this uniform colour space, colours with the same distance in any area of the space appear roughly the same colour difference. The uniformity of colour space was largely improved compared to CIEXYZ.

Metric Chroma C^*_{ab} is the distance from the centre (white point) on the a^*b^* plane and corresponds to the saturation component. Hue angle is defined by the rotating angle from the a^* axis. Metric Chroma and hue angle are defined by the following formula:

- CIE 1976 a^*b^* Metric Chroma C^*_{ab}

$$C^*_{ab} = \sqrt{\left(a^*\right)^2 + \left(b^*\right)^2}$$

(3.21)

- CIE 1976 a^*b^* hue angle h_{ab}

$$h_{ab} = \tan^{-1}\left(\frac{b^*}{a^*}\right)$$

(3.22)

Colour difference ΔE^*_{ab} is defined by Euclidean distance between colour 1 (L^*_1, a^*_1, b^*_1) and colour 2 (L^*_2, a^*_2, b^*_2), thus it is calculated by the following formula:

- Colour difference

$$\Delta E^*_{ab} = \sqrt{\left(\Delta L^*\right)^2 + \left(\Delta a^*\right)^2 + \left(\Delta b^*\right)^2}$$

(3.23)

where $\Delta L^* = L^*_1 - L^*_2$, $\Delta a^* = a^*_1 - a^*_2$ and $\Delta b^* = b^*_1 - b^*_2$.

Hue difference is calculated as follows:

$$\Delta H^*_{ab} = \sqrt{\left(\Delta E^*_{ab}\right)^2 - \left(\Delta L^*\right)^2 - \left(\Delta C^*_{ab}\right)^2}$$

(3.24)

where $\Delta C^*_{ab} = C^*_{ab1} - C^*_{ab2}$.

Therefore, colour difference ΔE^*_{ab} can be described using lightness difference ΔL^*, chroma difference ΔC^*_{ab} and hue difference ΔH^*_{ab}:

$$\Delta E^*_{ab} = \sqrt{\left(\Delta L^*\right)^2 + \left(\Delta C^*_{ab}\right)^2 + \left(\Delta H^*_{ab}\right)^2}$$

(3.25)

3.5.1.2 CIELUV

The other approach to improving the uniformity of the CIE 1931 xy is the development of a CIE 1976 $L^*u^*v^*$ (CIELUV) colour space. As shown in Figure 3.16, the CIE uv

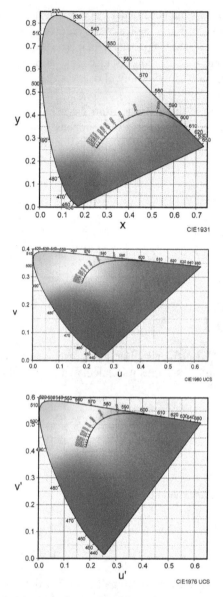

Figure 3.16   Development of CIE chromaticity diagram

uniform chromaticity diagram (1960) followed by the CIE *u'v'* uniform chromaticity diagram (1976) was developed. The chromaticity coordinates (*u*, *v*) of on the CIE *uv* chromaticity diagram are transformed from *x*, *y* chromaticity coordinates (or *XYZ* tristimulus values):

$$u = \frac{4x}{-2x+12y+3} = \frac{4X}{X+15Y+3Z} \tag{3.26}$$

$$v = \frac{6y}{-2x+12y+3} = \frac{6Y}{X+15Y+3Z} \tag{3.27}$$

Based on the CIE *uv* uniform chromaticity diagram, the lightness dimension was added, and the CIE 1964 $U^*W^*V^*$ uniform colour space was constructed in 1964. Here, (u, v) is normalised by subtracting the chromaticity coordinates of white under the same illumination (u_0, v_0). CIELUV is a value relative to the colour of the white surface under the same illumination, as in CIELAB:

$$W^* = 25Y^{1/3} - 17 \tag{3.28}$$
$$U^* = 13W^* (u - u_0) \tag{3.29}$$
$$V^* = 13W^* (v - v_0) \tag{3.30}$$

The CIE *u'v'* uniform colour diagram (1976) was proposed to improve uniformity. The only difference from the CIE *uv* uniform chromaticity diagram is the coefficient for calculating the v' component:

$$u' = \frac{4x}{-2x+12y+3} = \frac{4X}{X+15Y+3Z} \tag{3.31}$$

$$v' = \frac{9y}{-2x+12y+3} = \frac{9Y}{X+15Y+3Z} \tag{3.32}$$

Further, the three-dimensional CIE 1976 $L^*u^*v^*$ uniform colour space (CIELUV) was constructed by adding a lightness dimension to the CIE *u'v'* uniform chromaticity diagram. Here, the lightness dimension L^* is the same as that in CIELAB. The chromatic component is also normalised by white under the same illumination, by subtracting the chromaticity coordinates of white (u'_0, v'_0) from (u', v') of a colour:

$$L^* = \begin{cases} 116 \left(\frac{Y}{Y_0} \right)^{1/3} - 16 & \left(\frac{Y}{Y_0} > 0.008856 \right) \\ 903.3 \left(\frac{Y}{Y_0} \right) - 16 & \left(\frac{Y}{Y_0} \leq 0.008856 \right) \end{cases} \tag{3.33}$$

$$u^* = 13L^* (u' - u'_0) \tag{3.34}$$

$$v^* = 13L^* (v' - v'_0) \tag{3.35}$$

Metric chroma, hue angle, hue difference and colour difference are calculated in the same way as CIELAB.

- CIE 1976 u^*v^* metric chroma C^*_{uv}

$$C^*_{uv} = \sqrt{(u^*)^2 + (v^*)^2} \tag{3.36}$$

- CIE 1976 u^*v^* hue angle h_{uv}

$$h_{uv} = \tan^{-1}\left(\frac{v^*}{u^*}\right) \tag{3.37}$$

- CIE 1976 u^*v^* hue difference

$$\Delta H_{uv}^* = \sqrt{\left(\Delta E_{uv}^*\right)^2 - \left(\Delta L^*\right)^2 - \left(\Delta C_{uv}^*\right)^2} \tag{3.38}$$

- Colour difference ΔE_{uv}^*

$$\Delta E_{uv}^* = \sqrt{\left(\Delta L^*\right)^2 + \left(\Delta u^*\right)^2 + \left(\Delta v^*\right)^2} = \sqrt{\left(\Delta L^*\right)^2 + \left(\Delta C_{uv}^*\right)^2 + \left(\Delta H_{uv}^*\right)^2} \tag{3.39}$$

where $\Delta L^* = L_1^* - L_2^*$, $\Delta u^* = u_1^* - u_2^*$, $\Delta v^* = v_1^* - v_2^*$, and $\Delta C_{uv}^* = C_{uv1}^* - C_{uv2}^*$.

CIE recommended both the CIELAB and the CIELUV. CIELAB represents uniformity for the superthreshold colour difference such as the Munsell colour system, but CIELUV has better uniformity for stimuli with a threshold level. This is why two colour spaces co-exist in CIE. CIELAB is useful for dealing with uniform steps of colour appearance, whereas CIELUV would be useful for dealing with small colour differences such as colour discrimination. After CIELAB and CIELUV were recommended, the uv chromaticity diagram and CIE 1964 $U^*W^*V^*$ colour space have not used except for the calculation of correlated colour temperature and colour rendering index.

To obtain a more accurate colour difference and colour appearance, further improvement is continuing. Colour appearance models considering adaptation and brightness of the surrounding environment were proposed (CIECAM97, CIECAM02 and CIECAM16). Similarly, the colour difference formula (CIEDE2000) [18] and methods for evaluating colour differences in images [19] were also improved.

On the other hand, there are attempts to improve colourimetry from a different perspective. It is known that the basic sensitivity functions defined by CIE such as $V(\lambda)$ and the CIE1931 colour matching function themselves are not accurate enough to reflect our visual sensitivities. To establish a fundamental chromaticity diagram in which the coordinates correspond to physiologically significant axes, the CIE published two technical reports; CIE 170-1:2006 Fundamental Chromaticity Diagram with Physiological Axes—Part 1 and Part 2 [1, 20]. Part I of the report is limited to the choice of a set of colour matching functions and estimates of cone fundamentals for the normal observer, ranging in viewing angle from 1° to 10°. Part 2 aims at providing the user with practical colourimetric tools, in the form of chromaticity diagrams. Part 2 (CIE 170-2: 2015 Fundamental Chromaticity Diagram with

Physiological Axes—Part 2) proposes a fundamental chromaticity diagram. The new diagram is based on the cone fundamentals proposed by Stockman and Sharpe (2000). The sensitivity peaks of the cone fundamental $\bar{l}(\lambda)$, $\bar{m}(\lambda)$ and $\bar{s}(\lambda)$ are 570.2, 542.8 and 442.1 nm, respectively. The spectral sensitivity function is defined as $V_F(\lambda) = 0.68990272\bar{l}(\lambda) + 0.34832189\bar{m}(\lambda)$.

The colour notations based on the CIE 1931 chromatic values are widely used in the field of science, colourimetry and industry, so they would not be easily replaced, but it would be useful to consider these new metrics depending on the purpose.

3.5.2 Colour appearance models [21, 22]

Colour appearance is influenced by surrounding conditions and adapting conditions. Older colour spaces such as CIELAB and CIELUV do not take into account these factors. They have a perfect adaptation transform, but do not predict luminance-dependent effects, background or surround effects. CIECAM02 is a colour appearance model published in 2002 by the CIE Technical Committee 8-01 (Colour Appearance Modelling for Colour Management Systems) as a successor to CIECAM97s [21, 22]. CIECAM02 includes a chromatic adaptation transform, CIECAT02, and its equations for calculating the six technically defined dimensions of colour appearance: brightness, lightness, colourfulness, chroma, saturation and hue.

Brightness is the attribute of a visual perception according to which an area appears to emit, transmit or reflect, more or less light. Lightness is the brightness of an area judged relative to the brightness of a similarly illuminated area that appears to be white or highly transmitting. Colourfulness is the perceived colour of an area that appears to be more or less chromatic. Chroma is the colourfulness of an area judged as a proportion of the brightness of a similarly illuminated area that appears white or highly transmitting. This allows for the fact that a surface of a given chroma displays increasing colourfulness as the level of illumination increases. Saturation is the colourfulness of an area judged in proportion to its own brightness. Hue is the attribute of a visual perception according to which an area appears to be similar to one of the colours red, yellow, green, and blue, or to a combination of adjacent pairs of these colours considered in a closed ring. The colour appearance of the object surface is best described by lightness and chroma, and the colour appearance of the light emitted by or reflected from the object is best described by brightness, saturation and colourfulness.

As input CIECAM02 uses the tristimulus values of the stimulus X, Y, Z; the tristimulus values of an adapting white point X_w, Y_w, Z_w; adapting background (surroundings), and surround luminance information (adapting luminance, L_A), and information on whether observers are discounting the illuminant (colour constancy in effect). The model can be used to predict these appearance attributes or, with forward and reverse implementations for distinct viewing conditions, to compute corresponding colours.

When we observe a display, input data to the model would be the tristimulus values of the stimulus X, Y, Z, the tristimulus values of Display's white point (an adapting white point) X_w, Y_w, Z_w, and viewing parameters c, an exponential nonlinearity,

Table 3.1   Viewing parameters of CIECAM02

Viewing condition	c	N_c	F
Average ($S_R \geq 0.2$)	0.69	1.0	1.0
Dim ($S_R < 0.2$)	0.59	0.9	0.9
Dark ($S_R = 0$)	0.525	0.8	0.8

N_c, the chromatic induction factor, and F, the maximum degree of adaptation. The tristimulus values are the relative values normalised by $Y_w = 100$. Colour attributes such as lightness J and chroma C can be obtained as an output.

To apply CIECAM02, the adaptating field luminance L_A should be determined first. In the case of viewing a display, it is usually calculated as one-fifth of the absolute luminance of white on the display. In the case of viewing printed material or painted colour samples under a certain illumination, it is usually calculated as the illuminance divided by 5π. The surround S_R to determine the viewing parameters c, N_c and F is the ratio of the absolute luminance of surrounding white L_{SW} (cd/m²), and the absolute luminance of white on a display L_{DW} (cd/m²):

$$S_R = \frac{L_{SW}}{L_{DW}} \tag{3.40}$$

Viewing condition is defined as dark, dim and average for $S_R = 0$, $S_R < 0.2$ and $S_R \geq 0.2$, respectively. Viewing parameter is determined by the value of S_R as shown in Table 3.1.

CIECAM02 begins with a conversion from CIE 1931 tristimulus values X, Y and Z (scaled approximately between 0 and 100) to R, G and B responses based on the optimised transform matrix $\mathbf{M}_{CAT02}$:

$$\begin{pmatrix} R \\ G \\ B \end{pmatrix} = \mathbf{M}_{CAT02} \begin{pmatrix} X \\ Y \\ Z \end{pmatrix} \tag{3.41}$$

$$\mathbf{M}_{CAT02} = \begin{pmatrix} 0.7328 & 0.4296 & -0.1624 \\ -0.7036 & 1.6975 & 0.0061 \\ 0.0030 & 0.0136 & 0.9834 \end{pmatrix} \tag{3.42}$$

Then, the D factor, the degree of adaptation to the white point, is calculated as a function of the adapting luminance L_A and surround F. D ranges from 1.0 for complete adaptation to 0.0 for no adaptation. As a practical limitation, the minimum value of D would be about 0.6:

$$D = F\left[1 - \left(\tfrac{1}{3.6}\right) e^{\left(\frac{-(L_A+42)}{92}\right)}\right] \tag{3.43}$$

After D is established, the tristimulus responses for the stimulus colour are converted to adapted tristimulus responses R_c, G_c and B_c representing corresponding colours for an implied equal-energy illuminant reference condition. R_w, G_w and B_w are tristimulus responses for the adapting white. Y_w is a tristimulus value of the white of a display [6]:

$$R_c = \left[\left(\frac{Y_w D}{R_w}\right) + (1-D)\right] R \tag{3.44}$$

$$G_c = \left[\left(\frac{Y_w D}{G_w}\right) + (1-D)\right] G \tag{3.45}$$

$$B_c = \left[\left(\frac{Y_w D}{B_w}\right) + (1-D)\right] B \tag{3.46}$$

Next, coefficients dependent on the viewing condition are defined as follows:

$$k = \frac{1}{5L_A + 1} \tag{3.47}$$

$$F_L = 0.2k^4 \left(5L_A\right) + 0.1 \left(1 - k^4\right)^2 \left(5L_A\right)^{1/3} \tag{3.48}$$

$$n = \frac{Y_b}{Y_w} \tag{3.49}$$

$$N_{bb} = N_{cb} = 0.725 \left(\tfrac{1}{n}\right)^{0.2} \tag{3.50}$$

$$z = 1.48 + \sqrt{n} \tag{3.51}$$

where Y_b is the background relative luminance (generally, it is set to the value of the luminance of white on the display divided by 5). In other words, Y_b is Y_W (=100) /5 = 20).

In order to apply post-adaptation nonlinear compression, the adapted R, G and B values are converted from the M_{CAT02} specification to Hunt-Pointer-Estevez cone fundamentals that more closely represent cone responses:

$$\begin{pmatrix} R' \\ G' \\ B' \end{pmatrix} = M_{HPE} M_{CAT02}^{-1} \begin{pmatrix} R_c \\ G_c \\ B_c \end{pmatrix} \tag{3.52}$$

$$M_{HPE} = \begin{pmatrix} 0.38971 & 0.68898 & -0.07868 \\ -0.22981 & 1.18340 & 0.04641 \\ 0.00000 & 0.00000 & 1.00000 \end{pmatrix} \tag{3.53}$$

$$M_{CAT02}^{-1} = \begin{pmatrix} 1.096124 & -0.278869 & 0.182745 \\ 0.454369 & 0.473533 & 0.072098 \\ -0.09628 & -0.005698 & 1.015326 \end{pmatrix} \tag{3.54}$$

The post-adaptation nonlinearities are given in the following equations:

$$R'_a = \frac{400 \left(F_L R'/100 \right)^{0.42}}{27.13 + \left(F_L R'/100 \right)^{0.42}} + 0.1 \tag{3.55}$$

$$G'_a = \frac{400 \left(F_L G'/100 \right)^{0.42}}{27.13 + \left(F_L G'/100 \right)^{0.42}} + 0.1 \tag{3.56}$$

$$B'_a = \frac{400 \left(F_L B'/100 \right)^{0.42}}{27.13 + \left(F_L B'/100 \right)^{0.42}} + 0.1 \tag{3.57}$$

These values are used to obtain opponent-colour responses and formulate correlates of colour appearance.

Initial opponent-type responses, a and b axes and hue h are calculated as follows:

$$a = R'_a - \frac{12 G'_a}{11} + \frac{B'_a}{11} \tag{3.58}$$

$$b = \frac{1}{9} \left(R'_a + G'_a - 2B'_a \right) \tag{3.59}$$

$$h = \tan^{-1} \left(\frac{b}{a} \right) \tag{3.60}$$

An eccentricity factor e_t is calculated as follows:

$$e_t = \frac{1}{4} \left[\cos \left(h \frac{\pi}{180} + 2 \right) + 3.8 \right] \tag{3.61}$$

An initial achromatic response A is calculated by weighted summation of the nonlinear adapted cone responses modified with the brightness induction factor;

$$A = \left[2R'_a + G'_a + \tfrac{1}{20} B'_a - 0.305 \right] N_{bb} \tag{3.62}$$

Lightness J is calculated from an achromatic response A, that for white A_w, the surround factor c and the base exponent z:

$$J = 100 \left(\frac{A}{A_w} \right)^{cz} \tag{3.63}$$

Brightness Q is calculated from lightness J, the achromatic response for white A_w, the surround factor c, and the luminance level adaptation factor F_L

$$Q = \frac{4}{c} \sqrt{\frac{J}{100}} \left(A_w + 4 \right) F_L^{0.25} \tag{3.64}$$

A temporary quantity t is computed, and chroma C, colourfulness M and saturation s are calculated based on t:

$$t = \frac{50\,000}{13} \left[\frac{N_c N_{cb} e_t \sqrt{a^2 + b^2}}{R'_a + G'_a + \left(21/20 \right) B'_a} \right] \tag{3.65}$$

$$C = t^{0.9} \sqrt{\frac{J}{100}} \left(1.64 - 0.29^n\right)^{0.73} \tag{3.66}$$

$$M = CF_L^{0.25} \tag{3.67}$$

$$s = 100 \sqrt{\frac{M}{Q}} \tag{3.68}$$

Although CIECAM02 provides a viewing-condition-specific method for the transformation of the tristimulus values, X, Y, Z, to or from perceptual attribute correlates, it needed improvements. The CIE proposed the CIECAM16 model as the revision of CIECAM02 [23]. It is simpler than the original CIECAM02, but it maintains the same prediction performance for visual data as the original model.

3.6   Digital colour spaces (such as sRGB and HSV)

Digital colour spaces are defined by only 8-bit digits, and the actual colour exhibited by those values depends on the devices. Thus, it does not directly show colour appearance. However, their structures reflect the visual property. HSV corresponds to the three attributes of human colour perception: hue, saturation and value.

They can be transformed into the colourimetric values after adequate calibration of the devices.

Standard RGB (sRGB) is a standard RGB colour space for use for displays, printers and the Internet, and it is standardised by the IEC (International Electrotechnical Commission) IEC 61966-2-1:1999.

sRGB defines the colour appearance of an image on the defined 'standard' display in the defined 'standard' viewing condition. The luminance level (maximum luminance) of the display is assumed to be 80 cd/cm², and the gamma value is 2.2 with an offset of 0. The chromaticity of a display is defined by the CIE standard illuminant D65 as a white point ($x = 0.3127$, $y = 0.3290$) and three primaries R ($x = 0.6400$, $y = 0.3300$), G ($x = 0.3000$, $y = 0.6000$), B ($x = 0.1500$, $y = 0.0600$).

The sRGB standard viewing condition is defined as follows.

Reference background: 20% (16 cd/m²) of the luminance of the display. The chromaticity coordinates of D65 ($x = 0.3127$, $y = 0.3290$).

Reference surround: 20% (4.1 cd/m²) of ambient luminance (= 64 lx/π). The chromaticity coordinates of CIE standard illuminant D50 ($x = 0.3457$, $y = 0.3585$).

Reference ambient illuminance level: 64 lx

Reference ambient white point: D50 ($x = 0.3457$, $y = 0.3585$)

Reference veiling glare: 0.2 cd/m²

3.6.1   *Transformation from CIEXYZ to standard RGB*

Transform from RGB to sRGB is shown in the following formula. First, nonlinear 8-bit RGB values are transferred to linear RGB values:

$$A'_{sRGB} = A_{RGB} / 255 \tag{3.69}$$

if $A'_{sRGB} \le 0.04045$

$$A_{sRGB} = A'_{sRGB} / 12.92 \tag{3.70}$$

if $A'_{sRGB} > 0.04045$

$$A_{sRGB} = \left\{ \left(A'_{sRGB} + 0.055 \right) / 1.055 \right\}^{2.4} \tag{3.71}$$

where A denotes R, G or B.

Linear sRGB values are transformed to XYZ values is follows:

$$\begin{pmatrix} X \\ Y \\ Z \end{pmatrix} = \begin{pmatrix} 0.4124 & 0.3576 & 0.1805 \\ 0.2126 & 0.7152 & 0.0722 \\ 0.0193 & 0.1192 & 0.9505 \end{pmatrix} \begin{pmatrix} R_{sRGB} \\ G_{sRGB} \\ B_{sRGB} \end{pmatrix} \tag{3.72}$$

Transform from XYZ values to sRGB values is as follows:

$$\begin{pmatrix} R_{sRGB} \\ G_{sRGB} \\ B_{sRGB} \end{pmatrix} = \begin{pmatrix} 3.2406 & -1.5372 & -0.4986 \\ -0.9689 & 1.8758 & 0.0415 \\ 0.0557 & 0.2040 & 1.0570 \end{pmatrix} \begin{pmatrix} X \\ Y \\ Z \end{pmatrix} \tag{3.73}$$

Transformation from sRGB to 8 bit RGB is follows:
if $A_{sRGB} \le 0.0031308$

$$A'_{sRGB} = 12.92 \times A_{sRGB} \tag{3.74}$$

if $A_{sRGB} > 0.0031308$

$$A'_{sRGB} = 1.055 \times A_{sRGB}^{\left(\frac{1}{2.4} \right)} - 0.055 \tag{3.75}$$

$$A_{sRGB(8)} = 255 \times A'_{sRGB} \tag{3.76}$$

where

$A_{sRGB(8)}$: 8 bit $(0 - 255)$ RGB value

A'_{sRGB} : $0 - 1$ normalized RGB value

A_{sRGB} : sRGB value

sRGB is very useful for colour management that translates the colours of an object (images or graphics) from their current colour space to the colour space of the output devices such as displays and printers. It made managing colour possible so that it can be reproduced in a predictable manner across different devices and materials. It has to be described in a way that is independent of the specific characteristics of the mechanisms and materials used to produce it (e.g., colour LCD displays and colour printers use very different mechanisms for producing colour). Thus, a colour management system is required that colours described using device-independent colour coordinates such as sRGB are translated into device-dependent colour coordinates for each device.

There are other colour spaces commonly used such as Adobe RGB, which has a wider gamut than sRGB.

3.7 Colour signals for television systems

There is no single standard, and different broadcasting systems exist [24]. NTSC (National Television System Committee) and PAL (Phase Alternating Line) are two main types of colour encoding systems for analogue televisions. The NTSC standard is popular in the United States and Japan, while PAL is more common in countries in Asia, Africa, Europe, South America and Oceania. Furthermore, progressive systems such as high-definition television (HDTV) are in use. There is a third standard, called SECAM (Sequential Couleur Avec Memoire or Sequential Colour with Memory), that is used in France and Eastern Europe. There are some technical differences among systems. For example, while NTSC uses a frame rate of about 30 frames per second (fps), PAL uses a frame rate of 25 fps. The PAL system offers an automatic colour correction, but NTSC needs manual colour correction.

However, colour signals are transmitted in the same way. The original three colour signals (red, green and blue) are transmitted using three discrete signals (luminance and two subtractive chromatic channels) that are recovered as three separate colours to create a colour image. This reorganisation resembles the post-retinal colour processing in the human visual system and allows the chromatic channels to be compressed, saving transmission bandwidth.

3.7.1 Basic colour television systems
Here we briefly describe some examples of the basic television system.

3.7.1.1 EBU Y'U'V'
The EBU standard is used for PAL and SECAM colour encodings. It assumes CIE standard illuminant D65 as white point ($x = 0.3127, y = 0.3290$) and three primaries [R ($x = 0.6400, y = 0.3300$), G ($x = 0.2900, y = 0.6000$), B ($x = 0.1500, y = 0.0600$)]. Chromatic channels U' and V' are calculated by subtracting luminance component Y' from Blue (B') and Red (R'):

$$U' = 0.492111 \left(B' - Y' \right) \tag{3.77}$$

$$V' = 0.877283 \left(R' - Y' \right) \tag{3.78}$$

Thus, the transformation matrix becomes the following:

$$\begin{pmatrix} Y' \\ U' \\ V' \end{pmatrix} = \begin{pmatrix} 0.299 & 0.587 & 0.114 \\ -0.147141 & -0.288869 & 0.436010 \\ 0.614975 & -0.514965 & -0.100010 \end{pmatrix} \begin{pmatrix} R' \\ G' \\ B' \end{pmatrix} \tag{3.79}$$

The inverse is as follows:

$$\begin{pmatrix} R' \\ G' \\ B' \end{pmatrix} = \begin{pmatrix} 1.000000 & 0.000000 & 1.139883 \\ 1.000000 & -0.394642 & -0.580622 \\ 1.000000 & 2.032062 & 0.000000 \end{pmatrix} \begin{pmatrix} Y' \\ U' \\ V' \end{pmatrix} \tag{3.80}$$

3.7.1.2   NTSC Y'I'Q'

The American NTSC standard was introduced in the early 1950s, and this colour coding is not in wide use now. It assumes illuminant C as white point ($x = 0.3101$, $y = 0.3161$) and three primaries [R ($x = 0.6700$, $y = 0.3300$), G ($x = 0.2100$, $y = 0.7100$), B ($x = 0.1400$, $y = 0.0800$)]. Chromatic channels I' and Q' are calculated by subtracting luminance component Y' from Blue (B') and Red (R'):

$$I' = 0.27\left(B' - Y'\right) + 0.74\left(R' - Y'\right) \tag{3.81}$$

$$Q' = 0.41\left(B' - Y'\right) + 0.48\left(R' - Y'\right) \tag{3.82}$$

Thus, the transformation matrix becomes the following:

$$\begin{pmatrix} Y' \\ I' \\ Q' \end{pmatrix} = \begin{pmatrix} 0.299 & 0.587 & 0.114 \\ 0.596 & -0.275 & -0.321 \\ 0.212 & -0.523 & 0.311 \end{pmatrix} \begin{pmatrix} R' \\ G' \\ B' \end{pmatrix} \tag{3.83}$$

The inverse is as follows:

$$\begin{pmatrix} R' \\ G' \\ B' \end{pmatrix} = \begin{pmatrix} 1.000 & 0.956 & 0.621 \\ 1.000 & -0.272 & -0.647 \\ 1.000 & -1.107 & 1.704 \end{pmatrix} \begin{pmatrix} Y' \\ I' \\ Q' \end{pmatrix} \tag{3.84}$$

The gamut of some digital colour spaces is shown in Figure 3.17. There can be different objectives in colour reproduction depending on a situation, such as spectral colour reproduction (equal spectral reflectances or relative spectral power distributions), colourimetric colour reproduction (equal chromaticities and relative luminances), exact colour reproduction (equal chromaticities, relative luminances and absolute luminances), equivalent colour reproduction (equal chromaticities, relative luminances and absolute luminances to ensure equal colour appearance), corresponding colour reproduction (chromaticities and relative luminances to ensure equal colour appearance when the original and reproduction luminance levels are the same) and preferred colour reproduction (a more pleasing appearance) [25].

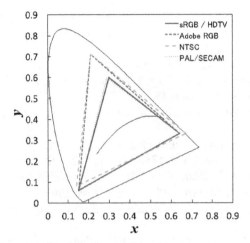

Figure 3.17   CIE chromaticity diagram showing the colour gamuts for various colour spaces

Spectral colour reproduction would be ideal, but almost impossible to achieve in present displays and television systems, and also colourimetric colour reproduction and exact colour reproduction would be difficult because of the limited gamut of a display, as shown in Figure 3.17. However, equivalent colour reproduction, corresponding colour reproduction and referred colour reproduction would be possible considering the property of colour vision.

3.8   Conclusion

It is important to consider and understand the characteristics of colour vision and how to describe colour properly when dealing with digital televisions. The human colour vision mechanism is very sophisticated. It has been developed to achieve efficient and adaptable coding. However, it makes the prediction of colour appearance difficult as these could be largely influenced by the surroundings, environment and viewing conditions. It is necessary to consider these factors in designing digital television systems.

References

[1]   *CIE 170-1:2006 fundamental chromaticity diagram with physiological axes-part 1*; 2006.
[2]   Hering E. *Outlines of a theory of the light sense.* ed. L.M.J. Translated by Hurvich D. Cambridge: Harvard University Press; 1878/1964.
[3]   Hurvich L.M., Jameson D. 'An opponent-process theory of color vision'. *Psychological Review.* 1957, vol. 64, Part 1(6), pp. 384–404.

[4]   Mullen K.T. 'The contrast sensitivity of human colour vision to red-green and blue-yellow chromatic gratings'. *Journal of Physiology*. 1985, vol. 359, pp. 381–400.

[5]   Van Nes F.L., Koenderink J.J., Nas H., Bouman M.A. 'Spatiotemporal modulation transfer in the human eye'. *Journal of the Optical Society of America*. 1967, vol. 57(9), pp. 1082–88.

[6]   De Valois R., Valois K.D. *Spatial Vision*. New York: Oxford University Press; 1988.

[7]   Kelly D.H. 'Luminous and chromatic flickering patterns have opposite effects'. *Science*. 1975, vol. 188(4186), pp. 371–72.

[8]   Berlin B., Kay P. *Basic Color Terms: Their Universality and Evolution*. Berkeley, CA: University of California Press; 1969.

[9]   Katz D. *The World of Colour*. London: Kegan Paul, Trench, Trubner & Co; 1935.

[10] Webster M.A. 'Adaptation and visual coding'. *Journal of Vision*. 2011, vol. 11(5), pp. 1–23.

[11] Brown R.O., MacLeod D.I. 'Color appearance depends on the variance of surround colors'. *Current Biology*. 1997, vol. 7(11), pp. 844–49.

[12] Foster D.H. 'Color constancy'. *Vision Research*. 2011, vol. 51(7), pp. 674–700.

[13] Mizokami Y., Ikeda M., Shinoda H. 'Color constancy in a photograph perceived as a three-dimensional space'. *Optical Review*. 2004, vol. 11(4), pp. 288–96.

[14] Fozard J.L. 'Vision and hearing in aging'. *Handbook of the Psychology of Aging*. 1990, vol. 3, pp. 143–56.

[15] Wuerger S. 'Colour constancy across the life span: evidence for compensatory mechanisms'. *PLoS One*. 2013, vol. 8(5), e63921.

[16] Werner J.S., Schefrin B.E. 'Loci of achromatic points throughout the life span'. *Journal of the Optical Society of America. A, Optics and Image Science*. 1993, vol. 10(7), pp. 1509–16.

[17] Enoch J.M., Werner J.S., Haegerstrom-Portnoy G., Lakshminarayanan V., Rynders M. 'Forever young: visual functions not affected or minimally affected by aging: a review'. *The Journals of Gerontology. Series A, Biological Sciences and Medical Sciences*. 1999, vol. 54(8), pp. B336–51.

[18] Pub C. *CIE 142-2001 Improvement to Industrial Colour-Difference Evaluation*. Vienna: Central Bureau; 2001.

[19] Pub C. *CIE 199:2011 Methods for Evaluating Colour Differences in Images*. Vienna: Central Bureau; 2011.

[20] *CIE 170-2:2015 fundamental chromaticity diagram with physiological axes-part 2*; 2015.

[21] Fairchild M.D. *Color Appearance Models*. John Wiley & Sons; 2013.

[22] *CIE 159:2004 a colour appearance model for colour management systems: CIECAM02*; 2004.

[23] *CIE 248:2022 The CIE 2016 colour appearance model for colour management systems: CIECAM16*. 2022.

[24] Reinhard E., Khan E.A., Oguz Akyuz A., Johnson G. *Color Imaging*. AK Peters Ltd; 2008. Available from https://www.taylorfrancis.com/books/9781439865200

[25] Hunt R.W.G. *The Reproduction of Colour*. John Wiley & Sons; 2005.

Chapter 4

Digital TV system approach

Marcelo Knörich Zuffo[1], Roseli de Deus Lopes[1], and
Laisa Caroline Costa de Biase[1]

4.1 Historical Perspective

Since the very early beginning of the human story, humans had developed the ability to share knowledge through immersive audiovisual narratives. An early conceptual model, of what nowadays we call television (TV), is a probable scenario 300 000 thousand years ago when humans invented the idea of joint together around a fire, where someone described some community event: a hunt, a tale from the tribe or any other casual life event relevant to the community at that time. Dynamic pictorial representations were necessary in the process and maybe the rupestrian paintings in caves around the world are media fossil remains of very early TV technology. It took centuries for humans to improve narrative techniques, by text, voice, pictures, music, fireworks, mimics, and other analog media.

The invention of photography, and immediately after the cinema, was a significant technological step in how humans communicate with each other through dynamic audiovisual narratives; the cinema brought audiovisual narratives to the masses, with significant large groups of people sharing an immersive audiovisual experience.

The invention of the vacuum tube [1] in the first years of the 20th century, and later an improved vacuum tube to display images on a phosphor screen, the cathode ray tube (CRT), [2] in conjunction with improvements in radio technology, led to the initial TV broadcasting in the first decades of last century. The broadcasting of the 1936 Summer Olympic Games in Berlin is another significant step, and the games were broadcast by the first TV station, the Fernsehsender "Paul Nipkow" (TV Station Paul Nipkow) [1].* At this moment, the commercial model was consolidated by mass propaganda associated with audience measurement concepts. Such

[1]Dept. Of Electronics Systems Enginnering, Escola Politécnica Universidade de São Paulo, São Paulo, SP, Brazil
*Vacuum tube or valves, are electrical circuits confined on near vacuum glass involvement, its functional principle is based on the photoelectric effect.

commercial models are a natural evolution of radio stations already consolidated during such period.

The emergence of electronics also was a significant step in TV evolution. Vacuum tube displays or CRTs were much smaller than huge cinema screens, and more than that: cheap and small enough for common citizens to buy and place them in the center of their family unit, the living room (Figure 4.1). The introduction of low-cost CRT technology was another significant step to widespread TV to the masses. Figure 4.1 despicts the cover of the August 1928 issue of Radio News magazine, published by Hugo Gernsback, featuring a couple watching "radio movies" at home. In the terminology coined by editor Hugo Gernsback, a "radio film" meant the transmission of a prerecorded film image, while "television" meant the transmission of a live image. The paper describes pioneering technologies employing disk scanning to transmit images from a conventional film strip. On the transmission side, strong light is thrown onto a scanning disk with 48 lenses. Light passes through pinholes in the disc, traveling through frames of film. The image is thus scanned in 48 separate horizontal "strips" during every fifteenth of a second. Each scanned image is received by a tube photocell and then transmitted.

Instant relevant audiovisual information becomes widely and massively spread on the globe by broadcasting TV, and a significant event in this technological cycle was the Apollo 11 live broadcasting of man landing on moon[†] on 20th July 1969 [3], where approximately 600 million people watched this event.[‡] Nowadays, TV broadcast can reach almost everyone in the planet earth in real time.

Last century, in the late 70s, the continuous need for audiovisual quality and the invention of the personal computer pushed the bounds of TV technology, which was the first step in the digital era. A high-definition TV (HDTV) trial demonstration was performed at the Los Angeles Olympics, introducing early digital technology concepts such as multimedia coding and signal multiplexing. The advent of the Internet was the last step to lead TV technology to all humans anywhere at any time, at incredibly low costs.

4.1.1   TV System

Fundamentally, a TV system's purpose is to remotely transmit an audiovisual narrative, usually designated by content, to a group of people or audience (Figure 4.2). The audience can range from a small group of people on a neighborhood community to billions of people on world-class mega events.[§] TV systems were a natural evolution of radio systems that have been invented just before.

A TV system can be decomposed into three major functional blocks: production, distribution and reception, according to Figure 4.3.

[†]How NASA broadcast Neil Armstrong live from the Moon, http://www.popsci.com/how-nasa-broadcast-neil-armstrong-live-from-moon.
[‡]http://www.telegraph.co.uk/news/science/space/5852237/Apollo-11-Moon-landing-ten-facts-about-Armstrong-Aldrin-and-Collins-mission.html.
[§]On 2014, the Brazilian Olympics Games have been watched by 5.00 billion people (https://www.reuters.com/article/olympics-rio-ioc-broadcast-idUKL1N1AY19I).

Figure 4.1   *Cover of Radio News magazine published in August 1928. Early TV Receivers Prototypes begin to become popular*

*http://www.vintageadbrowser.com/electronics-ads-1930s/5

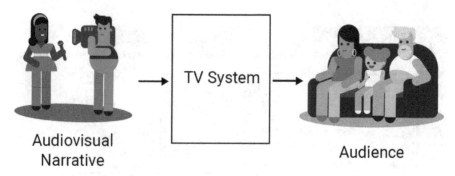

Figure 4.2   Purpose of a TV system

Production encompasses all activities related to the development of the narrative and its associated audiovisual content. From a technical perspective, production considers technologies for capturing, editing, processing, and storing audiovisual content that is ready to be distributed. Production includes another activity beyond technology such as financial, legal, regulatory, administrative, rights, journalism, history telling and artistic aspects. Productions can be real time or life when they are captured and transmitted immediately, or non-real time, when content is produced and delivered to the audience posteriorly.

Distribution is related to how audiovisual content is distributed to the audience; it considers the multimedia coding, modulation and signal transmission. The distribution also considers the infrastructure necessary for signal transmission including the TV station and the distribution network.

Since its origin, two major approaches for distribution technology have been considered: wired by the means of cables as a medium to propagate signals or wireless by the means of air propagation using terrestrial antennas and later, in the 1970s, using satellites in space.

The most common infrastructure considers wireless **terrestrial** or **over-the-air** broadcasting, but there are many other distribution technologies such as cable, both copper coaxial cable and fiber optics cable, satellite and more recently the Internet.

Terrestrial TV, broadcast TV or over-the-air TV is the most common TV distribution approach in which the television signal is transmitted by radio waves over the air from the terrestrial or Earth-based transmitter of a TV station to a TV receiver

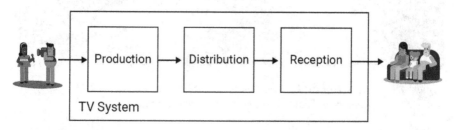

Figure 4.3   First-level block diagram of a TV system

Figure 4.4   TV reception

having an antenna at the audience location. The term "terrestrial" is used to distinguish this type of content distribution from modern distribution technologies such as satellite TV or direct broadcast satellite, in which the TV signal is transmitted to the receiver from an overhead satellite, and finally, we have cable TV, in which the signal is carried to the receiver through a cable that could be a coaxial metallic cable or a fiber optics cable.

Reception is related to how signals are received, processed and displayed to the final user, by the TV set, or TV receiver. Reception includes channel tuning, signal receiving, signal processing, audio and video decoding, display, front panel buttons, remote control and user interface.

From a very simple perspective, reception could be represented by a spectator in front of a TV set, with some limited control upon it, such as equipment on/off, channel selection and volume control, such an environment is presented in Figure 4.4. Over the years, TV receivers have become incredibly sophisticated, significantly increasing the interactivity between the viewer and the television. Manufacturers have improved their products in order to provide a better user experience when watching TV. Most of these improvements are related to the increased immersion and interactivity of the narrative. Modern televisions incorporate a variety of features, including 3D visualization, multi-channel audio, Internet connectivity, built-in cameras and adaptive light sensors, among others.

4.1.2   Analog TV

Analog TV is the original TV technology that uses analog electrical signals to transmit video and audio. In analog TV broadcast, information such as the brightness, synchronization, colors, and sound are represented by signal modulation or rapid variations of the amplitude, frequency or phase of the electrical signal.

Figure 4.5 represents an analog signal modulated in amplitude on the right and a digital signal with levels representing bits 0 or 1 on the left.

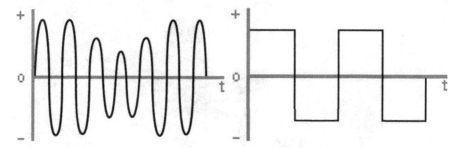

Figure 4.5   Analog signal (left) and digital signal (right)

Analog electrical signals can be modulated in amplitude (amplitude modulation [AM]), frequency (frequency modulation [FM]) and phase (phase modulation), and then transmitted using radio frequency (RF). Digital electrical signals are represented in levels; in such cases, electrical signals are discreet and voltage levels represent bits 0 or 1. Digital signals are also transmitted by RF; however, usually with the use of more sophisticated modulation algorithms that prevent and correct interference and errors. The binary representation of information can be enhanced with redundant information that can be reconstructed if some information is missed during transmission due to electromagnetic interference.

In the beginning, TV systems are fully analog based on electromechanical systems and later CRTs. Signals transmitted by RF were images in black and white transmitted in AM and audio was a single channel or mono transmitted in FM. Analog TV systems transmit RF signals in different carriers, so each one for different signal black and white images, audio, and synchronization signals. Later analog systems were improved to transmit additional color information and stereo or a dual channel audio.

Analog TV uses RF modulation to transport signals onto a very high frequency (VHF) channel or ultra-high frequency (UHF) channel. Each frame of a TV image is composed of lines drawn on the screen. The lines are an analog signal representing the varying brightness; the whole set of lines is drawn quickly enough that the human eye perceives it as one image, and this perceptual phenomenon of the human visual system is named image persistence. The next sequential frame is displayed, allowing the depiction of motion. The analog TV signal contains timing and synchronization information, so that the receiver can reconstruct a two-dimensional moving image from a one-dimensional time-varying signal.

Early analog TV systems are black and white, so the only relevant signals transmitted are the image luminance or black and white signal levels, organized in lines and frames. Spatial resolution was low with 480 lines by 640 columns. Image frame rates are displayed at 60 Hz[¶] frequency, just because that was the cheapest and

---

[¶]Some countries such as the Great Britain, alternate current is 50 Hz

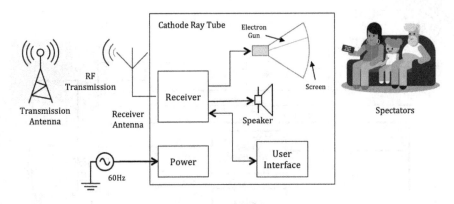

Figure 4.6   Analog TV set block diagram

widely available periodic frequency, which is the frequency from the (alternating current) from the power plug.

A practical analog TV system needs to take signals such as luminance (black and white), chrominance (in a color system), synchronization (control of horizontal and vertical resolution), and audio signals; these signals are broadcast over an RF transmission.

The TV set must receive this RF signal and properly display it on the CRT and speaker. The TV set needs to include a user interface panel for TV channel tuning or selection, sound volume control, contrast and brightness and power on and off. Figure 4.6 represents a high-level TV set block diagram.

An analog TV channel consists of two separated analog signals: the moving pictures are transmitted using AM on one frequency and the sound is transmitted with FM at a fixed offset frequency (typically 4.5–6 MHz) from the picture signal. Figure 4.7 presents the electronics block diagram of an early analog TV system, implemented with valve tubes.

The chosen channel frequencies represent a compromise between allowing enough bandwidth for video, satisfactory picture resolution and allowing enough channels to be packed into the available frequency band. In practice, a technique called vestigial sideband is used to reduce the channel spacing, which would be nearly twice the video bandwidth if AM was used.

Signal reception is invariably done via a superheterodyne receiver: the first stage is a tuner which selects a TV channel and frequency shifts it to a fixed intermediate frequency. The second stage is a signal amplifier that performs amplification of the intermediate frequency stages from the microvolt range to fractions of a volt.

4.1.3   TV Standards

The technology innovations on analog TV such as the introduction of color TV and technical specification discrepancies among manufacturers such as screen resolution led to different standardization efforts worldwide. Three main analog TV systems have been developed pushed by the color introduction in TV: National Television

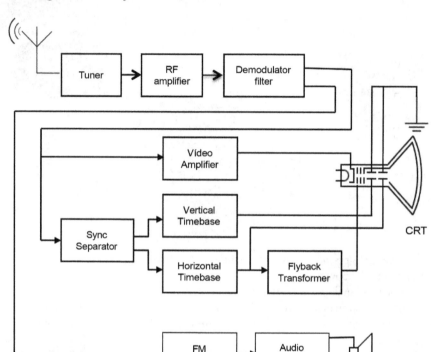

*Figure 4.7   Detailed block diagram of an analog TV set based on CRT
technology*

System Committee (NTSC), phase alternating line (PAL) and Séquentiel Couleur À
Mémoire (SECAM).

The NTSC is a standardization effort led by the FCC (Federal Communications
Commission), the US Goverment Telecommunications Regulatory Agency, prior
to the color introduction, aiming to prevent technology conflicts among analog TV
manufacturers in the USA, because NTSC efforts had started in the 1940s, further
in 1953 NTSC standardized color encoding. The PAL was a standard originally
developed by Telefunken in Germany and widely adopted in Europe; it has slightly
better performance in terms of resolution and color robustness compared to NTSC.
France decided to design its analog TV standard the SECAM, this was adopted by
some of its former colonies. Figure 4.8 shows the worldwide Analog TV distribution
by regions in the world.

4.1.4   Image Generation, Transmission and Display

At the early stage of TV systems, a fundamental problem was generating, transmitting, and displaying images in motion. Again the CRT technology was used both for
capturing and displaying images. To capture images early video cameras need to be

Figure 4.8   Analog TV systems standards

invented, they are CRTs with an inverse purpose: capturing an image that resulted in the **flying-spot scanner** (FSS). The FSS is a primitive video camera based on CRT technology, the output of the FSS-CRT was then transmitted by RF means. The image was just a number of analog signals from scan lines plus the synchronization signals to define the end of a line, a horizontal synchronization signal, and a synchronization signal to define the end of a frame, the vertical synchronization signal. Since the acquisition camera and the display are both analog CRTs with controlled electron beam, appropriate timing for vertical and horizontal repositioning or retrace is needed. Figure 4.9 shows the raster-scan image with respective signals.

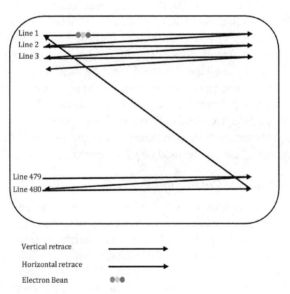

*Figure 4.9   Raster-scan image, in this case, we have the **progressive scan**,
            synchronization signals are used to define the end of a line and the
            end of a frame. The electron beam will excite the red, green, and
            blue phosphor.*

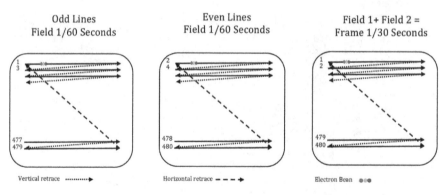

Figure 4.10   Interlaced scan lines are organized in even fields and odd fields that are scanned individually and interlaced on a final frame

Figure 4.9 presents the organization of the raster-scan image, in such case, lines are scanned sequentially or progressively, so such technique is named **progressive scan**. The whole image is scanned in 1/60 seconds or 60 Hz, so 60 images are displayed per second.

For years, a consensus has been established around a convenient number of lines and display form factors. The NTSC standardized the total number of lines transmitted to be 480 lines and also the frame form factor of 4:3 (columns:lines) resulting in a frame resolution of 640 columns and 480 lines.

Total transmission bandwidth becomes an intrinsic problem since the information in high-resolution frames does not fit in the available bandwidth. With this limitation, engineers invented the interlaced scan. An interlaced scan is a technique for doubling the perceived frame rate of a video display without consuming extra bandwidth. Figure 4.10 presents the interlaced scan lines of a TV image frame. The interlaced signal contains two fields of a video frame captured sequentially at two different times. This enhances motion perception to the viewer and reduces flicker by taking advantage of the eye persistence phenomenon.

4.1.5   The composite Video Signal

The analog video signal also named video carrier is demodulated after transmission to give the **composite video signal**. The composite video is an analog signal that combines luminance, chrominance and synchronization signals both for vertical and horizontal retracing. This video signal format is widely used by analog video systems until the introduction of digital TV (DTV). Figure 4.11 shows: a) the structure of a composite video signal with luminance, and chrominance signals and horizontal synchronization pulses (H-sync), b) the composite video signal in grey scale (luminance only), and c) the composite video signal of a color bar where chrominance and luminance signals are mixed, the white bar on the screen is the peak signal (0.7Vpp).

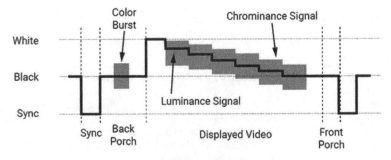

a) Structure of a Composite Video Signal

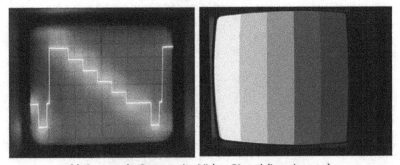

b) Greyscale Composite Video Signal (Luminance)
and Equivalent TV Greyscale bar

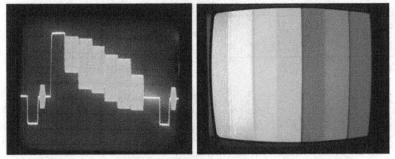

c) Color Composite Video Signal (Luminance + Chrominance)
and Equivalent TV Color bar

Figure 4.11   Structure of a composite video signal

4.1.6   Display technology and resolution

The **display resolution** of a TV is the number of distinct pixels in each dimension that can be displayed in time. The display resolution can consider its spatial density on a single frame or **spatial resolution**, and also the **time resolution**, which means the number of frames displayed per second. Display resolution can be an ambiguous

Table 4.1   Spatial resolution of TV systems

Spatial resolution	Description
Analog	**Analog and early digital systems**
352 × 240	Video CD (Compact Disk)
333 × 480	VHS (Video Home System), Video8, and Umatic
350 × 480	Betamax
420 × 480	Super Betamax and Betacam
460 × 480	Betacam SP, Umatic SP, and NTSC (National Television Standard Committee)
580 × 480	Super VHS, Hi8, and LaserDisc
700 × 480	Enhanced Definition Betamax and analog broadcast limit (NTSC)
768 × 576	Analog broadcast limit (PAL and SECAM)
Digital	**Modern digital systems**
500 × 480	Digital8
720 × 480	D-VHS, DVD (Digital Video Disc), miniDV, and Digital Betacam (NTSC)
720 × 576	D-VHS, DVD, miniDV, Digital8, and Digital Betacam (PAL/SECAM)
1280 × 720	D-VHS, HD DVD, Blu-ray, and HDV (miniDV)
1440 × 1080	HDV (miniDV)
1920 × 1080	HDV (miniDV), AVCHD (Advanced Video Codec High Definition(, HD DVD, Blu-ray, and HDCAM SR
1998 × 1080	2K flat (1.85:1)
2048 × 1080	2K digital cinema
3840 × 2160	4K UHDTV and Ultra HD Blu-ray
4096 × 2160	4K digital cinema
7680 × 4320	8K UHDTV
15360 × 8640	16K digital cinema
61440 × 34560	64K digital cinema

term especially as the displayed resolution is provided by either analog or digital technologies. Early analog TV systems using CRT technology consider their resolution in lines. Modern DTV systems using flat displays consider their resolution in pixels.

Spatial resolution is usually quoted as width × height, with the units in pixels: for example, "1024 × 768" means the width is 1024 pixels and the height is 768 pixels. The spatial resolution also considers the **picture form factor,** where the ratio between columns and lines observe a fixed proportion such as 4 × 3 or 16 × 9. Time resolution is usually quoted in frames per second. Table 4.1 presents the spatial resolution in pixels for several different analog and digital systems. Figure 4.12 presents the spatial resolution evolution for comparision purposes.

Table 4.1 is a list of traditional, analog horizontal resolutions for various media. The list only includes popular formats, not rare formats, and all values are approximate because the actual quality can vary from machine-to-machine or tape-to-tape.

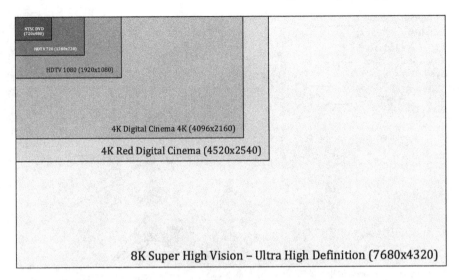

Figure 4.12   Spatial resolution evolution

For ease of comparison, all values are for the NTSC system. (For PAL systems, replace 480 with 576). Analog formats usually had less chroma resolution.

4.1.7   Flat display technology

Imaging technologies with electronic screens have evolved enormously in recent years. An electronic screen is a technological interface with information capable of stimulating the human visual system through the accurate reproduction of colors and images. Until the 1990s, it was relatively easy to buy a TV set, at that time the only technology available was the CRTs. Despite the advanced stage of technological maturity, the main disadvantage of CRT TVs was their large volume and weight, a 29-inch TV could weigh up to 50 pounds, which made it impossible to develop larger TVs and picture resolution, thus the weight and volume of CRT TVs drove the development of TVs using flat panel monitor technologies.

In the beginning, two flat panel display technologies emerged: plasma technology, or plasma display panel (PDP), and liquid crystal display (LCD) technology. Essentially the PDP and LCD technologies are very different, although both are considered flat panel technologies as is also the case with OLED (Organic Light-Emitting Diode) technology.

In PDP TV, the image formation is based on millions of colored microlamps of plasma or ionized gas. Thus, the flat panel using PDP technology is a sandwich composed of two sheets of glass, and in the middle, there is a mesh of RGB (Red, Green, Blue) pixels, each pixel is a cell or fluorescent lamp where there is a confined gas, when gas is electrically stimulated or ionized, it becomes a plasma that in interaction with the colored phosphor emits light. The chemical composition of phosphorus corresponds to the emission of light in blue, green or red.

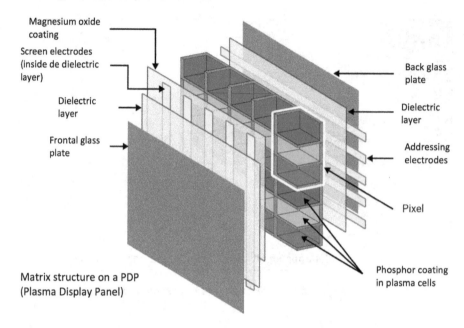

Magnesium oxide coating

Screen electrodes (inside de dielectric layer)

Dielectric layer

Frontal glass plate

Back glass plate

Dielectric layer

Addressing electrodes

Pixel

Phosphor coating in plasma cells

Matrix structure on a PDP (Plasma Display Panel)

Figure 4.13   Schematic structure of a plasma display panel

Figure 4.13 shows the PDP technology flat panel sandwich structure, as each point of light on the PDP screen emits light, we call this technology also as the emissive flat panel technology. An excellent quality of PDP technology is the viewing angle, as each point of the screen emits light, the viewing angle of the PDP screen is virtually 180°. A disadvantage of PDP technology is the difficulty of implementing smaller micro lamps, so this limitation makes plasma screens usually large and with not so high resolution.

In LCD TV, the image formation is based on millions of colored micro-windows that open and close by filtering a white backlight, the principle of opening and closing the micro-windows is based on the polarization of light using polarizing filters combined with electrically actuated liquid crystal dot arrays also polarize light, allowing light to pass through each point of the screen. Figure 4.14 shows the working principle of the liquid crystal dot matrix screen.

The LCD technology necessarily needs a white backlight unit. In the early days of early LCD TVs, white light was generated from fluorescent lamps, most recently white light is generated from an array of white light emitting diodes. Since every dot or pixel, on the LCD screen does not emit light and does not let light through, so we call LCD technology non-emissive or transmissive technology. An intrinsic problem with backlight displays is the leakage of this light (light leakage). This problem exists because LCD crystals cannot fully block light when pixels are black, light leakage directly influences the contrast of non-emitting flat screens. Another problem with LCD screens is that the color response depends directly on the backlight. Because the backlight is artificial, it is not considered completely white, so

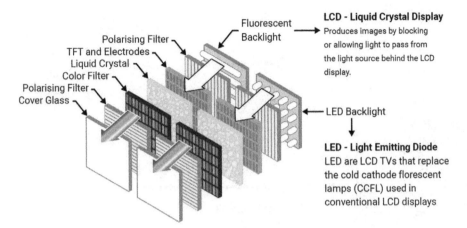

Fluorescent Backlight

Polarising Filter
TFT and Electrodes
Liquid Crystal
Color Filter
Polarising Filter
Cover Glass

LCD - Liquid Crystal Display
Produces images by blocking
or allowing light to pass from
the light source behind the LCD
display.

← LED Backlight
↓

LED - Light Emitting Diode
LED are LCD TVs that replace
the cold cathode florescent
lamps (CCFL) used in
conventional LCD displays

Figure 4.14   Schematic structure of an LCD display panel (a) LCD with fluorescent lamps and (b) LCD with LEDs

there are color distortions associated with the type of backlight adopted, in this case, fluorescent lamps or LEDs (Light-Emitting Diode).

Figure 4.14 shows the mounting structure of an LCD TV, both the LCD TV using fluorescent lamps (Figure 4.14a) and the LCD TV with LEDs (Figure 4.14b), which we will henceforth call LCD-LED.

When LCD screens first appeared in the 1980s, backlight technology based on fluorescent lamps was used. Subsequently, with the development of ultra-bright white LEDs, it was possible to incorporate these LEDs into the background of LCD TVs, allowing the construction of even thinner TVs.

4.2   DTV systems

DTV is a natural evolution of analog TV, where content is fully digital from production, distribution and receiving. When emerged in the late 1980 decade, it was a disruptive innovative audiovisual service that represents a significant evolution in TV technology since the introduction of analog color TV. Several regions of the world are in different stages of adaptation and are implementing different DTV systems standards.

4.2.1   Transport streams and multiplexing

A key concept in DTV systems is multiplexing. On a DTV system, the distribution of digitally processed audio, video signals and data could be multiplexed on a transport stream. The transport stream is a sequence of serial bits encoded and it is measured by bits/seconds, a sequence of images evolves some Megabytes of data, so normally the transport stream is quantified by Mbits/s. Figure 4.15 presents the multiplexing

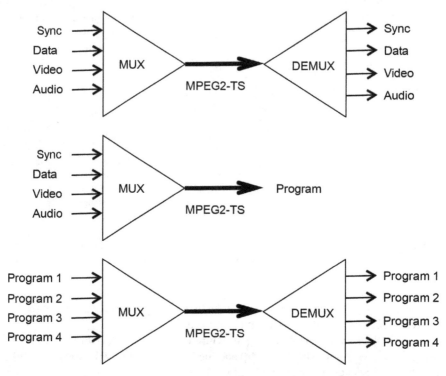

Figure 4.15   *DTV system multiplex coded content sources in a MPEG2 transport stream*

concept, in such a case a number of individual transport streams of video, audio and metadata are multiplexed in another transport stream and later de-multiplexed.

Multiplexing gives a lot of flexibility in terms of signal processing, distribution and storage on DTV systems. For example, the multiplexing approach allows DTV systems to support more than one program in the same channel bandwidth, this feature is called as multiprogramming. Eventually, multiple audio signals could be multiplexed on the same program to support multiple languages.

The advent of modern DTV systems arose with the development of multimedia standards to compress, encode, and transmit audiovisual content by digital means or transport streams. In 1988, the moving picture experts group (MPEG) was established. MPEG [4] is a working group of experts that was formed by the International Standards Organization (ISO) and IEC (International Electrotechnical Commission) to set standards for audio and video compression, encoding and transmission. MPEG evolves a wide range of standards to compress, encode, transport, decode, and uncompress audiovisual content encompassing a range of state-of-the-art multimedia processing techniques and algorithms.

The standardized transport stream specifies a container format encapsulating packetized elementary streams, with error correction and stream synchronization

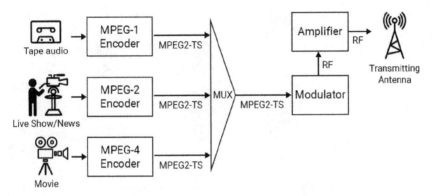

Figure 4.16   *Several programs in different formats and MPEG encoding standards can be transmitted at the same channel on DTV*

features for maintaining transmission integrity when the signal is degraded. MPEG transport stream (MPEG-TS, MTS or TS) is the MPEG standard digital container format for transmission and storage of audio, video, and program and system information protocol data. The transport stream is specified in MPEG-2 Part 1, Systems, formally known as ISO/IEC standard 13818-1 or ITU-T Rec. H.222.0. It is used in almost all DTV standards such as digital video broadcasting (DVB), advanced television systems committee and integrated services digital broadcasting terrestrial (ISDB-T) and Internet protocol (IP) broadband DTV systems (IPTV). Figure 4.15 presents a diagram how MPEG2-TS is generated, transmited and decoded.

The MPEG-TS transport streams differ from the similarly named MPEG program stream in several important ways: program streams are designed for reasonably reliable media, such as discs (like DVDs), while transport streams are designed for less reliable transmission, namely terrestrial or satellite broadcast. Furthermore, a transport stream may carry multiple programs. Figure 4.16 presents a diagram where multiple programs in different formats are multiplexed and streamed by MPEG-TS.

4.2.2 Multimedia compressing and encoding

MPEG committee also standardized the compressing and decompressing of multimedia information encompassing a range of encoding and decoding standards. These standards are well known as MPEG-1, MPEG-2, MPEG-4, and more recently MPEG-5.

Created in 1988, MPEG-1 is the first standard released by the MPEG committee and aims for lossy compression of video and audio. It was designed to compress VHS-quality raw digital video and CD audio down to 1.5 Mbit/s, with a video compression ratio of 26:1 and audio compress ratio of 6:1. The development of MPEG-1 ramped up a number of digital technologies such as CDs, digital cable, and satellite TV and the digital audio broadcasting service in the early 1990s. MPEG-1 is still the most widely adopted multimedia encoding standard, and its most well know part is its audio format the MP3 (MPEG-1 Audio Layer 3).

Figure 4.17   DTV system standards

MPEG-2 or H.222/H.262 as defined by the International Telecommunications Union (ITU), is an evolution of MPEG-1 audiovisual compression, it is an enhanced version of MPEG-1 supporting advanced features such as HDTV video format and multichannel 5:1 audio format. The advent of MPEG-2 made available the DVD format and the first DTV systems that adopted MPEG-2 for SDTV (Standard Digital Television) and HDTV (High Definition Digital Television) broadcasting.

MPEG-4 (H.264/AVC) and MPEG-5 (H.265/HEVC), are the newer and last state-of-the-art multimedia encoding standards proposed by MPEG.

4.2.3   DTV standards

DTV Standardization was a natural step with the introduction of DTV worldwide, interoperability, local regulation, industry policies, and industry interests guided these effors in national and international terms [5]. Figure 4.17 presents the global distribution of Digital TV standards adopted in different regions and countries of the world.

Below are the different widely used DTV broadcasting standards worldwide:

- DVB uses coded orthogonal frequency-division multiplexing (OFDM) modulation and supports hierarchical transmission [6]. This standard has been adopted in Europe, some countries in Latin America, Africa, Singapore, Australia, and New Zealand.
- Advanced TV system committee uses an eight-level vestigial sideband for terrestrial broadcasting. This standard has been adopted by six countries: the US, Canada, Mexico, South Korea, Dominican Republic, and Honduras.
- ISDB is a system designed to provide good reception to fixed receivers and also portable or mobile receivers. It uses OFDM and two-dimensional interleaving. It supports hierarchical transmission of up to three layers and uses MPEG-2 video and advanced audio coding. This standard has been adopted in Japan and the Philippines. ISDB-T International is an enhanced version of ISDB developed in Brazil using H.264/MPEG-4 AVC that has been widely adopted in South America and is also being embraced by Portuguese-speaking African countries.

- Digital terrestrial multimedia broadcasting adopts time-domain synchronous OFDM technology with a pseudorandom signal frame to serve as the guard interval of the OFDM block and the training symbol. The digital terrestrial multimedia broadcasting standard has been adopted in the People's Republic of China, including Hong Kong and Macau.

4.2.4 IPTV Internet Protocol Digital Television

In recent years, with the massive and global expansion of the Internet, DTV has become widely distributed by the Internet, for such purpose a specific protocol has been developed namely IPTV (Internet Protocol Digital Television. IPTV is a term that encompasses a set of Internet technologies where it is possible to deliver television services using the Internet protocol suite over a packet-switched network such as LAN (Local Area Network, WAN (Wide Area Network) or Internet, instead of being delivered via satellite signal. commercial terrestrial, cable television formats. In recent years IPTV become highly interactive, since receivers could be any computer connected to internet such as smartphones, tablets, laptops, desktop computers and recent TV receivers that embed advances computers on it.

4.2.5 Spectrum management and regulatory aspects

The terrestrial TV system for each country will specify the spectrum allocation in segments or bands or the number of TV channels within the UHF or VHF frequency ranges. The channel bandwidth varies from country to country ranging from 5MHz to 8MHz. The channel allocation is standardized by ITU. Spectrum is a scarce resource and strategic for military, commercial, and political mass communication purposes, so governments around the globe tend to make strict and complicated regulations around that. These regulations, local requirements and industry interests made efforts to standardize TV systems with a large fragmentation around the globe.

Proper regulation to allocate channels on the available spectrum needs to be conceived in order to avoid disputes among TV service providers and allocate spectrum to public services such as emergencies, military, and public TVs. The regulation efforts had started by 1940, and slight change from country to country. Figure 4.18 presents the channel allocation plan for some countries worldwide.

4.3   Interactive broadband and broadcast TV

In conjunction with the development and evolution of DTV transmission standards, a number of complementary solutions were developed to enable the provision of additional interactive content together with linear audio-visual content. A major step in Europe was the development of the interactive TV platform named multimedia home platform (MHP). The MHP applications are transmitted within the broadcast signal and are executed mostly by a Java engine in the end-users device. While MPH was predominantly successful in Italy, it failed in other European countries.

The reasons are manifold but patent issues played an important part in the non-success and also the fact that the knowledge of complex programming languages was required in order to write MHP applications. A similar approach was implemented in Brazil. Research institutes were assigned by the Brazilian government to develop a DTV standard complemented by a facility that improves the access to information to cope with and overcome the digital divide in Brazil. The result of these efforts was the ISDB-Tb (Integrated Services Digital Broadcasting Terrestrial Brasil) standard with its Ginga middleware – a more advanced system that has certain technical advances in relation to its precursors and is royalties-free.

Ten years after the development of MHP, Europe faced a turning point regarding the concept for the provisioning of interactive TV content. With the increasing availability of broadband connections in European households, Consumer Electronics manufacturers started to equip their devices with web browsers. Since implementations of browsers in these so-called Connected TVs or Smart TVs vary substantially between devices of different brands, the development of interoperable applications became very cumbersome. For the same functionality, a new application would have to be developed for each (proprietary) Smart TV system. This motivated the compilation of a standard aiming to define a common vocabulary to be understood by all Connected TVs. Moreover, the standard was to define a link between the broadcast signal and the broadband applications. With the implementation of these requirements, hybrid broadcast-broadband (HbbTV) was born.

Similar to the systems in Europe and Brazil, a number of other interactive TV systems came up all over the globe – aligned to the specific regional circumstances and conditions. Two major consequences follow from this technological diversity: the technical implementation efforts which are required to support a wide range of different specifications are very high for manufacturers and consequently, the quality and the precision of implementation often suffer under time pressure. Hence, interoperability of applications based on a certain standard is

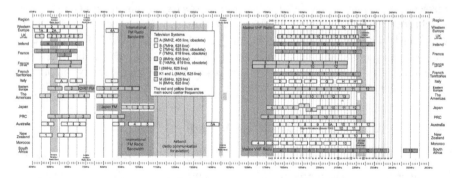

Figure 4.18   **VHF Channel Allocation Plan in different countries for Television, FM radio, amateur radio, marine radio, and aviation*

difficult to ensure across all available devices which in theory should support this standard. Another consequence is the increased complexity of developing globally available services in a fragmented market environment – unlike it has evolved on the Web which builds its success on global standards. Developers of TV applications must cope with all the different peculiarities specific to the different TV systems.

Increasing broadband availability has changed the game since it can be expected that prerequisites and requirements (including the availability of broadband access) will further converge in near future on a global level, there is no motivation to maintain this cumbersome diversity of technologies between the existing ITV (interactive Television) systems. At the same time, the possibilities coming with the emerging web technologies enabling personalized and social experience will be almost unbridled. An ITV system designed to have an impact in the future must consider these technologies and harvest their potential.

4.3.1   Connected TV or Smart TVs

In the past years, the world has witnessed a rapid market penetration with different flavors of Connected TV devices (also known as Smart TV). All consumer electronics manufacturers are now offering TV sets and set-top boxes which offer a connection to the Internet over Ethernet or WiFi. All these devices are capable of accessing the Internet-based service – yet limited to very different kinds of specific offerings. The market fragmentation is immense and while most devices can be used to download and install applications and surf the web – there are increasing problems due to vendor lock-in. These issues are more and more evident to the end-users. Manufacturers have become gatekeepers in a complex market of media, applications and national rights. A typical example is Amazon's video-on-demand service "LoveFilm": while in the UK, the service can (only) be accessed from Connected TVs of three major manufacturers, it is even more restricted in countries like Germany, where exclusive contracts are limiting the use in one manufacturer's device only. Examples like this demonstrate the non-transparent situation and have the potential to endanger the success of Connected TV.

Thus, while Connected TVs are based on well-established Internet technologies, the solutions we mostly see today contradict the Web paradigm and its openness. Even though some manufacturers include accessible Web browsers in their devices, they hardly offer any alternative, mainly due to the fact that regular websites are not really navigable on a TV by means of conventional remote control. Also, current TV sets and set-top boxes are limited and restricted in technology (e.g. low-performance CPU) as well as in format support and extensibility – i.e. they mostly do not (fully) support HTML5, Flash, Silverlight, Java etc. This very commonly leads to odd user experiences. In addition, a Connected TV set is not necessarily apt for truly hybrid interactive viewing experiences. While, in general, all Connected TV sets have two inputs – one for the broadcast signal (DVB or ISDB-T TV tuner) and one for the Internet (Ethernet/WiFi) connection – they do not necessarily offer converged services by making use of both distribution paths. In addition, add-on solutions like Apple TV, Boxee, and Google TV which do not offer a TV tuner at all are also pushing into the market.

In summary, these Connected TVs and add-on devices are usually only equipped for accessing proprietary portals for content and applications via the Internet and are under the control of the device manufacturers.

They either operate as a (limited) Internet terminal or as a broadcast device. This cannot be considered as converged equipment and the user eventually ends up with a (collection of) multi-purpose device(s) that just allow the viewing of broadcast TV content or using separated and limited add-on functionalities through the Internet connection on the same screen.

For the truly hybrid services enabling a seamless user experience, an "engine" is required that links the broadcast content offered via satellite, terrestrial, cable or IPTV networks and the Internet content offered via the IP channel. Only truly hybrid systems provide such an engine that is activated via appropriate signaling within the broadcast transport stream.

4.3.2   Ginga in Brazil

The Brazilian government opted to implement digital terrestrial TV as a derivation of the Japanese DTV standard, ISDB-T. The Brazilian derivate ISDB-Tb, also called SBTVD (Brazilian System on Digital Television), basically differs from the original Japanese ISDB-T by using H.264/MPEG-4 AVC as the video compression standard (ISDB-T uses H.262/MPEG-2), a presentation rate of 30 frames per second even in portable devices (ISDB-T, one seg, uses 15 frames per second for portable devices) and an interaction engine based on the middleware "Ginga" [8], composed by two programming flavours Ginga-Nested Context Language NCL (declarative programming) and Ginga-Java (procedural programming) modules. In the Japanese market, ISDB-T is complemented by the XML-based BML (broadcast markup language).

Ginga is an intermediate software layer between the operating system and applications (comparable to MHP). It has two main functions: one is to make applications independent of the operating system on the hardware platform used and the other is to better support the development of applications. Figure 4.19 presents the Ginga Middleware high level architecture with all modules relationships.

The "Ginga" specification is responsible for supporting the interactivity. Ginga is divided into two major integrated subsystems: the common core (Ginga-CC) and the application execution environment. The execution environment supports the execution of declarative NCL applications (Ginga-NCL) integrated into the runtime environment of Java applications (Ginga-J). Both environments are required by the Brazilian Terrestrial DTV for fixed terminals, while the Ginga-J is optional in the case of portable terminals. Also for IPTV DTV systems, according to ITU-T recommendation H.761 only the Ginga-NCL is mandatory.

The HbbTV is a hybrid TV framework proposed in Europe Built upon the DVB broadcasting standards [9], HbbTV provides an engine for interlinking broadcast and broadband content. The HbbTV encompasses the necessary signaling and specifies a CEHTML-based browser that has combined access to both, the data in the broadcast stream as well as to the services, applications and content provided via the

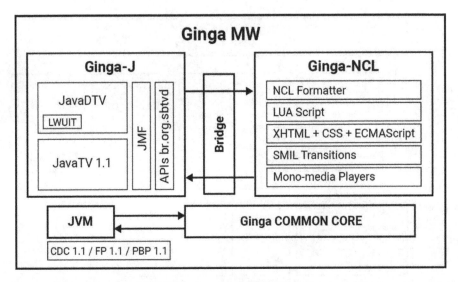

Figure 4.19   High-level architecture of the Ginga middleware

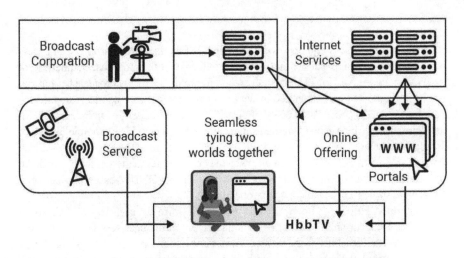

Figure 4.20   The Hybrid TV paradigm as implemented in HbbTV

Internet. Figure 4.20 presents the HbbTV paradigm where both broadcast services and internet are combined to deliver TV services to the home.

The HbbTV [10] specification is based on existing standards and Web technologies including open IPTV forum (OIPF), CEA (Consumer Electronics Association), DVB (Digital Video Broadcasting Association) and W3C (World Wide Web Consortium). Currently, requirements are collected and discussed to jointly define the next HbbTV specification, version 2.0, which is expected to align further with successful technologies from the Web like HTML5.

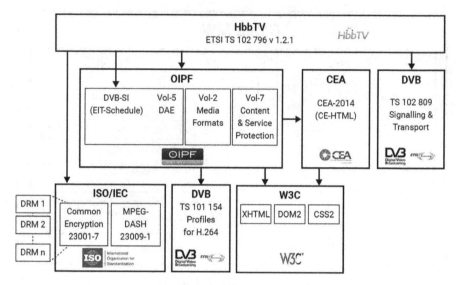

Figure 4.21   HbbTV 1.5 (ETSI TS 102 796 v1.2.1) and how it references to other standards

On a TV set that is equipped with the HbbTV function, the consumer can access interactive services, applications and content provided over the broadcast channel and/or the IP connection. All HbbTV services can be offered directly by a broadcaster, but can also be implemented and integrated by service providers, network operators and device manufacturers. This opens a diversity of possible business models, even independent of broadcasting, to which HbbTV remains agnostic as a platform. For example, a service channel may be used to signal an application which, by means of the HbbTV auto-launch functionality, is immediately kicked off when the device tunes to the service channel. An example is Eutelsat's commercial "Kabelkiosk choice," an HbbTV-based service portal that is pre-configured for cable network operators.

Generally, HbbTV is being used to build a platform for a variety of TV broadcasters, be it big or small, and have national, regional or local service areas. Some smaller TV broadcasters of them may not have the financial resources to pay for a 24/7 broadcast channel but share it with others or even offer Internet streaming only. The HbbTV-based portal allows for providing a harmonized view of all offered services and automatically redirects the consumer to the selected service regardless of whether it is delivered via broadcast or the Internet.

However, as stated above, HbbTV is not limited to broadcast-centric services. It can likewise be used to build application portals and platforms which provide Internet-only services.

Stakeholders are manufacturers but also independent application portals. Currently, the great majority of HbbTV services are broadcast-centric and thus link a given TV program with the portal of the program provider. Figure 4.21 presents the relationship among HbbTV and other relevant multimedia standards. Other MHP services are still in operation in Italy while the situation in the UK remains to be highly fragmented: With MHEG-5 service in operation on free TV channels, proprietary OpenTV middleware usage on Pay TV and the roll-out of the proprietary approach YouView, many market participants are searching for harmonization. DTG's D-Book 7 addresses this issue in Part B where it extends the interoperability requirements of Part A into the world of Connected TV. The D-Book defines a platform for signaling, transport, and presentation of interactive Web applications within the UK digital terrestrial TV platform. This document makes reference to ETSI TS 102 796 1.2.1 (HbbTV) as a baseline minimum requirement with additional features to meet the UK service provider requirements, which have been specified through profiling specifications from OIPF, W3C, MPEG, and DVB. In order to encourage harmonization, the DTG is working towards the adoption of these additional requirements within appropriate international specifications. Further requirements on buffering and seamless advert insertion are currently being investigated and will be included in a subsequent release of this document.

4.4 Conclusion

Considering the evolution of Humanity, Television is still a young technology with about 80-90 years of development. Essentially a technology to present immersive and interactive audiovisual narratives to remote audiences. The advent of highly integrated microelectronics and the digital computer in the 1980s has boosted technology considerably in recent years. Image spatial, color and frequency resolution are all significantly increased, delivering ultra-realistic images for every home. The advent of Digital TV in the 1990s triggered a process of innovation in the television industry that lasted almost 30 years. Recent advances are pushing for fully immersive TVs such as the introduction of 8K Super Retina displays, virtual reality streaming and flexible displays. Some recent advances in TV technology, such as holography, could have been considered science fiction a few years ago. In essence, Television is a technology for telling stories at a distance, as they provide great pleasure, and it is a necessity for human beings, a universal platform for storytelling that will continue to incorporate narrative and technological innovations as long as humanity persists.[**]

References

[1] *The TV station Paul Nipkow*. Available from https://en.wikipedia.org/wiki/Fernsehsender_Paul_Nipkow [Accessed 9 Jul 2017].
[2] *Vacuum tubes*. Available from https://en.wikipedia.org/wiki/Vacuum_tube
[3] Bird T.S. 'Role of the parkes radiotelescope in the first moon landing'. Presented at 2019 IEEE International Symposium on Antennas and Propagation and USNC-URSI Radio Science Meeting; Atlanta, GA, USA.

[**]https://en.wikipedia.org/wiki/Olympics_on_television

[4]   Chiariglione L. 'MPEG: achievements and future projects'. *ICMCS99*; Florence, Italy, 1999. pp. 133–38.

[5]   Ong C., Song J., Pan C., Li Y. 'Technology and standards of digital television terrestrial multimedia broadcasting [topics in wireless communications]'. *IEEE Communications Magazine*. 2010, vol. 48(5), pp. 119–27.

[6]   Bensberg G. 'Digital satellite transmission standards: an overview of the DVB satellite specification and its implementation' in *IEE colloquium on digitally compressed TV by satellite*. London, UK; 1995. p. 8.

[7]   Laflin N. 'Frequency planning for digital terrestrial television' in *IEE colloquium on digital terrestrial television*. London, UK; 1993. p. 10.

[8]   Gomes Soares L., Moreno M., Salles Soares Neto C., Moreno M. 'Ginga-NCL: declarative middleware for multimedia IPTV services'. *IEEE Communications Magazine*. 2010, vol. 48(6), pp. 74–81.

[9]   Merkel K. 'Hybrid broadcast broadband TV, the new way to a comprehensive TV experience'. *2011 14th ITG Conference on Electronic Media Technology*; Dortmund, Germany, 2011. pp. 1–4.

[10]   Kunic S., Sego Z. 'Analysis of the hybrid broadcast broadband television standard – HbbTV'. Presented at 2019 International Symposium ELMAR; Zadar, Croatia. Available from https://ieeexplore.ieee.org/xpl/mostRecentIssue.jsp?punumber=8911288

Chapter 5

Digital television basics

Konstantin Glasman[1]

5.1 Signals and systems

5.1.1 Analog signals and systems

Transmission and reception of television programs and exchange of music clips are examples of information communications carried out with the help of communication systems. A general model of the communication system (Figure 5.1) assumes the existence of a source of information that sends a message to the destination [1]. A message is a generalized concept that covers all types of transmitted information in any form – from texts of letters and numerals to speech, music and television images. A transmitter converts the message into a signal suitable for transmission over the communication channel. A signal is a physical representation of a message, and it displays a sequence of values of a certain message parameter. The message arriving at the system input can be called the primary signal directly representing the transmitted message. The receiver processes the received signal and reconstructs the transmitted message from it. Direct representation of the message received by a person or thing at the destination can be called the output signal of the communication system.

In analog communication systems, the signal displaying the message is a continuous sequence of values of a certain message parameter. The electrical signal can display fluctuations in the amount of pressure that generate a sound wave in the air. Such an electrical signal is called analog audio signal, it is a copy, or analog of a sound wave. The waveform of the audio signal displays the message being transmitted.

Images transmitted by television systems are multi-dimensional continual processes. For example, a static black and white image is described by a 2D function of two spatial coordinates, which determines the brightness at each point of the image. Dynamic image is a 3D brightness function that depends on two spatial coordinates and time. The 3D image of the object is described by a function of three spatial coordinates and time. The color image is described by three functions of spatial coordinates and time. The electrical signal displaying in some way such a multi-dimensional function is called analog video signal.

[1] Television Department, St. Petersburg Electrotechnical University (LETI), Russia

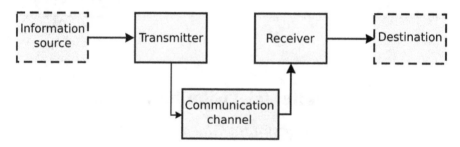

Figure 5.1   General communication system

The electrical signal can be transmitted to the receiver via a communication channel in the form of a wire or cable line. For communication at a distance, different physical processes can be used. For example, the transmission of messages in free space is carried out by means of electromagnetic oscillations – radio waves emitted by means of antenna systems. In order for radio waves to become carriers of the primary electrical signal, they are modulated. The modulation process consists in the fact that the parameters of the carrier electromagnetic oscillation, capable of propagating over long distances, vary according to the primary electrical signal. The modulator is included in the transmitting part of the communication system. On the receiving side, the restoration of the primary signal is performed using a demodulator. Optical lines can be used in communication channels. In this case, the modulation process consists in changing the parameters of the carrier wave, the frequency of which belongs to the optical range.

The advantage of the analog method of representing and transmitting a message is that an analog signal can in principle be an absolutely exact copy of a message. Disadvantages of the analog method are, as often happens, a continuation of its merits. The analog signal can be of any shape, so if, e.g., noise is added to the signal during the transmission, then it is very difficult and often impossible to isolate the original or transmitted signal against the noise background. The analog method is characterized by the effect of accumulation of distortions and noise, which can limit the expansion of the functionality of analog systems. Analog communication technology has gone a long way of improvement and has reached a high level. However, further expansion of functionality and improvement in the quality of analog equipment is associated with costs that can make new equipment inaccessible to the mass consumer audience. Now analog technology is giving way to digital systems.

5.1.2   Digital signals and systems

5.1.2.1   Digital signals

In digital communication systems, discrete signals that take a finite set of values are used for message transmission. A discrete representation is natural when displaying messages in the form of sequences of discrete symbols chosen from a finite alphabet. Examples of such discrete messages, called data, are telegrams, letters, sequences of numbers or machine codes.

But messages such as music, speech, or image are inherently continuous, or continuous processes that are naturally displayed using continuous analog signals. It should be noted that the division of messages and signals into continuous and discrete is not a question of terminology, but a problem of practical limitations. One can imagine the transmission of discrete messages displayed by discrete signals using an analog communication system. However, direct transmission of continuous messages through a digital system is not possible.

The need to use discrete signals is predetermined by the fact that all digital systems have limited speed and can only work with finite volumes of data represented with finite accuracy. To transmit continuous messages using a digital communication system, analog signals that display continuous messages must be converted into discrete signals that take a finite set of values. This operation is performed in the process of analog-to-digital conversion, the main stages of which are sampling, quantization and coding. Sampling or discretization means the transformation of a signal as a continuous function of time and spatial coordinates into a discrete sequence of values. The goal of quantization is the replacement of a continuous range of signal values by a finite number of discrete levels called quantization levels. In the process of quantization, signal values are rounded to the nearest quantization level. The values of the sampled and quantized signal at each moment of time coincide with one of the quantization levels whose number is encoded, e.g., using a binary code. A signal carrying a sequence of received numbers, code words, is digital. The analog-to-digital conversion can be interpreted as the transformation of a continuous message into a discrete data stream. The conversion of a digital signal into a continuous signal, the form of which is a transmitted message, is performed in a digital-to-analog converter (DAC).

5.1.2.2 Digital systems

From the transmitter and receiver, which are present in the generalized scheme of the communication system (Figure 5.1), it is possible to single out explicitly the blocks that solve the most important tasks of coding discrete messages: source coding and channel coding (Figure 5.2), which will be discussed in the next section. In the generalized scheme of a digital communication system, the purpose of the transmitter is to convert the encoded message into a signal suitable for transmission over a communication channel. The most important task solved by the transmitter is modulation. If the communication channel is made as an electrical or optical cable line, the modulator converts each symbol of the code word into a corresponding electrical or optical pulse signal. To transmit messages in free space, modulated electromagnetic oscillations or radio pulses are used. Then, a sequence of electrical, optical or electromagnetic pulses is transmitted over the channel. The main purpose of the receiver in this scheme is to demodulate or convert the sequence of impulse signals at the output of the communication channel into code words to be decoded and converted into a message.

The fact that a sequence of pulses from a finite set of allowable signals is transmitted via the communication channel determines additional possibilities

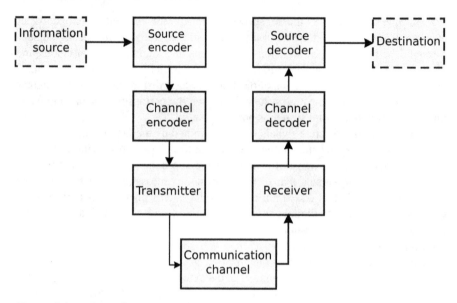

Figure 5.2   Digital communication system for transmission of discreet messages

for increasing noise immunity in digital systems in comparison with analog ones. With a relatively small amount of noise introduced by the channel, it is possible to regenerate weak, distorted and disturbed impulse signals at the input of the receiver, restoring their original form by using threshold schemes (such as the Schmitt trigger type). The greatest interference immunity is provided by the binary signals that always take one of two levels. At significant noise levels and significant distortions, the exact regeneration of the received pulse signals due to threshold schemes becomes impossible. Correction of errors in the communication channel is achieved through channel coding.

5.1.2.3   Source coding

In the digital communication system, the general scheme of which is shown in Figure 5.2, the analog-to-digital conversion of a continuous primary signal, mapping the continuum messages to a discrete sequence of numbers specified with finite accuracy, is performed in the source encoder. The source decoder performs digital-to-analog conversion and restores a continuous signal interpreted by the receiver as a message. The reconstructed signal displays some approximation of the original continuous signal. At the output of the digital audio communication system, a person hears the approximation of the transmitted sound and sees the approximation of the transmitted image. This approximation, the accuracy of which depends on the sampling interval and the number of quantization levels, is always associated with the appearance of noise and the appearance of distortions (frequency, nonlinear, and also some specific distortions). However, the analog-to-digital conversion is performed only once in the digital communication system.

The sound and the image in the digital form can then undergo any number of processing and transformations, and thus no additional distortions and noises are introduced. It must also be borne in mind that the accuracy of the approximation of continuous processes in digital systems can continuously increase. Moore's Law, according to which the computing resources of digital systems (processor speed and memory capacity) are doubled every 18 months, remains valid for several decades, so the thesis of practically unlimited increase in the quality of transmission of continual processes (such as sound and image) in digital systems does not seem too weird.

However, the introduction of digital technologies raises new problems. The frequency bandwidth of digital signals is much larger than the bandwidth of their analog "precursors." For example, the bandwidth occupied by a digital audio signal can reach several megahertz. The bandwidth occupied by a video signal in digital form can reach hundreds of megahertz. The use of broadband channels that have the necessary bandwidth can be technically impossible. This can also be economically unprofitable, as the cost of the communication channel increases with increasing bandwidth. An effective way to solve such a problem is encoding, which aims at compact representation of a digital signal by compression.

Compression involves packing data, such that cheaper low-bandwidth transmission channels can perform the task of exchanging television programs in digital form, provided that the quality of the reproduced image meets the specified requirements. The transformation of a signal into a digital form and its compact representation using compression of a digital data stream are two main tasks that are solved in the process of source encoding in the transmission of continuous signals (Figure 5.3). The output signal of the analog-to-digital converter (ADC) is subjected to special encoding with a compression encoder, which reduces the data bit rate and makes it possible to transmit digital data in a channel with lower bandwidth. The inverse transformation that restores the complete digital image signal is performed in the compression decoder included in the source decoder.

5.1.2.4 Channel coding

Source code words can be fed directly to the channel input of the communication system, provided that the communication channel does not introduce noise and distortion. However, the use of digital signals as an information carrier does not yet guarantee a high quality of transmission in real communication channels, in which there are noises and frequency and nonlinear distortions appear. At significant noise levels and significant distortions in the communication channel, errors may occur and the code words at the channel output may be different from the input code words. In this case, the recovery of code words transmitted through the communication channel is achieved by a special coding, called a channel coding, which allows detecting and correcting errors.

The source code words are processed by the channel encoder (Figure 5.3), in which special check symbols added to the input code words are calculated. The

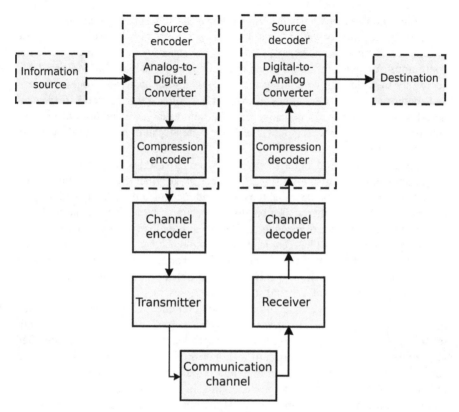

Figure 5.3   *Digital communication system for transmission of continuous messages*

words at the output of the channel encoder are called channel code words. The fact that the channel code words contain more symbols than the source code words means that the channel encoder introduces some redundancy into the data stream. The channel decoder uses the redundancy of the channel code words to detect and correct errors in the received code word. To date, a large number of effective codes have been developed to detect and correct various errors that appear in communication channels.

Additional objectives achieved in the channel coding process are to match the parameters of the digital signal with the properties of the communication channel and to provide self-synchronization. An example of matching signal parameters and channel properties is special coding, eliminating the constant component of the signal. Self-synchronization, which provides the ability to extract clock pulses directly from the channel output signal, is also solved by special coding, which makes regular changes to long sequences of the same symbols occurring in the source code words.

5.2 Digital representation of audio signals

5.2.1 Analog-to-digital and analog-to-digital conversions: principles

5.2.1.1 Analog-to-digital conversion

5.2.1.1.1 Sampling

Sampling or discretization is the first of three basic operations (Figure 5.4) that you need to perform to convert an analog signal into a digital form. Discretization is the representation of a continuous analog signal by a sequence of its values (samples). These samples are taken at times, separated from each other by an interval, which is called the sampling interval. The reciprocal of the interval between samples is called the sampling frequency. The sampling frequency should have a certain minimum value, depending on the width of the spectrum of the analog signal itself.

Figure 5.5 shows the initial analog signal (Figure 5.5a) and its sampled version (Figure 5.5b). Discretization can be considered as a process of amplitude modulation of equidistant pulses of short duration. The magnitude of the pulse is equal to the value of the signal at the time of sampling. These pulses can be considered samples or signal samples. With proper selection of the sampling interval, all the information contained in the continuous analogue signal is stored. In the example shown in Figure 5.5, a harmonic waveform having a frequency of 1 kHz (a period of 1 ms) is sampled at a frequency of 12 kHz. For the period of the signal at the ADC input, there are 12 samples, which allow to accurately reconstruct the original harmonic signal.

5.2.1.1.2 Quantization

Quantization is the replacement of the signal value by the closest value from the set of fixed values called quantization levels. In other words, quantization is rounding off the value of the sample. The quantization levels divide the entire range of possible changes in signal values by a finite number of intervals – quantization steps. The location of the quantization levels is determined by the quantization scale. They are used as uniform (the quantization step does not depend on the value of the signal) as well as nonuniform scales. The sampled and quantized signal obtained using a uniform quantization scale is shown in Figure 5.5c. In this example, the quantization step is 0.33 V 7 quantization levels can be found in the range of the input signal from −1 to + 1 V.

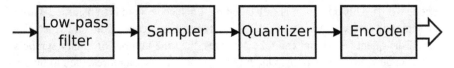

Figure 5.4   Schematic diagram of DAC

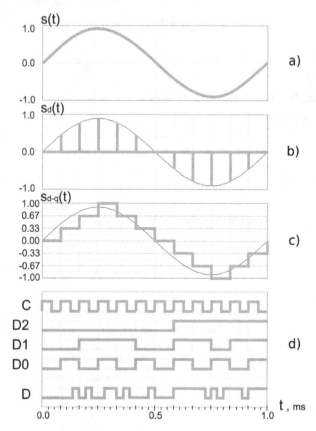

Figure 5.5   *Digital-to-analog conversion of a signal (a: analog signal; b: sampled signal; c: sampled and quantized signal; d: digital data stream in parallel form [D2-D0] and serial form [D])*

5.2.1.1.3   Coding

A quantized signal can only take a finite number of values. This allows us to represent it within each sampling interval by a number equal to the ordinal number of the quantization level. In turn, this number can be expressed by a combination of some characters or symbols. The set of characters (symbols) and the system of rules by which data are represented as a set of symbols is called a code. A sequence of code symbols is called a code word. A quantized signal can be converted into a sequence of code words. This operation is called coding. Each code word is transmitted within one sampling interval. A binary code is widely used to encode audio and video signals. If the quantized signal can take N values, then the number of binary symbols in each code word is $n \geq \log_2 N$. One bit, or symbol of a word represented in binary code, is called a bit. If the number of quantization levels is often equal to an integer power of 2, then $N = 2^n$.

Variants of binary coding of the numbers of the quantization levels shown in Figure 5.5c are given in Table 5.1. To number the levels, positive and negative numbers are used. At seven levels of quantization and binary coding, it is required to use code words of three binary symbols. The relationship between levels and three-digit code words depends on the code selected. Direct code assumes the representation of the sign in the highest bit of the code word (the bit of the sign is zero for positive numbers and one for negative ones). After the sign bit, the binary code of the absolute value of the quantization number follows. Direct code is natural for humans, but for processing data in the processor, the one's complement and two's complement codes turn out to be more convenient. Positive numbers are displayed identically in all three codes listed in Table 5.1. The code words of negative numbers in the one's complement code can be obtained from the words of the direct code by replacing the zeros by ones and ones by zeroes in two digits that represent the absolute value. To represent negative numbers in the two's complement code, add one to the words of the one's complement code. The main advantage of the two's complement code is that addition and subtraction of numbers can be performed in digital processors using the same digital device, the arithmetic adder. The operands must be represented as numbers with a sign, and subtraction is considered as the addition of the positive minuend and the negative subtrahend in the two's complement code.

5.2.1.1.4   Methods of transmission

Code words can be transmitted in parallel or serial forms (Figure 5.5d). For transmission in parallel form, n communication lines should be used (in the example shown in the Figure 5.5, $n = 3$), for each of which one bit of the code word is transmitted. The code word symbols are simultaneously transmitted along the lines within the sampling interval. For transmission in a serial form, the sampling interval must be divided into n subintervals – cycles. In this case, the word symbols are transmitted sequentially along one line, and one clock cycle is assigned to the transmission of one word symbol.

Table 5.1   Coding of quantization levels

Quantization level	Number of the level	Direct code	One's complement code	Two's complement code
1.00	3	011	011	011
0.67	2	010	010	010
0.33	1	001	001	001
0.00	0	000	000	000
−0.33	−1	101	110	111
−0.67	−2	110	101	110
−1.00	−3	111	100	101

Each word symbol is transmitted by one or more discrete pulse signals, so the conversion of an analog signal into a code word sequence is often called pulse-code modulation. The form of representing words by certain signals is determined by the format of the code. It is possible, e.g., to set a high signal level within the clock if the binary symbol 1 is transmitted in this clock cycle, and low if the binary symbol 0 is transmitted. This way of representation, shown in Figure 5.6d, is called the NRZ format – nonreturn-to-zero. In the example in Figures 5.5d and 5.3, bit binary words are used (this allows to have a maximum of 8 quantization levels). In a parallel digital stream, 1 bit of the 3-bit word is transmitted on each line within the sampling interval (the most significant bit [MSB] of the word is transmitted using the discrete signal D2, the least significant bit [LSB] is D0). In the serial stream (signal D), the sampling interval is divided into 3 cycles in which the bits of the 3-bit word are transmitted (beginning with the most significant one).

Operations related to the conversion of an analog signal into digital form (sampling, quantization and coding) can be performed using a separate integrated circuit. These operations can also be a task that is solved by a higher-level processor, e.g., the source coder. The low-pass filter shown in Figure 5.4 is an ADC element, and its purpose is discussed in the next section.

5.2.1.2   Digital-to-analog conversion

Recovering a continuous signal from a sequence of code words is performed in a DAC (Figure 5.6). The DAC decoder converts the sequence of numbers into a discrete quantized signal, which is similar to the signal shown in Figure 5.5c. Analog devices must be used to convert this signal into continuous. The problem consists in obtaining the envelope sequence of rectangular pulses forming the signal in Figure 5.5c. This can be done with the help of the Low Pass Filter (LPF) (Figure 5.6), obtaining the constant component of the sequence, which is proportional to the amplitude. In order to smooth the signal of the pulse sequence and obtain its constant component, the LPF must be sufficiently inert, its boundary frequency must be significantly less than the repetition rate of the pulses, i.e., sampling rate. And in order for the output of the low-pass filter to keep track of the amplitude of the pulses, it should not be too inert, and its boundary frequency must be higher than the highest frequency of the original continuous signal. Requirements for the filter, which plays an important role in the DAC scheme, are discussed in more detail in the next section.

The sampling of a continuous signal cannot introduce any distortion. With certain restrictions imposed on the sampled signal and proper selection of the sampling

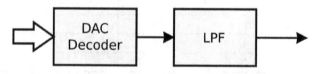

Figure 5.6   Schematic diagram of ADC

frequency, the original analog signal can be accurately reconstructed. However, quantization is always associated with irreversible distortions, and the signal reproduced at the output of the DAC is some approximation of the original continuous signal. Distortions and quantization noise are discussed in the next section.

5.2.1.3 Parameters

Digital signals can be described using parameters typical for analog signals, e.g., such as the width of a spectrum or a frequency bandwidth. But such a description would not be sufficient. An important indicator characterizing a digital stream is the transmission rate or the data bit rate. If the word length is n, and the sampling frequency is F_D, then the data transfer rate expressed in binary symbols per unit time (bit/s) is found as the product of the word length by the sampling frequency: $C = nF_D$. The bandwidth that is necessary for the transmission of the digital stream depends on the shape of the pulses used to transmit the code word symbols. Pulses satisfying the so-called Nyquist condition allow the data stream to be transmitted in binary form at a rate of 2 bits/s in a bandwidth of 1 Hertz (the specific bit rate is 2 [bit/s]/Hz). The required frequency band can be reduced if multi-position signals are used (but at the expense of reducing noise immunity). The use, e.g., of 8-position pulses allows to increase by three times the specific transmission rate (up to 6 [bit/s]/Hz) and to reduce by three times the required communication bandwidth.

5.2.2 Sampling: details

5.2.2.1 Spectrum of sampled signal

Within the theory of systems, sampling/discretization can be considered as a process of amplitude modulation of pulses of short duration (Figure 5.5b). Is it possible to fully preserve the information contained in the original continuous signal? What should be the sampling frequency in order to accurately reconstruct the signal shape at the output of the DAC? It is intuitively clear that the smaller the sampling interval and, correspondingly, the higher the sampling frequency, the smaller the differences between the original signal and its sampled copy. To quantify the requirements for the sampling frequency, it is advisable to consider the transformations to which the frequency components of a continuous signal of a complex shape undergo in the process of discretization. The study of these transformations allows us to formulate the requirements for the sampling frequency.

Figure 5.7a shows a periodic triangular waveform signal having a 2 V peak-to-peak span and a repetition rate of 1 kHz. As is known, any periodic signal can be represented as a sum of harmonic components with frequencies that are multiples of the repetition frequency, i.e., the spectrum of a periodic signal is discrete. Figure 5.7b shows the amplitude spectrum or distribution of the amplitudes of the harmonic components of the triangular signal. The triangular signal has certain symmetry properties; therefore, its spectrum contains only odd harmonics. The amplitudes of the components are expressed in dB (1 mV is used as the reference level). Theoretically, the spectrum has an infinitely large bandwidth, but the amplitudes of the harmonic components of the triangular signal fade out with increasing frequency

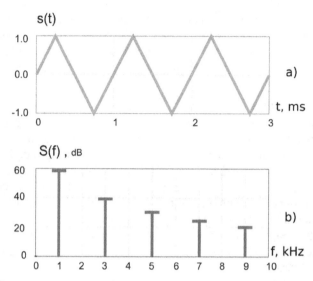

Figure 5.7   Periodic triangular waveform signal (a) and its spectrum (b)

rather quickly (inversely proportional to the square of the harmonic number). If in practice we limit ourselves to harmonics whose amplitude is not less than one hundredth of the first harmonic, then the ninth harmonic can be considered the highest frequency component.

The triangular waveform signal is sampled at a frequency of 11 kHz (Figure 5.8a), which exceeds the bandwidth of the signal (9 kHz). In the frequency spectrum of the sampled signal (Figure 5.8b), there are components of the original continuous signal (they are shown in red). As a result of the digitization in the frequency spectrum of the signal, additional components appear around the harmonics of the sampling frequency. The amplitudes of the harmonics themselves of the sampling frequency, as shown in Figure 5.8b, are zero, which is due to the zero DC component of the given sampled signal. This signal is chosen as an example because the audio signals do not contain a DC component. The components around each harmonic form a structure that corresponds to an amplitude-modulated oscillation.

The frequency band occupied by each amplitude-modulated oscillation is equal to twice the width of the spectrum of the original continuous signal. The components of the amplitude-modulated oscillation in which the carrier frequency is equal to the sampling frequency are selected in green. The amplitude-modulated oscillation has two side bands. The components of the upper side represent a copy of the spectrum of the original signal, shifted to the region of higher frequencies by an amount equal to the sampling frequency. The components of the lower side are the mirror image of the components of the upper side.

The appearance in the spectrum of a discretized signal of groups of frequency components that constitute an amplitude-modulated oscillation is natural, because in the course of discretization, the amplitude of the pulses of a short duration

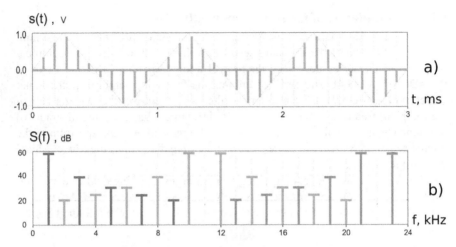

Figure 5.8   *Sampled triangular waveform signal (a) and its spectrum*
                       (b), sampling frequency of 11 kHz

following the sampling frequency occurs. In the spectrum of sampling pulses, there is not only a DC component and the fundamental harmonic but also higher harmonic components. Therefore, the components of the amplitude-modulated oscillations with carrier frequencies, which are equal to multiple sampling frequencies, are also present in the spectrum of the sampled signal (components of the amplitude-modulated oscillation with a carrier equal to twice the sampling frequency are in blue color).

The spectrum of the original signal and the side spectra corresponding to the amplitude-modulated oscillations with the carriers, multiples of the sampling frequency, overlap, so it is impossible to isolate the components of the original signal by frequency selection. But if the width of the spectrum of the original continuous signal did not exceed half the sampling frequency, then the spectra would not overlap. Elimination of overlapping spectra would be of great importance – it would be possible to isolate the components of the original signal with the help of a low-pass filter, whose bandwidth is equal to half the sampling frequency, and, therefore, to restore the original signal from its samples. However, if you pass through a filter with a bandwidth of 5.5 kHz (i.e., half of the sampling frequency) the sampled signal shown in Figure 5.8a, then the reconstructed signal will be a distorted copy of the original signal. The resulting distortions are nonlinear, and they are associated with the transformation of the components of the original signal. Instead of a component with a frequency of 7 kHz, a harmonic component with a frequency of 4 kHz will be reproduced. Instead of a component of 9 kHz, a component with a frequency of 2 kHz will be reproduced. Oscillations with frequencies of 4 and 2 kHz were not present in the original signal. They are a kind of mirror reflection of the components of the original signal with frequencies exceeding half of the sampling frequency, and the reflecting "mirror" is exactly half of the sampling frequency.

5.2.2.2   Limitation of the signal bandwidth

In order to prevent nonlinear transformations of the signal, it is necessary to limit the frequency band of the original signal to a value equal to half of the sampling frequency. This can be done with the help of a low-pass filter, which is a part of the ADC (Figure 5.4). The result of limiting the frequency band of the triangular waveform (Figure 5.7) using a low-pass filter whose cut-off frequency is half of the sampling frequency (5.5 kHz) is shown in Figure 5.9a. The time diagrams of the components (with frequencies 1, 3 and 5kHz) remaining in the spectrum of the filtered signal are shown in Figure 5.9b, and the time diagram of the filtered signal is shown in Figure 5.9c.

Apparently, low-frequency filtering has led to a smoothing of the source signal form. This is a charge for the possibility of signal recovery in a DAC without any nonlinear sampling distortions. The sampling of a signal with a limited band is

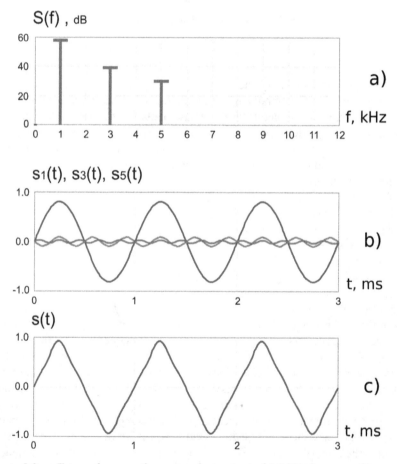

Figure 5.9   *Triangular waveform signal at output of 5.5 kHz low-pass filter (a: spectrum; b: harmonics, time diagram)*

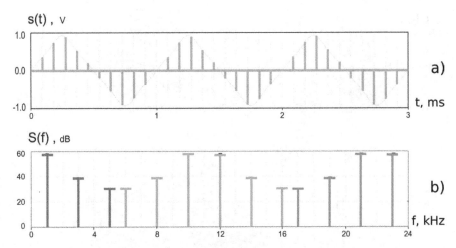

Figure 5.10   Sampling of triangular waveform signal with finite bandwidth (a, sampled signal; b, spectrum of sampled signal)

illustrated in Figure 5.10. The ratio of the sampling frequency and the bandwidth of the signal meets the requirements – the spectrum width does not exceed half the sampling frequency. There is no overlap of the spectra, and the signal sampled in the ADC can be reconstructed in a DAC without distortion. The recovery device is a low-pass filter (Figure 5.6), the passband of which is again half the sampling frequency. Figure 5.10 is an illustration of the sampling of an analog signal, which is performed in accordance with the provisions developed by Nyquist. The signal undergoing sampling can be reconstructed without distortion only if it has a limited spectrum, and the sampling rate is not less than twice the width of the spectrum of the signal being sampled.

5.2.2.3   Distortions of sampling

Discretization is an amplitude-pulse modulation or modulation of the amplitude of pulses of small duration (Figure 5.5b). Between the pulse sequence obtained during the sampling process and the signal recovered by the low-pass filter of the DAC (Figure 5.6), there is a one-to-one relationship. But the same pulse sequence can be generated by sampling different signals. There is no one-to-one correspondence between the input and output signals of the sampler. If the signals are harmonic, then the same pulse sequence is generated by sinusoidal signals whose frequencies are located symmetrically with respect to half of the sampling frequency (Figure 5.11). For example, a sinusoid with a frequency of 5 kHz (Figure 5.11a) and a sinusoid with a frequency of 6 kHz (Figure 5.11b) generate identical pulse sequences at a sampling frequency of 11 kHz. The sinusoid with the frequency of 5 kHz has an amplitude of 32 mV (this is the fifth harmonic of the triangular signal [Figure 5.10], the sampling of which was analyzed earlier). The amplitude of the signal with the frequency of 6 kHz is set to be the same −32 mV.

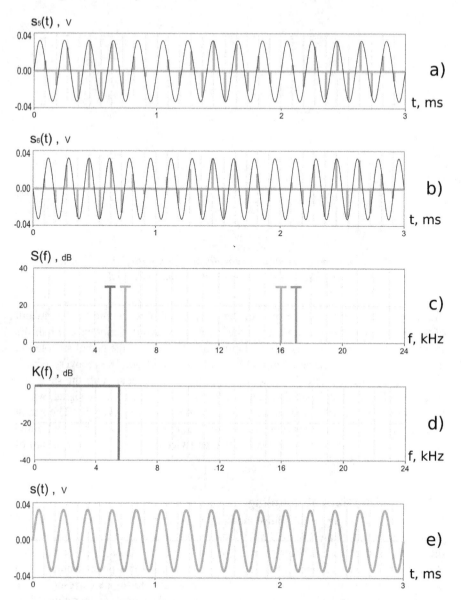

Figure 5.11   *Sampling distortions when sampling frequency is not high enough*
            (a: 5 kHz harmonic waveform signal sampled with the rate 11 kHz;
            b: 6 kHz harmonic waveform signal sampled with the rate 11 kHz;
            c: spectrum of sampled signals; d: frequency response of DAC low-
            pass filter; e: time diagram of restored signal)

Naturally, the frequency spectra of the two sampled signals are the same (Figure 5.11c). But the interpretation of the origin of the individual components of the frequency spectrum will be different. When sampling oscillations with a frequency of 5 kHz, the component of the spectrum of the sampled signal with a frequency of 5 kHz (it is shown in red in Figure 5.11c) refers to the spectrum of the original signal, and the components with frequencies of 6 and 16 kHz (shown in green) are components of the amplitude-modulated oscillation with the carrier, equal to the sampling frequency (11 ± 5 kHz). Blue shows the lower side frequency of the modulated oscillation with the carrier frequency equal to the second harmonic of the sampling frequency ($22 - 5 = 17$ kHz). The main and side spectra do not overlap, so the original signal can be extracted with a low-pass filter with a band of 5.5 kHz (Figure 5.11d). The result of reconstructing the sampled signal is shown in Figure 5.11e.

When sampling oscillations with a frequency of 6 kHz, the component of the spectrum of the sampled signal with a frequency of 5 kHz is the lower side frequency of the modulated oscillation with a carrier frequency equal to the sampling frequency ($11 - 6 = 5$ kHz), and the spectrum component with a frequency of 6 kHz refers to the spectrum of the original signal. The main and side spectra has overlapped (higher-order spectra of the first and second, second and third spectra overlap, etc.), so it is impossible to isolate the component of the original signal by frequency selection even in the case of using an ideal low-pass filter (Figure 5.11d). At the output of the low-pass filter of the DAC, a harmonic signal (with a frequency of 6 kHz) is reproduced with a false frequency (5 kHz). It can also be said that a component with a frequency of 5 kHz is the result of a mirror reflection of the original signal (6 kHz) from half of the sampling frequency (5.5 kHz).

Figure 5.11 illustrates the discrepancy distortions that arise when the sampling frequency is less than twice the width of the spectrum of the signal being sampled. Such distortions also occur when the frequency band of the signal being sampled exceeds half of the sampling frequency.

Even when the width of the spectrum of the original signal does not exceed half of the sampling frequency, the reconstruction can be associated with distortions due to insufficient suppression of the side spectra in the low-pass filter of the DAC. Such distortions are illustrated in Figure 5.12. A harmonic waveform with a frequency of 5 kHz is sampled at a frequency of 11 kHz (Figure 5.12a). The frequency ratio satisfies the requirements; the spectra do not overlap (Figure 5.12b). However, the low-pass filter (Figure 5.12c) does not have the frequency response of an ideal low-pass filter.

The filter used (Figure 5.12c) does not introduce any attenuation in the passband from 0 to 5 kHz, and a further increase in frequency leads to attenuation, which increases by 20 dB with increasing frequency by 1 kHz. The lower side frequency of the first side spectrum (the spectrum of the amplitude modulated oscillation with a carrier equal to the sampling frequency) equal to 6 kHz is attenuated by a factor of 10 (Figure 5.12d), and the remainders of the remaining components of the spectrum of the sampled signal can obviously be neglected. The reconstructed signal (Figure 5.12e) differs from the original one. In the spectrum of the original signal,

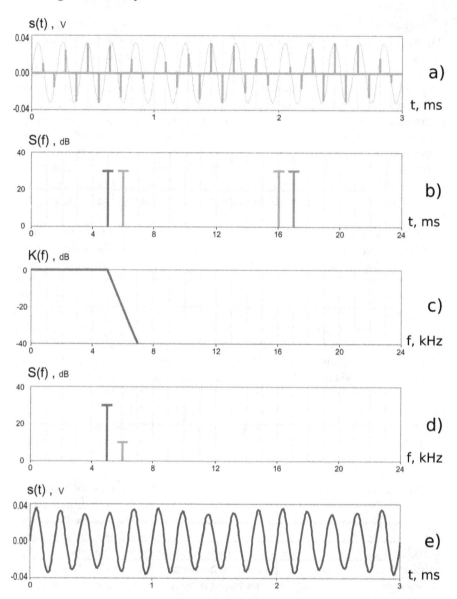

Figure 5.12   Sampling distortions when DAC low-pass filter is not good enough
(a: 5 kHz harmonic waveform signal sampled with the rate 11 kHz;
b: spectrum of sampled signal; c: frequency response of DAC
low-pass filter; d: spectrum of signal at the output of DAC low-pass
filter; e: time diagram of restored signal)

there is one harmonic component with a frequency of 5 kHz; in the spectrum of the reconstructed one, there are two (with frequencies of 5 and 6 kHz). Their addition gives an oscillation with a variable amplitude (the frequency of the amplitude change is equal to the frequency difference of the components of 6 and 5 kHz). This gives a well audible distortion.

Interestingly, an increase in the sampling rate by only 1 kHz (from 11 to 12 kHz) would significantly increase the suppression of the side spectra with the same low-pass filter. The lower side component of the first aliased spectrum would have a frequency of 7 kHz. At the output of the low-pass filter, it would have a level of −40 dB with respect to the signal at a frequency of 5 kHz, and the distortions of the reconstructed signal were 10 times smaller. Creating filters with a steep slope of the amplitude–frequency characteristic is a difficult technical problem if additional restrictions are imposed on the unevenness of the amplitude–frequency characteristic and the linearity of the phase characteristic in the passband. This example shows that in order to reduce the requirements for a low-pass filter included in the ADC and DAC, it is advisable to increase the sampling frequency in comparison with the theoretical limit – twice the width of the spectrum of the sample being sampled.

What is the sampling rate for analog-to-digital conversion of real audio signals? The maximum frequency of audible sounds is 20 kHz, in accordance with the theory stated; the sampling frequency should be greater than 40 kHz. At a sampling frequency of 44.1 kHz, low-pass filters in the ADC and DAC should pass without distortion of the frequency bandwidth to 20 kHz and significantly weaken the frequency components at frequencies of 22.05 kHz and higher. At a sampling frequency of 48 kHz, low-pass filters should significantly attenuate frequency components at frequencies of 24 kHz and higher. But other factors, described in the following sections, affect the selection of the sampling frequency.

5.2.3 Quantization: details

5.2.3.1 Quantization noise

Features of quantization, which is not preceded by sampling, are illustrated in Figure 5.13a using the example of a triangular waveform. In the range of signal values (from −1 to + 1 V), the quantization levels (−1.00,−0.67, ..., 0.67, 1.00) are determined using the quantization step equal to 0.33 V. This procedure determined the quantization scale. The scale of this example is uniform, since the magnitude of the quantization step is independent of the level. A quantized signal can only take values equal to the quantization levels. In the middle between the levels of quantization are the threshold levels, shown in the figure with dotted lines. The quantization procedure, illustrated in Figure 5.13, involves rounding the original signal to the nearest quantization level. When the original continuous signal passes through some threshold level located between adjacent quantization levels, the value of the quantized signal changes abruptly. Quantization is a nonlinear process. The transfer characteristic of the quantizer, which expresses the dependence between the values of the input and output signals, has a step-like character corresponding to the established quantization scale (Figure 5.14).

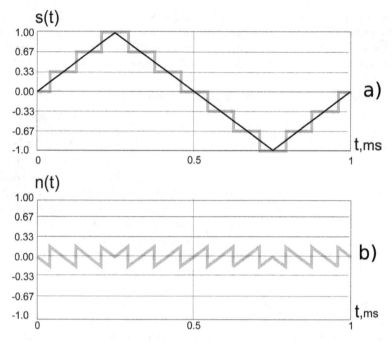

Figure 5.13   Quantization (a: quantized signal; b: quantization noise)

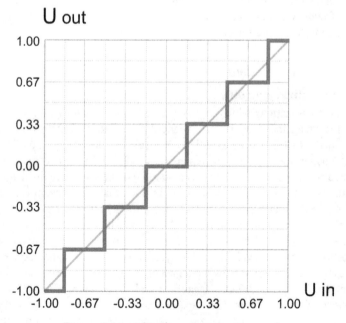

Figure 5.14   Uniform quantization scale

The characteristic of the quantizer can be considered as an approximation of the ideal characteristic shown in Figure 5.14 in green. Quantization always introduces errors into the digital representation of the signal. These distortions can be interpreted as some kind of interference, or quantization noise, added to the original signal. In such an interpretation, the quantization noise can be found as the difference between the quantizer's output and input signal. The quantization noise $n(t)$ introduced in the quantizer during the transformation of $s(t)$ signal of a triangular shape is shown in Figure 5.13b. It has the form of sawtooth pulses with a span equal to the quantization step. The value of the sawtooth signal – quantization noise – goes to zero when the source signal is equal to some level of quantization, and it changes abruptly when the quantized signal passes through the threshold levels.

As a measure of quantization noise, the power or root mean square (rms) noise is often used. Calculating the average value of the integral of the noise square $n(t)$, we can find that the signal power in the form of sawtooth pulses with amplitude $d/2$ is $d^2/12$ (here, d is the quantization step size). This is an important result. The power of the quantization noise depends only on the quantization step. The root-mean-square value of the quantization noise is $d/(\sqrt{(12)}) = d/(2\sqrt{3})$, i.e., half of the quantization step divided by the square root of 3. You can calculate the signal-to-noise ratio (the ratio of the effective signal value to the root-mean-square value of the quantization noise). The effective value of the triangular waveform is also equal to the amplitude divided by the square root of 3. Consequently, the signal-to-noise ratio is equal to the ratio of the signal amplitude to the half of the quantization step (or the ratio of the peak-to-peak signal span to the quantization step).

5.2.3.2 Dynamic range

An important indicator of digital sound systems is the magnitude of the dynamic range, which is equal to the ratio of the maximum value of the signal to the value of the smallest signal discernible in the noise. This ratio is the maximum value of the signal-to-noise ratio. The maximum signal amplitude can be found as the product of the number of quantization levels per quantization step. Therefore, the maximum signal-to-noise ratio is equal to the number of quantization levels N. Using the interrelation of the number of quantization levels with the code word length ($N = 2^n$) and passing to decibels, we can obtain a convenient formula for computing the dynamic range of a digital sound system: $20*\log(2^n) = 6n$ [dB], where n is the number of bits or bits in the code word at the output of the ADC for binary encoding. The physical meaning of the expression $6n$ [dB] is simple. With an increase in the number of digits, the number of quantization levels N rapidly increases, the quantization step decreases and, accordingly, the quantization noise level decreases.

The dynamic range, equal to $6n$ [dB], was found for a triangular waveform signal. However, this formula gives a good estimate for real audio signals. With a word length of 16 bits, the dynamic range is approximately 96 dB. Increasing the length of a word by 1 bit reduces the level of quantization noise and, accordingly, increases the dynamic range by 6 dB. With a code word length of 24 bits, the dynamic range increases to 144 dB.

Another important characteristic of noise is the spectral density, reflecting its frequency properties. When digitizing real wideband sound signals of large amplitude, a fairly good quantization noise model is a random process with a constant value of the spectral power density in the frequency band from 0 to the sampling frequency (this is a white noise equivalent to the fluctuation noise in analog systems). It is important to note that the power of quantization noise does not depend on the sampling frequency, and it is determined only by the quantization step and remains equal to $d^2/12$. This means that the value of the power density of the quantization noise is inversely proportional to the sampling frequency.

5.2.3.3 Distortions of low-level signals

In analog systems, nonlinear distortion of low-level signals is usually much less than distortions of high-level signals. In digital systems, the situation is fundamentally different. If the signal arriving at the input of the ADC has a small amplitude (comparable to the quantization step), the quantization errors correspond to the distortions of the complex type and are completely different from the noise existing in analog systems. The smaller the amplitude of the input signal, the stronger the distortion. The reason for this is that quantization is a nonlinear process with a stepped characteristic (Figure 5.14).

If the quantized signal of small amplitude has a low frequency (low in comparison with the sampling rate), the quantization leads to nonlinear distortions in the form of odd harmonics of the input signal. Figure 5.15 shows the discretization and quantization of a harmonic signal with a frequency (1/3) kHz (at a sampling frequency of 12 kHz) and an amplitude of 32 mV. When evaluating the quantization distortions, the signal is considered small if its range is comparable with the quantization step. Therefore, the diagram in Figure 5.15 is constructed under the assumption that the quantization step is 40 mV. The harmonic signal has turned into a pulsed signal. The quantized signal responds abruptly to an increase in the value of the input signal only in the fifth clock interval, when the value of the quantized signal becomes larger than the threshold level. In the spectrum of such an impulse signal one can find odd harmonics with rather significant amplitude.

Similar nonlinear distortions also occur in analog systems. However, the dependence of the amount of distortion on the signal level in analog and digital systems is fundamentally different. In analog systems, nonlinear distortions usually increase

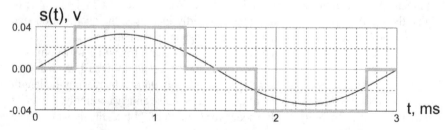

Figure 5.15   Sampling and quantization of low-level and low-frequency signal

with increasing signal level; for infinitesimal signals, any analog system can be considered linear. In digital systems, a decrease in the signal level leads to an increase in the harmonic coefficient of nonlinear distortions. This is understandable, because in the range of values corresponding to the amplitude of the signal, there are fewer quantization levels. The approximation of the signal in the quantizer becomes increasingly coarse, and the specific weight of the odd harmonics in the output signal of the quantizer increases.

If a weak signal has a high frequency approaching half of the sampling frequency, distortions occur that do not occur in analog systems. Figure 5.16 illustrates the discretization and quantization of a harmonic signal with a frequency of 5 kHz and amplitude of 32 mV (this is the fifth harmonic of the triangular signal

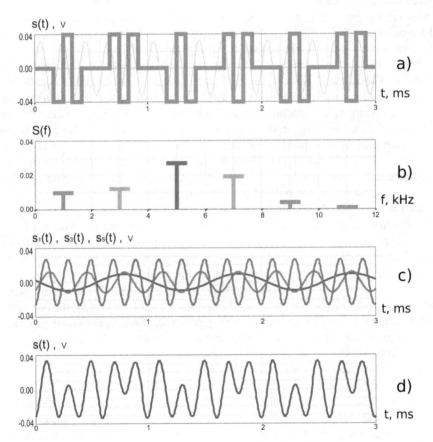

Figure 5.16   *Sampling and quantization of low-level and high-frequency signal (a: original and sampled and quantized signals; b: spectrum of sampled and quantized signal; c: frequency components of sampled and quantized signal in the band of DAC low-pass filter; d: time diagram of restored signal)*

[Figure 5.10], the sampling of which was analyzed earlier). The quantization step is 40 mV. The sampling is performed at a frequency of 12 kHz. The characteristics of the low-pass filters included in the ADC and DAC correspond to Figure 5.12c. As a result of discretization and quantization of the original continuous signal, shown in Figure 5.16a by a thin line, a three-level pulse signal (a thick blue line) is formed, which has a complex spectral composition. The harmonic components of the pulse signal, falling into the frequency band from 0 to 12 kHz, are shown in Figure 5.16b. In addition to the main component with a frequency equal to 5 kHz (shown by the red line), components with frequencies of 1, 3, 7, 9 and 11 kHz appeared.

The origination of the additional frequencies 1, 3, 7, 9 and 11 kHz is easy to explain if we mentally swap the processes of discretization and quantization in the ADC (Figure 5.4). Quantization of a signal with a frequency of 5 kHz and an amplitude not exceeding the quantization step leads to the formation of a bipolar three-level pulse signal, in the spectrum of which there are components with odd harmonics (5, 15, 25 kHz, ...). As was shown above, the discretization of such a wideband signal leads to the appearance of combinational components of the form $|m*f_D \pm (2n + 1)*f_S|$, where f_S is the fundamental frequency of the input signal equal to 5 kHz, and f_D is the sampling frequency equal to 12 kHz. For $m = 0$, this expression determines the spectrum of the sampled pulsed signal, and for $m = 1$, the first side spectrum, i.e., spectrum of amplitude-modulated oscillations with a carrier equals to the sampling frequency. The components with frequencies 3 and 7 kHz are components of the first spurious spectrum $|f_D \pm (2n + 1)*f_S|$ for $n = 3$ and 1 ($|12 - 3*5| = 3$, $|12 - 1*5| = 7$), which are shown in Figure 5.16b in green. Components with frequencies 1, 9 and 11 kHz are components of the second aliased spectrum $|2*f_D \pm (2n + 1)*f_S|$ for $n = 5$, 3 and 7 ($|2*12 - 5*5|=1$, $|2*12 - 3*5|=9$, $|2*12 - 7*5|=11$), which are shown in the Figure 5.16b in blue.

We already encountered analogous processes in analyzing the process of discretization of a signal having a complex spectral composition (Figure 5.8b), when the sampling frequency was not sufficiently big. Discrepancies were prevented by filtering the input signal with a low-pass filter included in the ADC (the bandwidth of the filter was set to half of the sampling rate). In the example under consideration (Figure 5.16), a similar low-pass filter does not allow to prevent the nonlinear interaction of the harmonics of the sampling pulses and the original signal, since the harmonics of the signal appear after the low-pass filter included in the ADC.

A low-pass filter with a 5 kHz band, included in the DAC, suppresses most of the combinational components. At the output of the DAC, components with frequencies of 1, 3 and 5 kHz remain (the diagrams of these components are shown in Figure 5.16c). As can be seen from the diagram of the reconstructed signal (Figure 5.16d), the combination frequencies 1 and 3 kHz significantly distorted the original signal with the frequency of 5 kHz. The presence of combinational components with frequencies below the frequency of the original signal creates an unpleasant sound. The sound is heard much better than the usual noise present in analog systems. Sounds like, e.g., the dying echoes or the fading sound of the percussion instruments, reproduced with the help of a digital system, do not fade away gradually but drown in well-audible noise.

There are ways to reduce the audibility of noise of crushing. A quantization error at a low signal level is an interference correlated with the input signal, which predetermines a good audibility of the noise. The quantization error can be de-correlated using broadband dithering noise added to the signal. The quantization error becomes a function of the dithering noise, not the input signal. With respect to the input signal, the system becomes linear but "noisy." Errors of quantization do not disappear, and their frequency transformation takes place. Subjectively unacceptable distortions are converted into broadband noise, which is less audible and annoying. But the main and radical way to deal with quantization noise is an increase in the number of quantization levels and a decrease in the quantization step at which the amount of quantization noise becomes below the threshold of audibility over the entire range of signal level variation. It can be assumed that when quantizing with the code word length $n = 24$, the noise practically disappears.

5.2.3.4 Distortions of high-level signals

The effects that occur in digital sound systems when the maximum allowable signal level is exceeded are different from the distortions that arise in a similar situation in analog systems. When the range of the input signal exceeds the allowable range of the quantizer, a clipping occurs. The diagram in Figure 5.17 illustrates the distortion of the shape at the frequency of the input signal of the ADC, which is much less than half of the sampling frequency (300 Hz at a sampling frequency of 12 kHz). In this example, the input signal is 3 V wide, i.e., the amplitude is 1.5 V, and the minimum and maximum quantization levels are, respectively, −1 and + 1 V. The source signal is limited to levels −1 and + 1 V. With such distortions, the appearance of odd harmonics of the signal is associated, and this is similar to the distortions that arise when limiting in analog systems. However, in digital systems, the limitation is absolute. Values exceeding the maximum level of quantization are not transmitted at

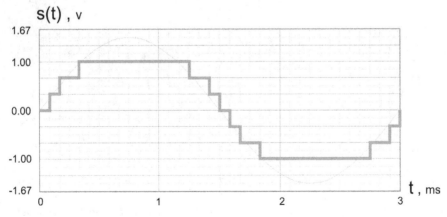

Figure 5.17   Clipping of sampled and quantized signal of low-frequency and high-level signals

all. When the amplitude of the input signal exceeds the maximum quantization level, the quantized signal becomes trapezoidal if the excess becomes significant, then the shape of the quantized signal approaches a rectangular signal.

At the frequency of the original signal approaching half of the sampling frequency, the quantization limitation leads to the appearance of interference, which from the point of view of the signal theory is similar to the noise of crushing. Figure 5.18 shows the clipping of a harmonic oscillation with a frequency of 5 kHz and amplitude of 1.4 V for the same as in the previous example (Figure 5.17), the minimum and maximum quantization levels and the sampling frequency. As a

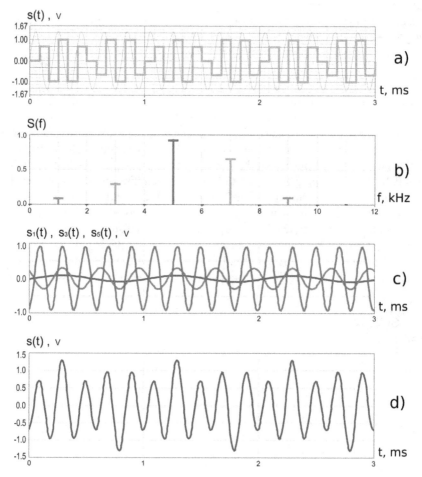

Figure 5.18   *Clipping of sampled (12 kHz) and quantized signal of high (5 kHz) frequency and high level (a: original and sampled and quantized signals; b: spectrum of sampled and quantized signal; c: frequency components of sampled and quantized signal in the band of DAC low-pass filter; d: time diagram of restored signal)*

result of nonlinear transformations during sampling and quantization, the waveform undergoes distortions (Figure 5.18a), which are fundamentally different from those associated with signal clipping in analog systems.

The interaction of odd harmonics of the input signal and harmonics of the sampling frequency leads to the appearance of a large number of combination oscillations. Harmonics of the input signal (Figure 5.18b) appear after the low-pass filter of the ADC, so low-frequency filtering of the input signal cannot prevent these distortions. A part of the combination oscillations falls into the pass band of the low-pass filter of the DAC. In this example, these are oscillations with frequencies of 1 and 3 kHz, which distort the main signal at a frequency of 5 kHz. The diagrams of all components of the signal at the output of the DAC filter are shown in Figure 5.18b, and the output signal of the DAC is shown in Figure 5.18c. It should be noted that although the quantized signal does not exceed one volt in absolute value, the signal at the filter output is outside the range (-1 and $+1$ V). The analog amplifier stages following the DAC must be designed for this overload to prevent further distortion. The frequencies of the combination oscillations caused by the limitation are much lower than the frequency of the main signal, and the amplitude of these oscillations is large and far exceeds the threshold of audibility, so they are usually not masked by a louder basic sound.

The quantizer output signal cannot be greater than the maximum quantization level and less than the minimum quantization level. Therefore, when measuring the level of a signal in digital systems, it is advisable to take as the reference value the largest value of the signal that can be transmitted in a digital system without limitation. This value is defined as 0 dB relative to Full Scale (0 dBFS). The level of 0 dBFS cannot be exceeded. It is clear that on this scale all other levels are negative in decibels.

Significant overload leads to audible distortions, which were described in the previous section. And how to detect a slight overload, analyzing a digital signal, which by definition cannot exceed the level of 0 dBFS? If the level is limited, several consecutive samples will have a zero value on the dBFS scale. This determines a simple algorithm for the operation of the measuring device, which signals the overload that occurred during the formation or processing of a digital signal. Such a device must count the number of zero values immediately following one another on the dBFS level scale. It is believed that overloading for 5–6 counts (at a sampling frequency equal to twice the maximum frequency of the audio signal) is quite difficult to hear in many musical works. If as a criterion to take 3 consecutive zero samples, then almost all audible overloads will be determined.

Digital and analog systems behave differently in relation to overload. In digital systems, the lower limit of the dynamic range is determined by the quantization noise. If we ignore the quantization errors, then the characteristic of the digital system in the entire range is a straight line (Figure 5.14). Exceeding the limits of the interval limited by the minimum and maximum quantization levels is associated with a complete clipping. The characteristics of analog systems are different. In the range of signal values from the noise level to some standard operating level, the system is linear. A further increase in the level does not immediately lead to clipping, only distortions gradually increase. Only when a certain maximum operational level is reached, a saturation or complete clipping occurs. Thus, in analog systems there is always some protective area between the boundary of the linear part of the amplitude characteristic and the saturation. In digital systems, there is no such area, so overloading is unacceptable. What should be the standard operating level of a digital audio signal to create a protective area? As an averaged value of the first approximation, the level −20 dBFS (Decibels relative to Full Scale) can be adopted.

5.2.4   One more time about sampling frequency

In accordance with the theory of discretization, the exact reconstruction of the sampled signal is possible if two conditions are met: the signal must have a frequency-limited spectrum and the sampling frequency should be no less than twice the width of the signal spectrum. The maximum frequency of audible sounds is 20 kHz, so the sampling rate should be at least 40 kHz. The low-pass filter of the ADC (Figure 5.4) should limit the bandwidth of the signals supplied to the input of the sampler to 20 kHz in order to prevent false frequencies from appearing due to overlapping spectra. The low-pass filter of the DAC (Figure 5.6) suppresses the side spectra and restores the sampled signal in the band up to 20 kHz. It may seem that a low-pass filter in the DAC is not needed if the sampling frequency exceeds 40 kHz, since all the aliased spectra are outside the frequency range of audible sounds. However, even a slight nonlinearity of the analog amplifier following the DAC will in this case lead to intermodulation distortions of the main signal. Therefore, the low-pass filter is an obligatory element of the DAC, suppressing the side aliased spectra.

ADCs and DACs are elements of a digital system that largely determine the quality of the reproduced sound. If they do not meet the specified requirements, then the entire digital system is discredited. Important elements of the ADC and DAC are low-pass filters. If you select a sampling frequency of 40 kHz, the filters must pass without distortion of the signals in the band to the upper frequency of the spectrum equal to 20 kHz and delay signals with frequencies exceeding half of the sampling frequency (40/2 = 20 kHz). Filters with the required properties are physically unrealizable, since they must in this case have an infinitely steep slope of the frequency response. At a sampling frequency of 44.1 kHz, the low-pass filters in the ADC and DAC must pass without distortion of the frequency bandwidth to 20 kHz and significantly attenuate the frequency components in the range from half the sampling frequency of 22.05 kHz and higher. The frequency gap between the main spectrum, which must be passed without distortion, and the first side of the spectrum, which should be substantially attenuated, is 2 kHz, which reduces the requirements for filters and makes them physically realizable. At a sampling frequency of 48 kHz, low-pass filters should significantly attenuate frequency components at frequencies of 24 kHz and higher, which further simplifies the creation of filters.

But even at a sampling rate of 48 kHz, low-pass filters of high order are required, which do not have the characteristics necessary for high-end audio systems. This refers to the linearity of the phase-frequency response and the uniformity of the amplitude–frequency characteristic in the passband. To improve the quality of digital audio systems, it is advisable to further increase the sampling frequency, since the requirements to the slope of the amplitude–frequency response of the filters are reduced. This makes it possible to use simpler low-order filters that have a better linearity of the phase characteristic and a smaller unevenness of the amplitude–frequency characteristic in the passband. If you increase the sampling rate 8 times (up to 384 kHz), you can use simple filters with a cutoff frequency of 30–40 kHz at −3 dB and an attenuation out of band pass at a rate of 12 dB per octave. With such filter parameters, the uniformity of the amplitude–frequency and the linearity of the phase-frequency characteristics, which are necessary for high-end systems, are easily ensured.

Increasing the sampling frequency by several times with respect to the theoretical value simplifies the implementation of low-pass filters – analog elements of ADC and DAC. However, at the same time, the bit rate of digital data representing the audio signal is also increased by the same amount. Processing of data with an increased frequency means unproductive consumption of digital computing resources (clock frequency, memory capacity, communication bandwidth, etc.). After all, the theory of discretization asserts that an arbitrary signal with a limited frequency band (audio signals is just such) can be uniquely represented by samples that follow a frequency equal to twice the width of the spectrum. The solution of the problem is possible by reducing the frequency of the sampling of the audio signal after the ADC. Decimation of the digital data stream at the output of the ADC using high-quality digital filtering allows to reduce the sampling frequency to some standard value (e.g., 32, 44.1 or 48 kHz) for processing or transmission over communication channels. A reverse procedure must be performed before the DAC. The sampling frequency is increased to the sampling rate in the ADC by calculating the intermediate samples by means of digital interpolation.

The choice of the sampling frequency is also affected by the tendency to reduce the audibility of quantization noise. The manifestations of quantization noise are manifold. If the amplitude of the input signal of the ADC is much larger than the quantization step, but less than the maximum quantization level, then the quantization noise is similar to the fluctuation noise existing in analog systems. But if the input signal of the ADC has a very small amplitude (comparable to the quantization step) or very large at which a clipping arises, the quantization noise can no longer be likened to a fluctuation process. For a harmonic signal of low frequency (small in comparison with half of the sampling frequency), the nonlinear distortions of the quantizer result in the appearance of harmonics.

At a big signal frequency comparable to half of the sampling rate, components that do not lie at the harmonic frequencies of the input signal appear as a result of quantization, which leads to specific and more audible noise. If you significantly increase the sampling frequency, then all audio frequencies will be much less than half of the sampling frequency. Therefore, nonlinear transformations of the quantizer for the entire frequency range will lead to nonlinear distortions similar to those in analog systems and not to unpleasant interference. Thus, an increase in the sampling frequency provides tangible benefits.

With analog-to-digital conversion of real wideband sound signals of large amplitude, the quantization noise can be described quite well by a random process with a constant value of the spectral power density in the band up to the sampling frequency. The quantization noise power does not depend on the sampling frequency and is determined only by the quantization step: it is equal to the square of the quantization step divided by 12. This means that the value of the spectral density of the quantization noise power is inversely proportional to the sampling frequency. If you increase the sampling frequency, the quantization noise bandwidth will increase, and the spectral density value will decrease. If a digital low-pass smoothing filter with a band corresponding to a theoretical value of 20 kHz at the output of the ADC with an increased sampling rate, then the quantization noise power will be reduced by eliminating the frequency components of the high-frequency noise lying outside the audible range of sounds. With an eightfold increase in the sampling rate, the noise power will decrease by a factor of 8, and the rms value will increase by 2.8 times. The signal-to-noise ratio will increase by 2.8 times. The gain can be even greater if you use special ADCs, which allow you to concentrate most of the power of quantization noise in the high-frequency region – outside the range of audible sounds. Digital filtering after the ADC eliminates high-frequency components of quantization noise and further increases the gain in the signal-to-noise ratio.

5.2.5   AES/EBU interface

Transmission of digital audio signals from one element of the system to another (Figure 5.1) requires a preliminary agreement on the parameters of the connection. Figure 5.5d illustrates the principle of parallel and serial transmission of audio data. But in practice, it is necessary to agree on the voltage values corresponding to the logical levels of the digital signal, the methods of synchronization of the transmitter and receiver at the level of clock pulses and data words, to agree on the length of the code word and the order of the word bits, to specify the requirements for the mechanical and electrical characteristics of the connectors, to select connector type, etc. A means of pairing two systems or parts of a system in which all parameters (physical, electrical, and logical) meet preliminary agreements are called an interface. The Society of Audio Engineering (AES)/European Broadcasting Union (EBU) interface (also known as the AES3 standard) regulates the serial transmission of two linearly quantized digital audio signals over a single communication link over distances of up to several hundred meters [2]. The interface specification was prepared by the AES and the EBU and submitted for approval in 1992.

5.2.5.1   Format

The code words carrying information about the audio signal samples are transmitted in a sequential form (the principle of serial transmission of code words is illustrated in Figure 5.5d). The value of the sound sample is represented in the format of an two's complement code, with positive numbers corresponding to the positive voltage of the input signal of the ADC. The number of word bits can be set in the range from 16 to 24, with two transmission options. In one of them, for the data word, 24 bits are assigned, and in the other – 20 bits. In the variant of a word with a length of 20 bits, the 4 LSBs are allocated for additional information. The transmission specification is independent of the sampling frequency; however, there are a number of frequencies specified explicitly: 48, 44.1, 32, 24, 96, 192, 22.05, 88.2, and 176.4 kHz. For studio use, the recommended frequency is 48 kHz [3]. If two audio signals are transmitted via the interface, the sampling rates of both signals must be the same.

The above parameters refer to the actual sound data. But to synchronize the transmitter and receiver at the level of clock pulses and data words, as well as to solve many other problems of receiving signals and interpreting the received data together with the audio data, additional information is transmitted. The packet in which the data word is transmitted on one sample of the audio signal and additional information is called the subframe (Figure 5.19).

Each subframe is divided into 32 parts (time slots), numbered from 0 to 31. Slots 0–3 are assigned to the preamble, which is the word for synchronization and data identification. In slots 4 through 27, the audio data words are transmitted, with the MSB of the word always in slot 27. The LSB is transmitted in slot 4 with 24-bit encoding and in slot 8 – at 20-bit. If the source creates audio data words with a shorter length (less than 24 or 20), then the unused LSBs are set to the logical zero state.

The validity bit V for which slot 28 is assigned indicates whether a subframe sample can be used to convert to an analog signal. If bit V is 0, then the audio sample data are valid and can be used for digital-to-analog conversion. In the slot number 29 (user data U), any information is transmitted: this is the user data channel. The format of user data is not regulated by the interface specification. The channel status data bit in slot 30 is one of the bits of the data packet that is transmitted in a sequence of 192 subframes and in a specific format carries information about the audio channel. The parity bit P in the last slot of the subframe with number 31 is used to detect an error (or an arbitrary odd number of errors) in a subframe.

A frame is a sequence of two subframes in which data of two audio channels are transmitted (Figure 5.20). Subframes 1 and 2 have different preambles. About 192 consecutive frames are combined into a group called a block. The frame rate is equal to the sampling frequency. The first subframe usually starts with a preamble of type

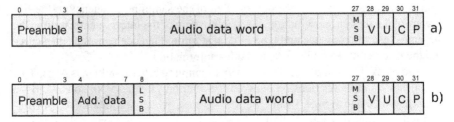

Figure 5.19   AES/EBU subframe format (a: 24 bit audio data word; b: 20 bit audio data word)

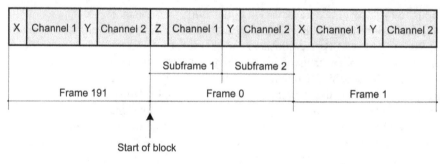

Figure 5.20   AES/EBU frame format

X, but at the beginning of the block (in the first subframe of the block), the preamble X is replaced with a preamble of type Z. The second frame subframe always starts with a preamble of type Y.

The AES/EBU interface assumes the transmission of audio data in several modes:

- Two-channel mode, in which subframes 1 and 2 are used to transmit data from two independent audio channels.
- Stereo mode in which the left channel (or channel "A") is transmitted in subframe 1, and the right channel (channel "B") is in subframe 2. In this case, the samples of the left and right channels should be taken simultaneously.
- Single-channel (mono) mode, in which the audio data are transmitted in subframe 1. In the second subframe, bits 4–31 may be identical to bits of subframe 1 or set equal to logical zero.

To properly interpret the data, the receiving side needs information about the operating mode that the transmitting side uses. This information is reported to the receiver using the C bits transmitted in each subframe. Data on the status of each channel are transmitted in 192-bit packets that are composed of 24 bytes. Individual bits of bytes are transmitted in C time slots of subframes. The first bit of the first byte of each block is transmitted in frame 0 marked with Z type preamble. In the data packet of the channel status data, various information are transmitted: the length of the audio data word, the number of channels, the transmission mode (e.g., stereo, mono), the presence or absence of a binding of the sampling frequency to the frequency of the reference oscillation, the identification of the linear quantization scale, the time code, the presence and type of predistortion, the code names of the source and the receiver of the audio data, and other information.

The block duration of 192 frames with a sampling rate of 48 kHz is 4 ms. This establishes a simple relationship between the sampling frequency of sound and the frequency of video frames (frame period: 40 ms), which simplifies the synchronization and transmission of digital video and audio signals along a single communication line. Therefore, the sampling rate of 48 kHz audio signals is the most convenient for television [3].

5.2.5.2   Channel coding

The AES/EBU standard prescribes channel coding in accordance with the two-phase code method (Figure 5.21). Each bit of audio data and bits V, U, C, P is transmitted using a two-bit binary symbol or a symbol consisting of two states. Thus, the symbol corresponds to a pulse signal occupying two clock intervals. The first state of the symbol is always different from the second state of the previous character. This means that a pulse signal corresponding to a sequence of symbols always changes its polarity at the symbol boundary. The second state of the symbol is identical to the first if the transmitted audio data bit is equal to logical zero. It is opposite to the first if the bit of the transmitted audio data is equal to the logical one (in this case, the pulse signal changes its polarity in the middle of the interval of one symbol as well). This channel coding eliminates the constant component of the transmitted signal and makes the interface "insensitive" to changing the polarity of the signal. The pulsed signal, the carrier of the data subjected to channel coding, always changes its polarity at the symbol boundary. That is why conditions are created for simple and reliable synchronization of the transmitter and receiver (it can be said that this property of the channel code ensures self-synchronization).

Features of the two-phase code channel coding method were used in developing the structure of the preamble – a packet for synchronization and identification of subframes, frames and blocks. Slots 0–3 are allocated at the beginning of each subframe for the preamble. It takes 8 clock intervals in which signals representing 8 consecutive logical states are transmitted. The first state of the preamble must be different from the second state of the preceding symbol, which always represents the parity bit (Figure 5.22). Therefore, there are two versions of each preamble. The X preamble corresponds to the sequence 11100010 if the previous state is 0 and

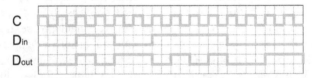

Figure 5.21   *AES/EBU channel coding (C – clock pulses; D_{in} – input signal of channel coder; D_{out} – output signal of channel coder)*

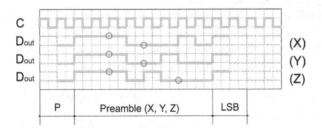

Figure 5.22   *AES/EBU X, Y and Z preambles*

00011101 if the previous state is 1. The Y preamble corresponds to the sequences 11100100 and 00011011. The Z preamble corresponds to the sequences 11101000 and 00010111, respectively. In Figure 5.22, the preambles are shown, provided that the second state of the preceding parity bit is equal to 0.

Like the audio data encoded with the two-phase code method, the preamble signals do not have a constant component and have the self-synchronization property, i.e., their shape provides a simple selection of clock pulses. However, the rules for channel coding do not apply to the preamble structure. In Figure 5.22, red circles indicate the points where there should be level differences for the encoded sequence of audio data (these points are on the boundaries of the symbols carrying the data bits). The preamble signals differ from any sequence of audio data encoded by the two-phase code method in at least two clock intervals. The preamble signals differ from each other in two clock intervals. These features provide reliable synchronization at the level of subframes, frames and blocks.

5.2.5.3   Electrical parameters

The bit rate of data passing through the AES/EBU interface is very high. With a sampling rate of 48 kHz and coding using a 24-bit word, the audio data rate is 24*48 kHz = 1152 kbit/s. To the 24 time slots allocated in each subframe to transmit bits of audio data in serial form, "overhead" is added in the form of 8 time slots for the preamble and bits V, C, U, P. As a result, the bit rate increases to 32*48 kHz = 1536 kbit/s. When transmitting the audio data of two channels, the total bit rate is 3072 kbit/s = 3.072 Mbit/s. The transmission of such a data stream in a serial form requires a significant frequency band.

The interface does not assume the use of any special coding, which allows to increase the specific transmission bit rate for bandwidth reduction. The transmission of the binary signal is performed by pulses whose shape is close to rectangular. A practical estimate of the frequency band occupied by rectangular pulses is the reciprocal of the pulse duration. The specific bit rate calculated for such a method of transferring binary data would be 1(bit/s)/Hz (1 bit per second in the frequency band equal to 1 Hz). This value is half the Nyquist 2(bit/s)/Hz limit, which can be achieved using encoding methods such as the partial sampling method. But the use of a two-phase code method requires the use of two pulses to transmit one bit, which reduces the specific rate to 0.5 (bit/s)/Hz. Therefore, the AES/EBU interface should be designed for bandwidth (3072 Mbit/s)/(0.5 [bit/s]/Hz) = 6144 MHz at a sampling rate of 48 kHz. If you do not explicitly include the sampling frequency in the above calculation, you can get an overall estimate – the bandwidth allocated for AES/EBU signal transmission should be 128 times the maximum frame rate, i.e., the maximum sampling frequency.

The interface specification establishes the use of a symmetrical communication line with 110 Ω impedance (ITU-T Recommendation V.11) at a signal span from 2 to 7 V. In professional applications, it is recommended to use an XLR connector (IEC 60268-12).

The lines used for general purpose data networks can also be used for the implementation of the interface. Experiments have shown that when using a Category 5 twisted pair and RJ45 connectors, the transmission range reaches 400 m without frequency correction and 800 m when using the frequency correction of the received signal (at a frame rate of 48 KHz).

5.3 Digital representation of video signals

5.3.1 Analog-to-digital conversion of video signals

5.3.1.1 Sampling

To convert an analog signal into a digital form, you need to perform three basic operations: sampling, quantization and coding. Sampling or discretization is the representation of a continuous analog signal by a sequence of its values (samples). These samples are taken at times, separated from each other by an interval, which is called the sampling interval. The reciprocal of the interval between samples is called the sampling frequency. Figure 5.23 shows the original analog signal and its sampled version. The pictures given under the time diagrams are obtained on the assumption that the original analog and sampled signals are television video signals of one line, the same for all the lines of the TV screen.

It is understood that the smaller the sampling interval and, correspondingly, the higher the sampling frequency, the less the difference between the original signal and its sampled copy. The stepped structure of the sampled signal can be smoothed using a low-pass filter. Thus, the analog signal is reconstructed from the sampled signal. But the recovery will only be accurate if the sampling rate is at least 2 times the bandwidth of the original analog signal, as shown in the previous section. If this

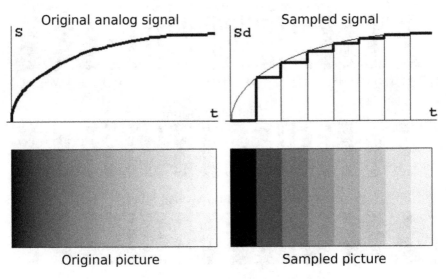

Figure 5.23   Sampling of video signal

condition is not met, then the discretization is accompanied by irreversible distortions. The matter is that as a result of discretization in the frequency spectrum of the signal, additional components appear around the harmonics of the sampling frequency in the range equal to twice the width of the spectrum of the original analog signal. If the maximum frequency in the frequency spectrum of the analog signal exceeds half of the sampling frequency, additional components fall into the frequency band of the original analog signal. In this case, it is impossible to restore the original signal without distortion.

The manifestation of the discretization distortions on the television image is shown in Figure 5.24. The analog signal (suppose again that this video signal is the signal of a TV line) contains a wave whose frequency first increases from 0.5 to 2.5 MHz and then decreases to 0.5 MHz. This signal is sampled at the frequency of 3 MHz. Figure 5.24 also shows images corresponding to the original analog signal, sampled and reconstructed after sampling the analog signals (the image in Figure 5.24 has standard definition). A low-pass filter has a bandwidth of 1.2 MHz. As you can see, low-frequency components (less than 1 MHz) are restored without distortion. A wave with a frequency of 2.5 MHz after the restoration turned into a wave with the frequency of 0.5 MHz (this is the difference between the sampling frequency of 3 MHz and the frequency of the original signal of 2.5 MHz). These diagrams and pictures illustrate the distortions associated with an insufficiently high frequency of spatial sampling of the image.

If the sampling frequency is set and fixed, then for the absence of sampling distortions, the bandwidth of the original signal must be limited and not exceed half of the sampling rate. Figure 5.25 illustrates the preliminary low-frequency filtering

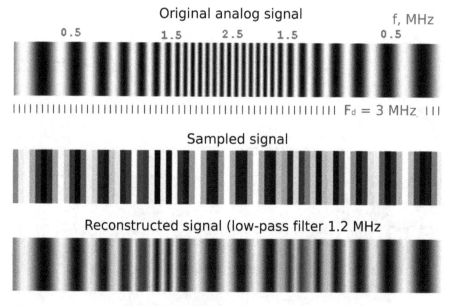

Figure 5.24   Sampling distortions when sampling frequency is not high enough

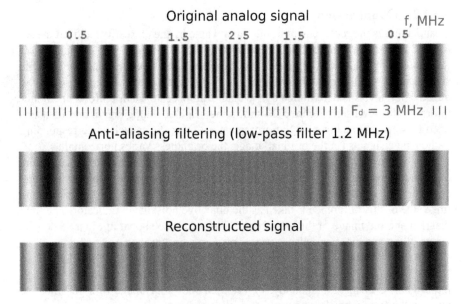

Figure 5.25   *Low-pass filtering of video signal to avoid sampling distortions*

of the sampled signal aimed at preventing the appearance of sampling distortions. The original signal (the upper image) is passed through a low-pass filter with a 1.2 MHz band. The output signal of the filter (middle image) is sampled at the frequency of 3 MHz. In the reconstructed signal (bottom image), discretization distortions are almost imperceptible. It should be noted that a low-pass filter of the first order was used to model the process of discretization and recovery, which does not completely suppress the wave with a frequency of 2.5 MHz. Therefore, in the reconstructed signal (bottom image), one can see a wave with a frequency of 0.5 MHz of small amplitude (the visibility depends, of course, on the characteristics of the printing process when printing a book).

If you require that in the process of digitization there is no distortion of a video signal with a boundary frequency, e.g., 6 MHz, then the sampling frequency should be not less than 12 MHz. However, the closer the sampling frequency to the doubled cut-off frequency of the signal, the more difficult it is to create a low-pass filter, which is used in the reconstruction, as well as with the preliminary filtering of the original analog signal. This is explained by the fact that when the sampling rate approaches the doubled cut-off frequency of the signal being sampled, restoration filters with a frequency response that is more and more approximates to a rectangular signal are required (a filter with a rectangular characteristic cannot be physically realized, and the theory shows that it would have to make an infinitely long delay of the transmitted signal). Therefore, in practice, there is always some interval between the doubled boundary frequency of the original signal and the sampling frequency.

5.3.1.2 Quantization

Quantization is the replacement of the signal value by the closest value from the set of fixed values, the quantization levels. In other words, quantization is rounding off the value of the sample. The quantization levels divide the entire range of possible signal values by a finite number of intervals – quantization steps. The location of the quantization levels is determined by the quantization scale. Both uniform and nonuniform scales are used. Figure 5.26 shows the original analog signal and its quantized version obtained using a uniform quantization scale, as well as corresponding to the image signals (in the original image, the brightness varies horizontally).

The distortions of a signal arising in the process of quantization are called the quantization noise. In the instrumental estimation of quantization noise, the difference between the original signal and its quantized copy is calculated, and, e.g., the rms value of this difference is taken as the objective noise level indicator. The time diagram and the image of the quantization noise are also shown in Figure 5.26 (the image of the quantization noise is shown on a gray background).

The manifestation of quantization noise depends strongly on the structure of the image. Figure 5.27 shows the quantization of the video signal of the image, in which the brightness changes diagonally. Figure 5.28 shows an image of a portrait type quantized into two levels, and the quantization noise corresponding to such a number of levels, in which it is not difficult to discern the outline of the original image. Figure 5.29 shows an image quantized into four levels and quantization noise corresponding to this number of levels. The quantization noise decreases with increasing

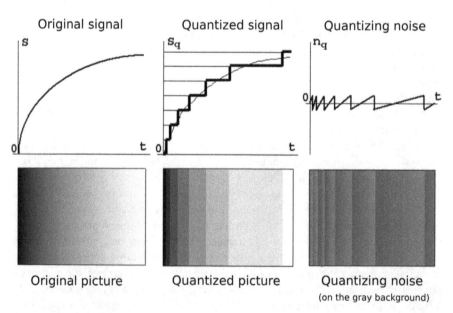

Figure 5.26   Quantization of video signal (image brightness is changing in horizontal direction)

Original picture   Quantized picture   Quantizing noise (on the gray background)

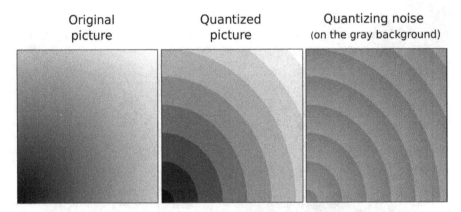

Figure 5.27   *Quantization of video signal (image brightness is changing in diagonal direction)*

Quantized picture (2 quantizing levels)   Quantizing noise (on the gray background)

Figure 5.28   *Quantized picture and quantization noise on gray background (two quantization levels for each component of color picture)*

number of quantization levels. Unlike fluctuation noise, the quantization noise is correlated with the signal, so the quantization noise cannot be eliminated by subsequent filtering. The image shown in Figure 5.30 is obtained using 128 levels. At such a relatively large number of levels, quantization noise is similar to ordinary fluctuation noise. The noise has dropped, so it was necessary to increase this noise values by 128

Quantized picture
(4 quantizing levels)

Quantizing noise
(on the gray background)

*Figure 5.29   Quantized picture and quantization noise on gray background (four
quantization levels for each component of color picture)*

Quantized picture
(128 quantizing levels)

Quantizing noise
(on the gray background, amplified)

*Figure 5.30   Quantized picture and quantization noise on gray background (128
quantization levels for each component of color picture)*

times when getting a picture of quantization noise so that the noise was noticeable. A few years ago, it seemed sufficient to use 256 levels for quantizing a television video signal. Now, it is considered the norm to quantize video signals at 1024 levels.

5.3.1.3   Coding

A quantized signal, in contrast to the original analog signal, can take only a finite number of values. This allows us to represent it within each sampling interval by a number equal to the ordinal number of the quantization level (Figure 5.31). In turn, this number can be expressed by a combination of some characters or symbols. The set of characters (symbols) and the system of rules by which data are represented as a set of symbols is called a code. The sequence of code symbols is called a code word. A quantized signal can be converted into a sequence of code words. This operation is called coding. Each code word is transmitted within one sampling interval. A binary code is widely used to encode audio and video signals. If the quantized signal can take N values, then the number of binary symbols in each code word is $n \geq \log_2 N$. One digit, or symbol of a word represented in binary code, is called a bit. Usually, the number of quantization levels is equal to an integer power of 2, i.e., $N = 2^n$.

Code words can be transmitted in parallel or serial forms (Figure 5.31). For transmission in parallel form, n communication lines should be used (in the example shown in Figure 5.31, $n = 4$). The code word symbols are simultaneously transmitted along the lines within the sampling interval. For transmission in a serial form, the sampling interval must be divided into n subintervals – cycles. In this case, the word symbols are transmitted sequentially along one line, and one clock cycle is assigned to the transmission of one word symbol.

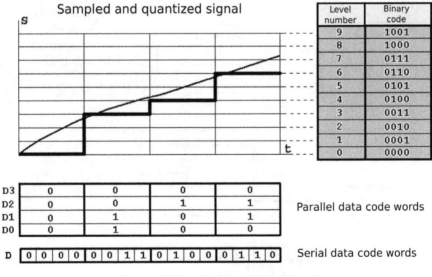

Figure 5.31   Coding of video signal

Each word symbol is transmitted using one or more discrete signals – impulses. The conversion of an analog signal into a sequence of code words is therefore often called pulse code modulation. The form of representing words by certain signals is determined by the format of the code. For example, you can set a high signal level within the cycle if the binary symbol 1 is transmitted in this clock cycle, and low if the binary symbol 0 is transmitted (this way of representation, shown in Figure 5.32, is called the NRZ format). The example in Figure 5.32 uses 4-bit binary words (this allows 16 levels of quantization). In a parallel digital stream, 1 bit of a 4-bit word is transmitted over each line within the sampling interval. In a serial stream, the sampling interval is divided into 4 cycles in which the bits of the 4-bit word are transmitted (beginning with the most significant one).

Operations related to the conversion of an analog signal into digital form (sampling, quantization and coding) are performed by one device – ADC. The ADC can simply be an integrated circuit. The reverse procedure, i.e., restoration of an analog signal from a sequence of code words is performed in a DAC. ADC and DAC are an integral part of digital television systems.

Digital video signals can be described using parameters typical for analog technology, e.g., such as the frequency band. An important indicator characterizing a digital stream is the data bit rate. If the word length is n, and the sampling rate is F_D, then the data transfer rate expressed in binary symbols per unit time (bit/s) is found as the product of the word length by the sampling frequency: $C = nF_D$.

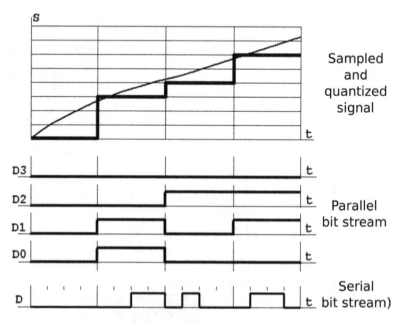

Figure 5.32   Parallel and serial data streams

5.3.2 *Digital representation of component video signals of standard definition*

5.3.2.1 Sampling and quantization

A component standard definition video signal can be digitized in accordance with Recommendation ITU-R 601, which was adopted in 1982 [4]. This recommendation establishes rules for the separate sampling, quantization and coding of the luminance signal Y and the two color difference signals R-Y (Cr) and B-Y (Cb). The sampling frequency for the luminance signal Y is set to 13.5 MHz, and for color difference signals – 6.75 MHz, i.e., the sampling frequency of the luminance signal is 2 times greater than the sampling frequency of the color difference signals (Figure 5.33, the color band signal).

If we take, as usual, the frequency of 3.375 MHz as a conventional unit, then the sampling frequencies of the luminance and two color difference signals will be in the ratio 4:2:2, which gives the commonly used name of the standard. With such sampling rates, you can practically convert the luminance signal in the band of 5.75 MHz without distortion into digital form, and the color difference signals in the band of 2.75 MHz (remember the "spare" interval between the cutoff frequency of the signal and half of the sampling frequency). Standard 4:2:2 is used as a "base" in evaluating other sampling options, and the value of 5.75 MHz is often referred to as the border of the full video signal bandwidth.

The length of the code word is 10 bits (in the original version – 8 bits), which allows to number 1024 quantization levels. However, the numbers 0..3 and 1020..1023 are reserved for digital synchronization signals. To quantize the

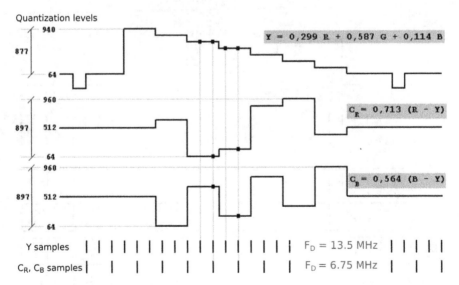

Figure 5.33   *Sampling and quantization of component video signal (color bar image)*

luminance signal, 877 levels are allocated (Figure 5.33). The black value in the video signal corresponds to the quantization level of 64 and the nominal value of the white value to the level of 940. To quantify the color difference signals, 897 levels are allocated, with the quantization level of 512 corresponding to the zero value of the analog signal. The gamma-corrected signals are coded.

The given ranges of quantization levels are often used when compared with other quantization options. In this case, they are often referred to as dynamic range or full resolution by the signal level, since the number of quantization levels determines the quantization noise and, accordingly, the dynamic range. As shown in section 1.2.3.2, the magnitude of the dynamic range, or the ratio of the signal peak-to-peak span to the effective value of the quantization noise, can be calculated as $20*\log(2^n) = 6n$ [dB], where n is the number of bits or bits in the code word. With a code word length of 8 bits, the value of the dynamic range can be estimated at 48 dB, and at a length of 10 bits – 60 dB.

Values of 13.5 and 6.75 MHz are integral multiples of both the horizontal scanning frequencies in the 625/50i scanning system and the scanning frequency in the 525/60i system. This allowed us to have an almost unified world standard for digital encoding of the component video signal, in which the active part of the line contains 720 samples of the brightness signal and 360 for each color difference signal (Figure 5.34). The difference in the systems 625/50 and 525/60 consists in a different number of lines and slightly different duration of the blanking interval. In the system 625/50, the total number of samples of the signal Y in the line is 864, in the system 525/60–858. The total number of samples of signals Cr and Cb in the system 625/50 is 432 in one line, in the system 525/60–429. The number of the active lines is equal to 576 in the 625/50 system and 480 in the 525/60 system. The total bit rate

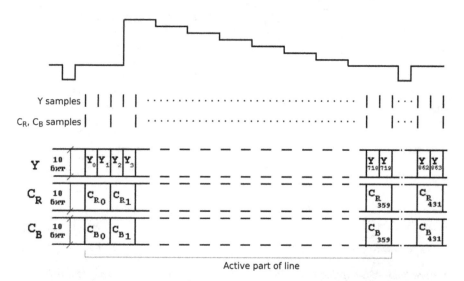

Figure 5.34   Samples of component video signal (625/50 system)

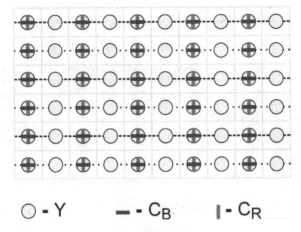

$\bigcirc$ - Y $-$ - C_B $|$ - C_R

Figure 5.35   *Sampling structure of TV picture (4:2:2 format, progressive capture)*

of the digital component video signal is $10*13.5 + 10*6.75 + 10*6.75 = 270$ Mbit/s at ten bits per one sample. The sampling frequencies represent the harmonics of the line frequency, which provides a fixed orthogonal frame of the TV image counts (Figure 5.35).

When interlaced, the frame of the TV image consists of two fields. The structure of the samples in the image frame and in two fields (upper and lower) is shown in Figure 5.36. The lines of the top field are shown as a dashed line, and the lines of the bottom field are shown in dotted lines.

There are other formats for representing the component signal digitally. Coding to the 4:4:4 standard assumes the use of a 13.5 MHz frequency for all three components: R, G, B or Y, Cr, Cb (Figure 5.37). This means that all components are transmitted in full bandwidth. For each of them, 576 lines of 720 elements are digitized in the active part of the frame. The bit rate of the digital stream with a 10-bit word is 405 Mbit/s.

The 4:1:1 format offers a twofold reduction in the sampling frequency of color-difference signals (compared to the 4:2:2 standard). The luminance signal Y is

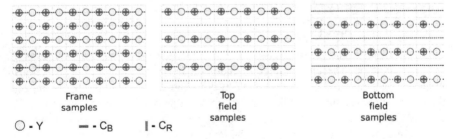

Figure 5.36   *Sampling structure of TV picture (4:2:2 format, interlaced capture)*

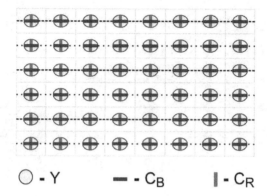

Figure 5.37   Sampling structure of TV picture (4:4:4 format)

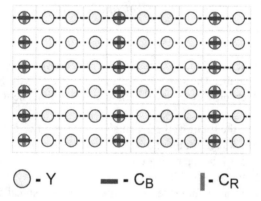

Figure 5.38   Sampling structure of TV picture (4:1:1 format)

sampled at the frequency of 13.5 MHz, and the color difference signals (Cr and Cb) are 3.375 MHz. This means a twofold reduction in the horizontal resolution in color. In the active part of the frame 576 lines, each of which contains 720 elements of the luminance signal and 180 – color-difference signals (Figure 5.38). Color difference signals of 4:1:1 format can be obtained from color difference signals of 4:2:2 format by discarding every second sample horizontally (in combination with preliminary low-frequency filtering of color difference components).

The 4:2:0 format offers an image in which the luminance component Y contains 576 lines of 720 samples in the active portion of the frame, and the color difference components C_R and C_B have 288 lines of 360 samples. This means a twofold reduction in resolution in color and in the horizontal and vertical directions in comparison with the 4:4:4 format. With the progressive scanning of the image, the color difference signals of the 4:2:0 format can be obtained from the color difference signals of the 4:2:2 format (Figure 5.35) by discarding every second sample vertically. However, when interlacing the image (Figure 5.36), this cannot be done. If you discard color difference samples in every second line of the TV frame, then one field will not participate at all in the formation of the color image. This means not only a twofold reduction in vertical color definition but also a twofold reduction in temporal resolution with respect to large color details of the image. The frequency of sampling in color for large details of the image will be halved (from 50 to 25 Hz). This can lead to noticeable distortions in the transmission of color dynamic images.

A possible variant of forming the structure of samples of the 4:2:0 format is shown in Figure 5.39. The matrix of samples of color-difference components can be obtained from the prototype structure 4:2:2 by alternately excluding one color difference component in every second line of each field. This is the 4:2:0 format option used in DV video recording system.

The 4:1:1 and 4:2:0 encoding options have the same data rate of 202.5 Mbps for the code word length of 10 bits and 162 Mbps for 8 bits per word. If you transfer only the active part of the image (without blanking time intervals), the value of the digital stream at 8 bits per word will be 124 MB/s. The digital signals of these two formats can be obtained from 4:2:2 signals by preprocessing and decimation of samples in order to reduce the bit rate. The 4:1:1 format is recommended for systems with the scanning standard of 525/60, and a 4:2:0 format for systems of 625/50. In order to understand the prerequisites for such a conclusion, it is necessary to evaluate the vertical and horizontal resolution in the digital image, encoded in accordance with Recommendation 601.

The format of the standard television image, or the ratio of the width of the image to its height, is 4/3. If we proceed from the requirement of the same resolution in the horizontal and vertical directions, then when the TV image is decomposed into 625 lines (576 active lines), each line should contain 576*4/3 = 768 pixels, and when decomposed into 525 lines (480 active lines) − 480*4/3 = 640 pixels. But the active part of the line in accordance with Recommendation 601 contains 720 pixels for both the 625/50 system and the 525/60 system. This means that the pixel is not square in either the 625/50 system or the 525/60 system, and that the resolution in the horizontal and vertical directions is not the same. In the 625/50 system, the actual pixel is stretched horizontally (its format can be estimated at 1.07) and the horizontal resolution is worse than the vertical. In the 525/60 system, the pixel is compressed horizontally (its format is 0.89), and the resolution in the horizontal direction is better than the vertical direction.

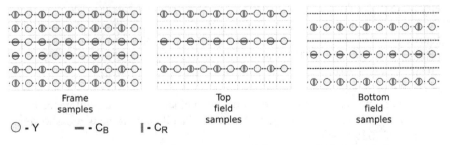

Figure 5.39   Sampling structure of TV picture (4:2:0 format)

The above calculations refer only to the luminance component of the image, and the results obtained do not take into account the effects of the interlacing of the television image and the differences in the properties of the human visual analyzer in the horizontal and vertical directions. However, from the above reasoning follows the main conclusion – in the system 625/50, there is a certain lack of horizontal definition (in comparison with the vertical one), and in the system 525/60 – vertical (in comparison with the horizontal). Therefore, a 4:1:1 sampling structure that degrades the horizontal clarity in color is less acceptable for the 625/50 system than the 4:2:0 structure. It can also be noted that the 4:2:0 structure, which degrades the definition in color along the vertical, is less suitable for the 525/60 system than the 4:1:1 structure.

The 3:1:1 format is used, in which the horizontal resolution for both the brightness component (from 720 to 540) and for the color difference (from 360 to 180) is reduced (in comparison with 4:2:2). The active part of the frame contains 576 lines with 540 luminance component samples and 180 samples for color difference ones. The data bit rate of 3:1:1 format is 135 Mbit/s at 8 bits per one sample.

To significantly reduce the bit rate, the resolution of the luminance component is reduced approximately twofold in the vertical and horizontal directions, and the color-difference component resolution is reduced by 4 times vertically and 2 times horizontally (in comparison with the 4:2:2 standard). This type of representation is described by the format Common Interchange Format. One frame of this format contains 352 samples for the luminance component in the active part of 288 lines and 144 lines of 176 samples for the color difference components. When transferring only the active part of the image, the bit rate is about 30 Mbit/s at 8 bits per sample.

Formats 4:2:2, 4:2:0, 4:1:1 and others associated with discarding part of the color difference samples are used in signal transmission to reduce the bit rate of the transmitted data. But during playback, the pixels of all components of a color image must be known at all points in the image. This means that in the receiver it is necessary to essentially go to the 4:4:4 format. It can be done by calculating the excluded samples, e.g., by interpolation.

The formats associated with the increase in the bit rate of the stream of transmitted data are also used. The 4:4:4:4 format describes the coding of four signals, three of which are components of the video signal (R, G, B or Y, Cr, Cb), and the fourth (alpha channel) carries information about signal processing, e.g., transparency foreground images when overlaying multiple images. An additional fourth signal may also be a luminance signal Y in addition to the signals of the primary colors R, G, B. The sampling frequency of all signals is 13.5 MHz, i.e., all signals are transmitted in full bandwidth. The data bit rate at 10 bits per word is 540 Mbps.

The relative sizes of the matrixes of the luminance and color-difference components of the digital color television image in different formats are shown in Figure 5.40 (625/50 system).

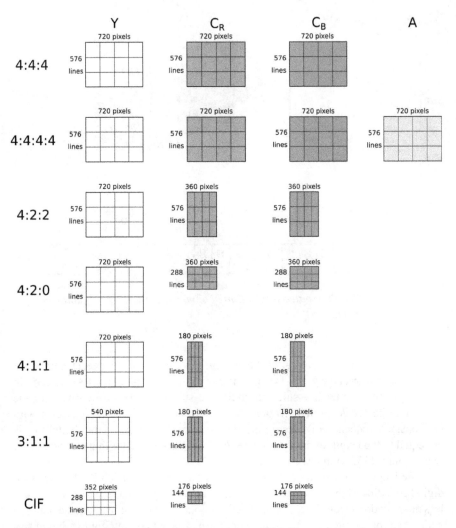

Figure 5.40 *Relative sizes of pixel matrixes of the luminance and color-difference components of the digital color television image in different formats*

5.3.3 *Digital representation of composite video signals*

The composite signals for PAL and NTSC systems are sampled at a frequency of 4 fsc, which is equal to the fourth harmonic of the color subcarrier. Figure 5.41 illustrates the sampling and quantization of the composite video signal (the signal of the color bars is shown). In the National Television System Committee (NTSC) system, the line contains 910 samples, of which 768 form the active part of the digital line.

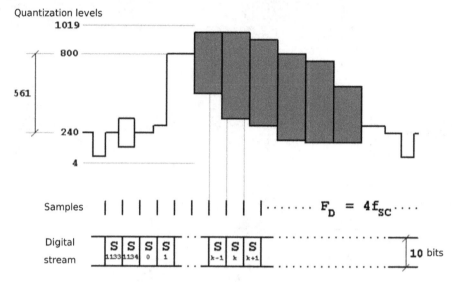

Figure 5.41   *Sampling and quantization of composite video signal (color bar image)*

In the PAL system, the interval of the analog line contains a noninteger number of samples with a frequency of 4 fsc. This is due to the fact that in the PAL system, in addition to the quarter-line shift, an additional subcarrier frequency offset for the frame rate 25 Hz is applied. To preserve a continuous digital sample stream with the constant frequency of 4 fsc, in the PAL system, the length of the digital line is not equal to the length of the analog line. All lines of the field (with the exception of two) contain 1135 samples, and two – for 1137.

The length of the code word is 10 bits (in the original version 8). The need for digital encoding of the sync pulses of a composite analog signal results in about 30% less quantization levels for the range from the nominal value of black to the nominal white than for the signal in the component form. The data bit rate for the digital signal in the NTSC system is 143 Mbit/s, and in the PAL system – 177 Mbit/s.

5.3.4   Digital video interfaces

5.3.4.1   Parallel interface

The standards for digital interfaces describe the connection between two digital devices. In this case, the formats of transmitted signals, type, number and purpose of connecting lines are clearly defined [5].

The parallel interface for the 4:2:2 standard offers the simultaneous transmission of binary code word symbols in NRZ format over separate trunks (Figures 5.31 and 5.32): the number of which is 10 (the length of the code word). Since three component signals must be transmitted digitally on 10 lines, the interface uses time multiplexing, or multiplexing (Figure 5.42). Words of video data are transmitted

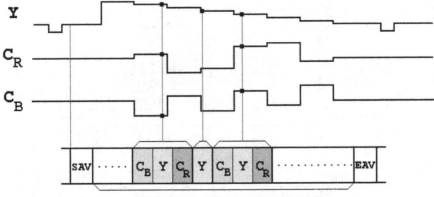

Y - 720 words; C_B - 360 words; C_R - 360 words

SAV - Start of Active Video

EAV - End of Active Video

Figure 5.42   *Parallel digital interface (multiplexing of components for 4:2:2 format)*

in the following order: [Cb, Y, Cr], [Y], [Cb, Y, Cr], [Y], ... Here, the three words [Cb, Y, Cr] describe the three coincided components of one picture element, and the word [Y] refers to one luminance component of a picture element. This is due to the fact that the frequency of the samples of the brightness signal is 2 times greater than the sampling frequency of the color difference signals. The words follow with a frequency of 27 MHz (with a period of 37 ns). The start of the active part of the digital string is preceded by a 4-word Start of Active Video (SAV) synchronization signal, and after the active part of the line the End of Active Video (EAV) reference signal, whose duration is also 4 words, is followed. The basis of SAV and EAV signals is words corresponding to the numbers 0 and 1023, which cannot be used when encoding video data. In the active part of a line, 1440 words (720 words for the luminance signal Y and 360 for the Cr and Cb) should be transmitted, and the total number of pixel cycles in the line is 1728 for the 625/60 system and 1716 for the 525/60.

In the parallel form, a digital composite signal can also be transmitted. Multiplexing is not required in this case, so the code words are transmitted with the frequency of the fourth harmonic of the color subcarrier 4 fsc. The reference time signal for the digital stream is the Timing Reference Signal (TRS) package of 3 words formed from the numbers 0 and 1023. The TRS signal follows the edge of the analog sync pulse of the lines. In the digital signal stream PAL, the duration of the digital line is not equal to the duration of the analog line, so the position of the reference signal TRS relative to the sync pulse of the analog line is gradually shifted. The TRS signal is corrected one time a field by changing the number of samples in the last line of the field. Following the TRS packet, the ID line number identification signal is transmitted in the digital line within the sequence of 4 (for the NTSC system) or 8 (for the PAL system) fields.

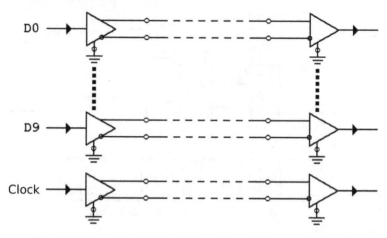

Figure 5.43   Parallel digital interface (schematic diagram)

The parallel interface specifies the transmission of binary bits of code words using symmetrical two-wire lines (Figure 5.43). Transmitters and signal receivers are compatible with integrated circuits of emitter-coupled logic. It is important to note that in addition to the 10 communication lines for 10 bits of the code words, another similar line is used, over which clock pulses are transmitted. Without clock pulses, digital signals in the NRZ format cannot be correctly decoded on the receiving side. Component 4:2:2 and composite 4 fsc parallel digital interfaces allow data transfer up to a distance of 50 m without the use of frequency correctors. With the help of frequency correctors, the transmission distance can be increased; however, for long connecting lines, errors are possible due to the time delay difference for the signals of different lines.

For comparatively small distances between connected devices, a parallel interface may be the most practical option for transmitting video data. However, the parallel form is not very convenient if digital signals need to be switched. The disadvantages are also relatively complicated and bulky connectors and cables.

5.3.4.2   Serial interface

The serial interface assumes the serial transmission of all the binary digits of each code word over one communication line (Figure 5.44). This can be achieved by time multiplexing; hence, the clock frequency must increase at least by the number of times equal to the length of the code word. When transmitting data in a parallel stream at a rate of 27 Mwords/s (4:2:2 standard), the transmission rate for one line is 27 Mbps. In a serial variant, the entire data stream with the bit rate of 270 Mbit/s, which corresponds to the clock frequency of 270 MHz and clock period of 3.7 ns, must be transmitted along one line. For a composite serial interface, the clock frequency is 173 MHz (PAL).

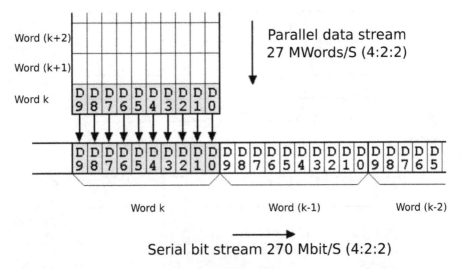

Figure 5.44   *Serial transmission of bits of code word over one communication line*

The transmission of such a broadband signal (hundreds of MHz), such as a serial stream of video data, is not the only difficulty on a communication line. It is also necessary to solve the problem of data synchronization during reception because clock transmission on a separate line does not correspond to the concept of a serial interface. Therefore, the clock pulses must be recovered on the receiving side from the video data signal rises and falls. If this is possible, then such a video data signal is self-synchronized. To make the video data self-synchronized, as well as to match them with the frequency response of the communication line, a special kind of coding is used in the channel coding process.

The simplest Non Return to Zero (NRZ) code, corresponding to the diagrams in Figure 5.32, does not have the self-synchronization property. It has a DC component, low-frequency components dominated in its spectrum. The modified Non Return to Zero Inverted (NRZI) code of differs from NRZ in that when the «1» data is transmitted, level rise or fall is formed. If you limit the number of the following continuously zero data, then the selection of clock pulses becomes possible. In the serial digital interface, the code of NRZI in combination with scrambling is used as a channel code. The procedure, called scrambling, involves adding data with a pseudo-random sequence.

The serial interface specifies the transmission of data over an unbalanced line with a wave impedance of 75 Ohm (coaxial cable and a BNC (initialism of "Bayonet Neill–Concelman") type connector). Transmitting and receiving processors are performed in the form of integrated circuits (Figure 5.45). They allow you to transfer video data of a component 4:2:2 or composite formats up to 300 m.

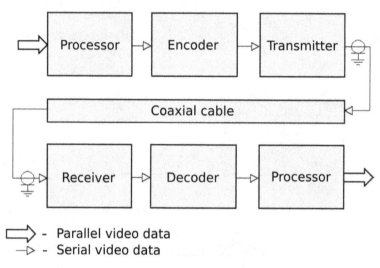

- Parallel video data
- Serial video data

Figure 5.45   Serial digital interface

5.3.4.3   Transmitting ancillary data

In addition to the video data over the video interface, additional information can be transmitted: audio signals, time code, error data and other information [6]. For its transmission, blanking intervals for the line and for the field can be used. Significant convenience gives the transmission of additional data within the serial interface.

The largest part of the space for placing ancillary data is provided by the line blanking interval. In the composite signal, you can use the horizontal sync area following the TRS-ID reference signals, which gives 64 (PAL) or 55 (NTSC) words per line. More space for ancillary data can be allocated in the component 4:2:2 digital signal. This is the area between the end and start reference signals of the digital line EAV and SAV (280 words for the 625/50 system and 268 words for the 525/60). Ancillary data can also be placed in certain areas of the field blanking interval. The share of ancillary data in the total digital stream can be significant – about 20% for a 4:2:2 component signal and about 7% for a 4 fsc composite signal. Information that may already be present in the video signal in special fields blanking intervals (e.g., test signals) is digitized together with the video signal, being considered a part of it.

The ancillary data is embedded in the video signal as a block, i.e., in the package form. In addition to the actual payload data array, the packet contains a header indicating the data type (e.g., AES/EBU audio data), the block number for sequential transmission of additional data of the same type in several blocks, the number of ancillary data words in this block and also the checksum for validation data on reception. The ancillary data are combined with the digital video stream by multiplexing performed by the transmitter processor and extracted by demultiplexing in the receiver processor. Processors should have buffer memory blocks to match the rates of digital video streams and ancillary data (e.g., sound).

The audio signals are included into the video stream in the form of groups; each of which consists of two AES/EBU format signals (1 group is 2 stereo or 4 mono channels). A single AES/EBU group of digital audio signals (2 stereo or 4 mono channels) can be included in the stream of the 4 fsc composite signal. Four AES/EBU groups can be included into the 4:2:2 component digital signal, which gives 8 stereo or 16 mono sound channels.

Combining sound with video and transmission over a single cable promises significant benefits, e.g., in the form of saving material resources (amplifying and distributing and switching equipment, cable equipment). When correcting temporal distortions, co-transmission can eliminate problems with nonsynchronous picture and sound. But if, e.g., independent processing of image and sound signals is required, then the separate transmission of digital streams can be the best way.

5.3.5 *Parameter values for high definition and ultra-high definition television systems*

Definition is one of the main qualitative indicators of a television image, showing how clearly the small details are visible in the image, and characterizing the resolution of the television system. Definition is determined by the maximum number of dark and light lines that are distinguishable in the image under given viewing conditions. The often used definition estimate in digital systems is the number of picture elements, or pixels, horizontally and vertically.

The idea of what is high definition has undergone significant changes during the existence of television. About 405 lines TV introduced in 1936 in the UK was officially referred to as high-definition television (HDTV). The new TV replaced the television system, in which the number of lines was 30. In 1950, the State Prize of the USSR was awarded "For the development of a new high-quality HD system for 625 lines." The history of the development of television technology reflects the creation of HDTV systems, although each of them, we can say today, was such only in comparison with the previous one.

A fundamentally new approach to the creation of HDTV systems as elements of the expandable hierarchy of digital formats was proposed and developed in the 1990s by the 11th Study Group on Television of the International Telecommunication Union under the chairmanship of Professor M. I. Krivocheev. The Group developed Recommendation ITU-R BT.709 (the first version was adopted in 1993), which established a standard for the production and international exchange of programs in the HDTV format and determined the values of key parameters of the television image, among which, first of all, the number of active lines equal to 1080 [7].

The number 1080 appeared as a result of applying the principle of doubling the parameters of the digital signal coded in accordance with Recommendation 601. Additional requirements: one and the same resolution in the horizontal and vertical directions and the transition to a wide image format 16:9.

As noted above, the image that is coded in accordance with Recommendation 601 has a different definition in the horizontal and vertical directions, i.e., a pixel of the image is not square but rectangular in both the 625/50 system and the 525/60 system. The parameter that is the same for both systems is the number of active elements in the line equal to 720. In order for the pixel to be square with an image format of 4:3, the number of lines should be 540 (720*[3/4] = 540). The number of elements per line required to obtain a wide format of the image while saving this number of lines (540) is 960 (530*[16/9] = 960). Thus, a widescreen image of standard definition with a square pixel should have a sample matrix of 960 × 540. The doubling of the obtained numbers gives the matrix 1920 × 1080, which determines the parameters of the image format for today's HDTV. This format was later called 2 K, since the maximum number of pixels along a coordinate is close to 2000.

In digital photography, an often used parameter is the total number of pixels in the image. In this case, the high-definition digital TV system of Recommendation 709 could be called 2 M, since the total number of pixels is close to two million. Standard definition digital television could get the names of 0.7 K (about 700 pixels per line) and 0.4 M (about 0.4 million pixels per frame), respectively.

It should be noted that we should not understand the doubling of the number of lines and picture elements as a primitive approach. Doubling allows the most simple way to implement the principle of scalability of a television image, according to which the image parameters in systems of different levels are among themselves in integer ratios.

Recommendation 709 determines the different frame rates for progressive and interlaced picture capture: 25, 30, and 60Hz. The total number of lines is 1125. Sample structures of 4:2:2 and 4:4:4 are defined. The sampling frequency of the luminance signal is 74.25 MHz in the system 30P and 60I (progressive picture capture with the frame rate of 30 Hz and interlacing with the frequency of 60 Hz, respectively) and in systems 25P and 50I. In systems 60P and 50P, the sampling frequency of the luminance signal is 148.5 MHz. Linear quantization is recommended at 8 or 10 bits per sample of each component.

The value of the frame rate of 24 Hz is also defined with the progressive capture of the image. The number 24 Hz is the frame rate in standard cinematography. The progressive capture of the image into 1080 active lines at a frame rate of 24 Hz is a universal format for the production of television programs and films. This format has become a radical solution to the problem of interaction between television and cinema areas. This format opened a new stage in the development of both television and cinema systems.

The creation of the HD format 1920 × 1080 was a huge step. However, the target viewing distance at which the HDTV system can be considered "transparent" is a value equal to three screen heights. But there are many applications in which the viewer should view the image on large screens and at a large angle, e.g., computer graphics, printing, medicine, video information systems, and cinema. The penetration of digital television systems into new areas is promising for the development of television itself because the increase in the angle at which the viewer watches the images enhances the effect of immersion in reality [8–10].

International standardization was extended to systems with a resolution of more than 2 K (1920 × 1080), chosen for HDTV systems. In 1995 the first world recommendation ITU-R VT.1201 [8] was developed, which defines the image formats of even higher definition, including 4 K (3840 × 2160) and 8 K (7680 × 4320), selected as ultra-high definition UHDTV formats. The total number of pixels in the frame for 4 K and 8 K formats is 8 M and 32 M, respectively. These formats correspond to the principle of building a hierarchy of scalable high-definition systems. The parameters of the UHDTV systems are given in Recommendation ITU-R BT.2020 [11]. A wide range of possible frame rates is defined (24, 25, 30, 50, 60, 100 and 120 Hz) in combination with a progressive picture capture. Three possible sampling structures are determined: 4:4:4, 4:2:2 and 4:2:0. Linear quantization is recommended at 10 or 12 bits per sample of each component.

The transition to the 4 K and 8 K UHDTV systems is not just a way to improve image definition. As noted above, the HDTV system 2 K 1920 × 1080 was developed based on the viewing distance, equal to 3 screen heights. The horizontal viewing angle is approximately 30 degrees. 4 K system allows you to reduce the viewing distance to one and a half heights of the screen and increase the angle at which you can comfortably watch the image, up to 60 degrees. 8 K system allows you to reduce this viewing distance to three-fourths of the screen height and increase the horizontal viewing angle up to 100 degrees. It should be noted that the viewing angle of the 8 K UHDTV system is a significant part of the natural field of view of a person, which provides almost the maximum possible effect of immersion in reality.

Both UHDTV formats can be used for the reproduction of TV programs on a home TV with a diagonal of one and a half meters or more to enhance the sense of the presence of the viewer in the depicted place. Of course, this will be attractive to the viewer at home. But even more advantageous is the use of these formats for demonstration on large and huge screens. These are screens of concert halls, scoreboards of sports arenas. UHDTV systems can have a huge effect in applications where ultra-high definition is required, e.g., in medicine and computer graphics. An immense sphere of application is video information systems. Large screens can be used to simultaneously display several different images.

But there is a sphere where UHDTV systems are simply necessary. This is the cinema. The 4 K UHDTV systems can be successfully used at all stages of digital film production, they satisfy all the requirements of the production process of films for theatrical display in conventional cinemas, in which 35-mm film was previously used. The parameters of the 8 K UHDTV systems approach the parameters of the IMAX system (70 mm film with a longitudinal position of the frame), which ensures the highest quality of the theatrical display. So, the use of ultra-high definition TV systems promotes the convergence of television and cinema and their transformation into important components of the country's information infrastructure.

References

[1] Shannon C.E. 'A mathematical theory of communication'. *Bell System Technical Journal*. 1948, vol. 27(4), pp. 623–56.
[2] EBU Tech Doc 3250: Specification of the digital audio interface (The AES/EBU interface).

[3] Recommendation ITU-R BS.646-1. Source encoding for digital sound signals in broadcasting studios.

[4] Recommendation ITU-R BT.601-7. Studio encoding parameters of digital television for standard 4:3 and wide-screen 16:9 aspect ratios.

[5] Recommendation ITU-R BT.656-5. Interface for digital component video signals in 525-line and 625-line television systems operating at the 4:2:2 level of recommendation ITU-R BT.601.

[6] Recommendation ITU-R BT.1305-1. Digital audio and auxiliary data as ancillary data signals in interfaces conforming to recommendations ITU-R BT.656 and ITU-R BT.799.

[7] Recommendation ITU-R BT.709-6. Parameter values for the HDTV standards for production and international programme exchange.

[8] Recommendation ITU-R BT.1201. Extremely high resolution imagery (Question ITU-R 226/11).

[9] Recommendation ITU-R BT.1680-1. Baseband imaging format for distribution of large screen digital imagery applications intended for presentation in a theatrical environment.

[10] Recommendation ITU-R BT.1769. Parameter values for an expanded hierarchy of LSDI image formats for production and international programme exchange.

[11] Recommendation ITU-R BT.2020-2. Parameter values for ultra-high definition television systems for production and international programme exchange.

Chapter 6

Error correction for digital signals

Konstantin Glasman[1]

6.1 Channel coding in digital television systems

6.1.1 Near Shannon limit performance of digital television systems

Digital television in accordance with the standards of the Digital Video Broadcasting (DVB) and Advanced Television Systems Committee (ATSC) families appeared over twenty years ago. But the second generation of the DVB-2 standards [1–3] and the third generation of the Advanced Television Systems Committee (ATSC 3.0) standards [4] are already being introduced everywhere.

The design goal of the second-generation DVB was to ensure quality at the level of quasi-error-free delivery of TV programs with greater efficiency, i.e., with an increase in the speed of the transmitted data. In DVB-T2 (terrestrial television), DVB-S2 (satellite television) and DVB-C2 (cable television) systems, this increase is about 40%. In the DVB project, the level of quasi-error-free delivery quasi error free corresponds to the packet error rate in the transport stream, measured at the input of the demultiplexer of the receiving side, equal to 10^{-7} (one of 10 million packets is affected by errors), if the signal-to-noise ratio is not lower than the threshold value [1–4]. This means less than one uncorrected error per hour of transmission at a digital data rate of 5 Mbps.

DVB is a transport system for delivering picture and sound data. Therefore, it is natural to evaluate its efficiency in terms of the transmitted data stream rate and the threshold signal-to-noise ratio. The DVB-T system was capable of transmitting a data stream at the maximum rate of 31.7 Mbit/s in the 8 MHz band with a C/N ratio of 20 dB (carrier-to-noise power ratio in 8 MHz band). The DVB-T2 system under the same conditions is capable of transmitting data at the rate of 45.5 Mbit/s. The threshold signal-to-noise ratio required for data transmission at 24 Mbit/s in DVB-T is 16.7 dB. In the DVB-T2 system, the threshold value of the signal-to-noise ratio required for the same speed is 10.8 dB [5].

The given data on the ratio of stream rates for DVB-T and DVB-T2 do not give a complete picture of the significance of the achieved increase in efficiency. To fully understand this significance, the capacity of the 8 MHz communication channel should be estimated. The throughput of a continuous communication channel with a frequency band W, in which there is white noise of power N, with an

[1]Television Department, St. Petersburg Electrotechnical University (LETI), Russia

average power of the transmitted signal P, is determined by the famous Shannon formula [6]:

$$C = W * log_2 \frac{P + N}{N}.$$ (6.1)

Substituting a bandwidth of 8 MHz and a signal-to-noise ratio of 20 dB into the formula, we obtain a channel capacity of 53.3 Mbps. The data rates achieved in DVB-T and DVB-T2 systems are 59% and 85% of the theoretical capacity, respectively. But when close to the limit, it is better to estimate the distance that separates the achieved level from the theoretical limit. From (6.1), you can find the theoretical minimum value of the signal-to-noise ratio, at which data are transmitted at a rate of 24 Mbit/s in the 8 MHz frequency band. The calculation gives a value of 8.5 dB. This estimate shows that the threshold value of the signal-to-noise ratio for the DVB-T system is 8.2 dB higher than the theoretical value, and for DVB-T2, it is only 2.3 dB. It should be borne in mind that the assessment of the throughput according to the (6.1) does not take into account the properties of real channels and gives a somewhat overestimated value. The achieved threshold signal-to-noise ratio is actually quite close to the theoretical limit. The research results presented in [7] show that in a wide range of operating modes the threshold value of the signal-to-noise ratio in the DVB-S2 system is 0.6–0.8 dB away from the theoretical limit. Similar indicators were obtained for the DVB-C2 system [8]. The given data allow us to evaluate the level of DVB-2 systems as 'Near Shannon limit performance'. This is a very significant result, exceeding the achieved level is unlikely in the foreseeable future.

Efficiency was also one of the most important concepts in the development of the third generation ATSC digital television system. New options allowed to bring the operation of the ATSC-3.0 closer to the theoretical Shannon limit.

How did one manage to reach a level close to the theoretical limit? The answer to this question is both simple and complex. The theoretical possibility of reaching this level was predicted by Claude Shannon many years ago, the practical possibility has appeared recently. The key to understanding how the transformation of DVB systems into 'Near Shannon' was achieved lies in the field of channel coding, one of the main tasks of which is error correction in digital communication systems.

The history of channel coding can be traced back to 1948, when an article by the outstanding mathematician and engineer Claude Shannon 'A mathematical theory of communication' was published [6]. In this work, Shannon proved that with the help of channel coding it is possible to provide data transmission with an arbitrarily small error probability, if the data transfer rate does not exceed the communication channel capacity. Shannon did not indicate how to find an algorithm for such coding, he only noted that it may be necessary to use fairly complex coding systems. The search for channel coding methods has become the subject of intense research.

In 1950, Hamming invented a class of codes for correcting single errors in a codeword (Hamming codes). These codes were weak against the background of the expectations generated by Shannon's theory, but this was the first major step. The breakthrough came in the late 1950s and early 1960s of the twentieth century. A large class of multiple error-correcting codes was found in 1959 and 1960 thanks to the works of Bose, Chaudhuri and Hocquenghem (BCH codes), Reed and Solomon (RS codes) [9, 10].

In 1960, a postgraduate student Robert Gallager proposed low-density parity-check (LDPC) codes in his dissertation work, written before he was 30 years old [11]. These codes had excellent characteristics, but their implementation was extremely time consuming and not realizable in real time, even on digital computers of that time. Gallager believed that in order to speed up decoding, it is necessary to create a special parallel computing device using analogue adders, adders modulo 2, amplifiers and non-linear devices [12]. Gallager's codes were so ahead of their time that they did not find an application and were forgotten for more than three decades. In 1995, they were reopened by MacKay and Neal [13], after which their triumphal march began. It is the LDPC codes that made it possible to approach the theoretical limit and achieve in the second-generation DVB systems the rate of transmitted data, which is very close to the theoretical limit – the channel capacity.

The history of the discovery of BCH and LDPC codes, in itself quite intriguing, describes the external outline of events and does not come close to understanding the significance of the results of the research performed for today and for the future. To understand, you need to dive into communication theory.

6.1.2   Digital television system as a communication system

The transmission and reception of television programs provided using the DVB and DVB-2 families is an example of information communications performed using communication systems. The generalized model of a communication system (Figure 6.1) assumes the presence of the information source that sends a message to

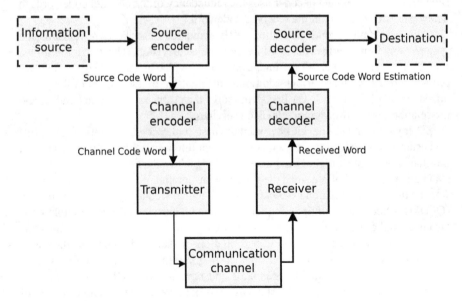

Figure 6.1   Generalized model of a communication system

the recipient [6]. A message is a generalized concept that covers all types of transmitted information in any form – from text letters and numerical measurements to speech, music and television images.

In digital communication systems, discrete signals that take a finite set of values are used to transmit messages. Discrete representation is natural when displaying messages as sequences of discrete symbols selected from a finite alphabet. Examples of such discrete messages, called data, are telegrams, letters, sequences of numbers or machine codes. But by their nature, they are continuous, or continuous processes that are naturally displayed using continuous analogue signals.

Source codewords can be applied directly to the input of a communication system channel, provided that the communication channel does not introduce noise and distortion. However, in real communication channels, there is noise and frequency and non-linear distortions appear. With significant noise levels and significant distortions in the communication channel, errors may occur and codewords at the channel output may differ from the input codewords. Recovery of codewords transmitted over a communication channel is achieved through a special coding called channel coding, which allows you to detect and correct errors.

Source codewords are processed by a channel encoder (Figure 6.1), in which special check symbols are calculated, added to the input codewords. The words at the output of the channel encoder are called channel codewords. The fact that channel codewords contain more symbols than source words means that the channel encoder introduces some redundancy into the data stream. The transmitter converts the message into a signal suitable for transmission over the communication channel. The receiver processes the received signal and reconstructs the transmitted message from it. The channel decoder uses the redundancy of the channel codewords to detect and correct errors in the received codeword.

The first- and second-generation DVB standards describe data transmission systems for digital terrestrial, satellite and cable television. The transmitted data represents information about the image and sound: video and sound signals in compressed form, as well as various additional information. In the context of the generalized model of Figure 6.1, DVB systems solve the problems of channel coding and modulation for multi-program television services.

There are many similarities between the first- and second-generation standards. To ensure interoperability between devices from different hardware manufacturers, the standards define a digital modulated radio signal. In the first-generation DVB-T standard, a novelty that provided the system with significant competitive advantages was the use of a modulation system called orthogonal frequency-division multiplex (OFDM) – frequency multiplexing with orthogonal carriers. OFDM modulation is also used in the second-generation systems.

DVB standards describe the transformation of data and signals in the transmitting part of a digital television broadcasting system. The signal processing in the receiver is not regulated by standards and remains open. Of course, this does not mean that the creators of the standards did not foresee the principles of building receivers, everything is quite the opposite. But the absence of a standard for receivers intensifies competition among receiver manufacturers and stimulates efforts to create high-quality, low-cost devices.

The use of any one channel coding system does not give the desired effect in television conditions, for which a variety of noise, interference and distortions are typical, leading to errors with different statistical properties. In such conditions, a more complex error correction algorithm is required. DVB and DVB-2 systems use a combination of two types of channel coding: outer and inner, designed to deal with errors of different structure, frequency and statistical properties and provide, when used together, almost error-free operation.

Data interleaving is a means of increasing the coding efficiency not related to decreasing the code rate. Coding can detect and correct errors, but interleaving increases coding efficiency because interleaving breaks up error bursts that the coding itself could not cope with into smaller chunks that are within the capabilities of the coding system. The role of interleaving in the digital television system is very significant.

In the first-generation DVB, the outer coding system uses the RS code. The inner coding is based on a convolutional code. In the second generation of DVB-2 television systems, BCH codes are used for outer coding and Gallager codes for inner. It was the use of Gallager codes, or codes with a low density of parity checks, that made a decisive contribution to increasing efficiency and allowed second-generation DVB systems to come close to the theoretical Shannon limit. BCH codes are used for outer coding and LDPC codes for inner coding system are used in ATSC 3.0 digital television systems [14, 15].

6.1.3 Error-correcting coding as a digital system design tool

The implementation of perfect algorithms for channel coding and error correction in the format of integrated circuit technology has become possible today thanks to the success of microelectronics. The number of gates per chip on an integrated circuit has doubled approximately every 18 months according to Moore's Law for more than 50 years. As a result, channel codecs, capable of processing huge amounts of data in real time, ensure almost error-free operation of television broadcasting systems. But the field of application of coding for error correction today is extremely wide.

The electromagnetic spectrum is increasingly filled with signals from different applications. Channel coding allows communication systems to work successfully in environments with strong interference and high levels of noise. In many systems, there are limits on the power of the transmitted signal. For example, in satellite relaying systems, mobile communication systems, in communication systems for the Internet of Things, increasing the power of the emitted signal is very expensive or almost impossible. The use of channel error-correcting codes allows data to be transmitted at high rates with a low signal-to-noise ratio.

LDPC codes have found application in second-generation DVB systems DVB-T2, DVB-S2, DVB-C2, in the third-generation ATSC television broadcasting system ATSC 3.0 and in the fifth-generation 5G mobile communications system. They are a prime candidate for communications applications for the Internet of Things.

The variety of applications described is associated with a very important result of theoretical research on coding for error correction. Channel coding techniques, which continue to evolve, are beginning to play a central role in the design of communications systems and are becoming an important factor in the design and optimization of various devices. The creation of extremely complex and, accordingly, very expensive channels and communication lines with a low noise level and high signal power can be opposed by the use of error-correcting channel codes with highly complex algorithms implemented in the form of integrated circuits.

Channel coding theory is becoming a tool for designing digital systems and a component of scientific and technical culture. Understanding of coding methods that allow to correct errors in communication channels becomes important for specialists of different professions in various fields of technology.

6.2   Error-correcting codes: basics

6.2.1   Error correcting

Transmitted messages in the form of a sequence of letters from the alphabet containing a finite set of letters are converted by the source encoder into binary code words consisting of binary symbols and called informational symbols. If, e.g., there are two such letters, then they can be encoded in the form of one-bit binary codewords. To encode samples of a video signal at the output of an analogue-to-digital converter using 256 quantization levels, information words must be eight binary symbols long. Information words u are fed to the input of the channel encoder. The task of the channel encoder is to form a codeword x of greater length according to certain rules, i.e., words with additional data called checks. The codeword is transmitted over a channel in which it can be corrupted. The introduced check data redundancy must be sufficient for the decoder to correct errors in the received word and then decide which channel word was transmitted.

A certain analogy that makes it possible to understand the basic principles of decision-making by a decoder can be seen in the recognition of the text of a book in which there are typos. Texts written in any language have a certain redundancy that allows you to understand what is written, despite a number of typos. This is due to the fact that the number of words that can be formed from the letters of the alphabet is much greater than the number of words in the dictionary. A typo usually does not result in one word from the dictionary being converted to another word from the dictionary. The misspelled word is not in the dictionary, but it looks like the original one, into which the typo has crept, which allows us to determine which word was garbled by the typo and to understand the text. It may turn out that a misspelled word looks like several words from a dictionary. In this case, when reading and recognizing text, you can evaluate which dictionary word is 'closer' to a word with a typo and which dictionary word it looks more like. Oral speech also has redundancy, which allows us to understand foreigners who speak with an accent or mistakes.

Returning to the channel code words, it can be noted that the coder's task is to make the code words as 'unlike' as possible. The more 'dissimilar' the codewords, the greater the 'distance' between them, the more errors the decoder can correct in order to recover the transmitted word.

6.2.2   *Block and tree codes*

There are two different channel coding rules. Block coding involves splitting a sequence of information symbols into fragments, or blocks, each containing k symbols. A set of n symbols $(n > k)$ is formed from each information block, which is called the channel codeword. In a channel decoder, channel codewords are processed independently of each other.

A block code over an alphabet of q symbols is defined as a set of M q-ary sequences of length n, which are channel codewords. If $q = 2$, then the symbols are called bits (binary digits), and the code is called binary. The number $M = q^k$ is called the cardinality or size of the code, which is denoted as an (n,k)-code. The ratio of the length of the information word to the length of the codeword is called the block code rate:

$$R = \frac{k}{n}. \tag{6.2}$$

Tree coding involves processing an information sequence without first dividing it into independent blocks. The encoder processes information continuously. Each long information sequence is associated with a code sequence consisting of a larger number of symbols.

Block codes have been better researched by now. This is due to the fact that block codes are built on the basis of well-studied mathematical structures. The channel codes currently used in television systems of the DVB-2 and ATSC 3.0 families are block codes.

6.2.3   *Repetition code*

The message is a sequence of two letters: a and b. They can be represented in the form of one-bit binary codewords u according to the rule: $u = 0$ when transmitting the letter a and $u = 1$ when transmitting the letter b. The word u goes to the input of the channel encoder. Output word x is a threefold repetition of each information word. So, the length of the codeword is $n = 3$, the size is $M = 2^k = 2^1 = 2$. This is a block code with parameters $(n, k) = (3,1)$. The coding process can be described using a coding table (Table 6.1).

Table 6.1   Coding table for $(n,k) = (3,1)$ repetition code

Information words u	0	1
Code words x	000	111

6.2.4   Hamming distance and minimum distance

In section 6.2.3, three parameters have already been introduced that are important for evaluating the code: code size M, block length n and information length k. Let us introduce one more parameter – the minimum distance d^*, which is a measure of the difference between the two most similar codewords. In general, in coding theory, Hamming distance is introduced as a measure of difference. The distance between two words is defined as the number of positions in which they are different. For codewords 000 and 111 of the repetition code (section 6.2.3), the distance $d(000, 111) = 3$. Since there are only two words, this distance is also the minimum one.

The minimum distance d^* is the most important parameter of the block code. If word x is transmitted over the channel and one error occurs in the channel, then the word y received by the channel decoder differs in one position, i.e., the Hamming distance between them is $d = (x, y) = 1$. If the minimum code distance is not less than 3, then the distance from y to any other codeword is not less than 2. This means that the decoder can correct the error, assuming that the codeword closest to y has been transmitted, which is at a distance of 1 from the received one, and make the right decision.

The minimum distance for the repetition code under consideration is 3, so the decoder of the $(3, 1)$-code can correct any single error.

In general terms, the relationship between the minimum Hamming distance and the number of errors to be corrected can be expressed as follows. If t errors have occurred in the channel, and the distance to any word that is not equal to the transmitted one is greater than t, then the decoder will be able to correct t errors by choosing the closest to the received word as the transmitted word. This means that the minimum distance between codewords must meet the condition:

$$d^* \geq 2t + 1. \tag{6.3}$$

6.2.5   Transmission errors

Possible 'typos' in the transmitted codewords due to errors for $(3, 1)$-repetition code (section 6.2.3) are shown in Table 6.2. The rows of the table show the code words (the first row of the table) and words received at the channel output with various errors. Errors are shown as three-digit error words. A single value in some bit of the error word indicates an error that occurs in the communication channel in the corresponding bit of the code word. In fact, this means that an additive model has been adopted to describe the effect of errors, within which the transmitted codeword is added to the error word, and the addition modulo 2 occurs ($0 + 0 = 0, 0 + 1 = 1, 1 + 0 = 1, 1 + 1 = 0$). In case of error 000, the received word is the same as the transmitted one (second row of the table). If the error is described by the word 001, then the error occurs in the last bit of the transmitted code word and instead of the word 000, 001 comes to the decoder input and instead of 111, 110 comes to the decoder input (the third row of the table). The next two rows describe the impact of single errors in the second and first bits of the codeword (error words 010, 100).

Table 6.2   Table of transformations of transmitted words because of errors for (n,k) = (3,1) repetition code

(3,1)-code		Received words	
Code words x		000	111
Errors	000	000	111
	001	001	110
	010	010	101
	100	100	011
	011	011	100
	101	101	010
	110	110	001
	111	111	000

If a word equal to the code word (e.g., 000) arrives at the input of the channel decoder, then it is natural to assume that it is this code word that was transmitted. If, e.g., word 110 is received, then we can assume that word 111 was transmitted, corrupted by one error 001. The explanation may be the fact that word 111 is at a Hamming distance of 1 from the received one, and the code word 000 is at a distance of 2 from received word. But the decisions described above, it might seem, contradict the last four lines of Table 6.2, which describe the effect on the transmitted codeword of two and three errors that can occur in the channel. Word 000 can be received in a situation when words 111 is transmitted, but three errors occur in the channel (the last row of Table 6.2). Word 110 can be received if codeword 000 is transmitted, and two errors occur in the channel (the penultimate row of Table 6.2). What decision should the decoder make?

6.2.6   Decoding as a statistical decision and the maximum likelihood method

The decision on the transmitted codeword made by the decoder is always statistical. It represents the best hypothesis on the basis of the information available and therefore may not be correct. The decoder cannot correct absolutely all errors. The challenge for developers is to find the code that makes a wrong decision much less likely than a correct one.

In order to fully assess the capabilities of the code to develop a decision-making procedure, it is necessary to obtain the probabilistic characteristics of the communication channel. Figure 6.2 shows a schematic model of the binary symmetric channel. The transition probabilities $P(y|x)$ of receiving the symbol y are set for the channel, provided that the symbol x has been transmitted: the probability that the received symbol coincides with the transmitted $(1 - \varepsilon)$, and the probability of receiving the opposite symbol ε. It is assumed that $(1 - \varepsilon) > \varepsilon$ and that the transition probabilities for each symbol are independent of the preceding symbols in the sequence

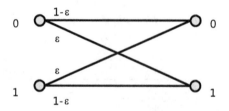

Figure 6.2   Model of the binary symmetric channel

(such channels are called memoryless channels). It should be borne in mind that the channel in Figure 6.2 includes a modulator, the actual communication channel and a demodulator (Figure 6.1). The channel is symmetric, therefore $P(0|0) = P(1|1) = (1 - \varepsilon)$ and $P(0|1) = P(1|0) = \varepsilon$. The binary symmetric channel model is used most often in research, although it does not accurately describe many real channels.

If the binary symbols are transmitted on a channel without channel coding, the error probability is ε. The value $\varepsilon = 0.1$ can be taken as some estimate. In 10% of cases, the transmission is accompanied by errors, and only 90% of the transmission occurs without errors. Let us estimate the result of channel coding using the code $(n, k) = (3, 1)$ with a minimum distance $d^* = 3$ (the repetition code considered in section 6.2.3)

If the codeword has length n, then the transition probabilities for words can be calculated as the product of the transition probabilities for individual symbols of the word:

$$P_n\left(y|x\right) = \Pi_{i=1}^{n} P_i\left(y_i|x_i\right).\tag{6.4}$$

A block of symbols equal to 000 can be received in two situations: if the codeword 000 passed through the channel without errors, and if the codeword 111 was corrupted with errors in 3 bits. The probability that word 000 will be received at the channel output if codeword 000 arrives at the channel input is $P(000|000) = (1 - \varepsilon)^3$. The probability that word 000 will be received at the channel output if the code word 111 arrives at the channel input is $P(000|111) = \varepsilon^3$. Since $(1 - \varepsilon) > \varepsilon$, receiving a word without errors at the channel output is more likely than receiving a word with three errors. If $\varepsilon = 0.1$, then $P(000|000) = (1 - \varepsilon)^3 = 0.729$, and $P(000|111) = \varepsilon^3 = 0.001$.

Word 110 can be received at the channel output also in two cases: if the transmitted word 111 was affected by one error and if the codeword 000 was corrupted by two errors. The probability that word 110 will be received at the channel output if the codeword 111 arrives at the channel input is $P(110|111) = (1 - \varepsilon)^2\varepsilon$. The probability that word 110 will be received at the channel output if codeword 000 is received at the channel input is $P(110|000) = (1 - \varepsilon)\varepsilon^2$. Since $(1 - \varepsilon) > \varepsilon$, receiving a single-error word at the channel output is more likely than receiving a two-error word. If $\varepsilon = 0.1$, then $P(110|111) = (1 - \varepsilon)^2\varepsilon = 0.081$, and $P(110|000) = (1 - \varepsilon)\varepsilon^2 = 0.009$.

The calculations show that it is more likely to receive any word without errors at the decoder input than to receive a word with errors. Getting any word with one mistake is more likely than getting a word with two or three mistakes. Therefore, the best solution on the receiving side will always be decoding into the codeword that differs from the received word in the smallest number of bits. This decoding is called maximum likelihood decoding. It should only be noted that this conclusion is valid provided that all codewords are transmitted with the same or close probabilities.

The decision-making process for decoding by the maximum likelihood method for the repetition (3,1)-code is described using a decoding table (Table 6.3).

The codewords form the first row of the decoding table. If a word is received that coincides with one of the codewords, then in accordance with the maximum likelihood method, it is decided that this particular codeword was transmitted. Decisions made for other possible words at the output of the receiving side demodulator are made in accordance with the word lists under each codeword. Words from the list are decoded into the codeword, which is at the top of the list. For a block (3,1)-code with a codeword length of $n = 3$, there are only eight variants of received words. Each word appears only once in the decoding table.

Using the data in Table 6.3, we can find the decoding error probability for the repetition (3,1)-code. The received words 000, 001, 010, 100 are decoded into codeword 000. The probability of correct decoding can be found as the sum of the transition probabilities for each word from the list:

$$P_{cor} = P\left(000|000\right) + P\left(001|000\right) + P\left(010|000\right) + P\left(100|000\right) =$$
$$= (1 - \varepsilon)^3 + 3(1 - \varepsilon)^2\varepsilon. \tag{6.5}$$

For $\varepsilon = 0.1$, the probability of correct decoding is 0.972. For $\varepsilon = 0.01$, the probability of correct decoding is 0.9997.

Received words 111, 110, 101, 011 are decoded into codeword 111. If word 000 was actually transmitted, then each such decision will be in error. The probability of decoding error can be found as the sum of the probabilities that words 111, 110, 101, 011 are received if word 000 was actually transmitted:

$$P_{err} = P\left(111|000\right) + P\left(110|000\right) + P\left(101|000\right) + P\left(011|000\right) =$$
$$= \varepsilon^3 + 3(1 - \varepsilon)\varepsilon^2. \tag{6.6}$$

For $\varepsilon = 0.1$, the error probability is $P_{err} = 0.028$. For $\varepsilon = 0.01$, the error probability is $P_{err} = 0.000298$.

Table 6.3   *Decoding table for (3,1) repetition code.*

Code words	000	111
	001	110
Other received words	010	101
	100	011

For a repetition code with the codeword length n, the minimum distance is equal to the length n. In general, an (n, k)-code with repetition can be written as an $(n, 1)$-code. If it is necessary to correct, e.g., two errors in a codeword, the length of the codeword should be increased to 5. If it is necessary to correct three errors, then the length should be $n = 7$. Repetition codes have good error correction capabilities, but the code rate is slow. For $n = 7$, the code rate $R = k/n$ is only 1/7.

6.2.7   One parity-check code

Let the message be a sequence of four letters: a, b, c, d. They can be encoded by the source encoder using two-bit binary information codewords u according to the rule: $u = 00$ when transmitting letter a, $u = 01$ when transmitting letter b, $u = 10$ when transmitting letter c, $u = 11$ when transmitting letter d ($k = 2$). The information word u is fed to the input of the channel encoder. The word at the output of the encoder x is formed by adding one check symbol p so that the number of ones in each codeword is even. The symbol p is called the parity bit. The process of adding a check bit is described in the coding table (Table 6.4). Each codeword contains an even number of ones (the number 0 is considered even). So, the length of the codeword is $n = 3$, the size is $M = 2^k = 2^2 = 4$. This is a block code with parameters $(n, k) = (3,2)$, which can be generally described as an $(n, n - 1)$-code or $(k + 1, k)$-code.

The distance between all words is the same: $d(000,011) = d(000, 101)... = d(101,110) = 2$. The minimum code distance d^* is 2, so no error can be corrected. A code with one parity check can only be used to detect one error in the communication channel (Figure 6.1). With one error, the total number of ones will become odd. Checking the number of ones in the received word in the decoder allows detecting the fact of an error, but not correcting it. But if two errors occur in the channel, then the number of ones will remain even and no errors will be detected. Correcting ability of the code is at the minimum level, but the code rate is maximum for a given codeword length n. In the example, it is equal to 2/3; in the general case, the code rate with one parity check is $R = k/(k + 1) = (n - 1)/n$. The code is used, e.g., when writing numbers to memory and in other cases where you need a small opportunity to detect errors with minimal hardware costs.

6.2.8   Systematic codes

In each codeword of the considered code with one parity check (section 6.2.3), first there are the symbols of the information word, and then the parity-check bit. This is an example of a systematic code. A code is called systematic if in each codeword of length n, first there are k symbols of the information word, and then $(n - k)$ check symbols.

Table 6.4   *Coding table for (3,2) one parity-check code*

Information words u	00	01	10	11
Code words x	000	011	101	110

6.2.9 Linear codes

6.2.9.1 Elements of algebra

6.2.9.1.1 Finite fields

Many good codes that are widely used in practice are based on algebraic structures. Such codes have special structural regularities that provide the possibility of practical implementation of encoding and decoding operations without compiling encoding and decoding tables. The first step to understanding such codes is to identify the arithmetic operations that can be performed on binary symbols.

The arithmetic systems introduced in algebra obey certain rules that are often applicable to ordinary number systems. They consist of sets and operations on elements of these sets. A group is a system in which one main operation is defined and an operation that reverses it. For example, these can be operations of addition and subtraction or multiplication and division. A ring is a system in which two basic operations are defined: addition and multiplication, and the inverse of addition – subtraction. A field is a system that defines two main operations and reverse operations for each of the main ones.

Real numbers form a field – a set of mathematical objects that can be added, multiplied, subtracted and divided according to the rules of ordinary arithmetic. This field contains an infinite number of elements. Arithmetic systems used in coding theory contain a finite number of elements. Such fields are called finite fields. The rules of ordinary arithmetic for real numbers do not apply to finite fields.

The minimum number of elements that form a finite field is 2. This is due to the fact that the field must contain two unit elements: 0 relative to addition and 1 relative to multiplication. The rules for addition and multiplication in a field of two elements are given in Tables 6.5 and 6.6.

The operations in Tables 6.5 and 6.6 are addition and multiplication modulo 2. The result of the operation is the remainder of the division by two of the operation result in accordance with the rules of the field of real numbers, i.e., the rules of ordinary arithmetic. From the equality $1 + 1 = 0$, it follows that $-1 = 1$, i.e., subtraction is equivalent to addition. From the equality $1 \times 1 = 1$, it follows that $1^{-1} = 1$, i.e.,

Table 6.5   Addition table for GF(2)

+	0	1
0	0	1
1	1	0

Table 6.6   Multiplication table for GF(2)

×	0	1
0	0	0
1	0	1

the reciprocal of 1 is 1. Thus, subtraction and division are always defined, except for division by zero. The set of two symbols together with addition and multiplication modulo 2 is called the Galois field of two elements $GF(2)$. Many codes are based on parity-check ideas. Section 6.2.7 uses one parity check. The word at the output of the encoder x is formed by adding one check symbol p to the information symbols so that the number of ones in each codeword is even. The added symbol p has been called the parity bit. If we use the rules of arithmetic for symbols from the field $GF(2)$, then the p bit can be entered as the sum of the characters of the information word: $p = u_1 + u_2$.

6.2.9.1.2   Vector spaces

An example of a vector space is given by three-dimensional Euclidean space. A vector value is a directional segment that depends on two elements of different nature: an algebraic element – a number that measures the length, or the modulus of the vector, and a geometric element – the direction of the vector. A scalar value is completely characterized by one number. A vector a in three-dimensional space (Figure 6.3) can be specified using three algebraic elements, or scalar values – the projections of the vector on the coordinate axes $a = (a_x, a_y, a_z)$.

Above, the Galois field of two elements $GF(2)$ was introduced. Elements from this field 0 and 1 are scalars. We have already looked at sets or sequences of symbols called words. These are ordered sequences of n field elements denoted as $(a_1, a_2, ..., a_n)$. The sequences are vectors, and the set of all sequences forms a vector space if the addition operation is defined for the pairs of sequences, and the operation of the multiplication of sequences by a scalar is defined for the sequence and field element. A mandatory requirement – the result of operations is an element from the set of sequences. Addition of sequences of length n is defined as componentwise addition, or bitwise addition:

$$(a_1, a_2, ..., a_n) + (b_1, b_2, ..., b_n) = (a_1 + b_1, a_2 + b_2, ..., a_n + b_n)$$

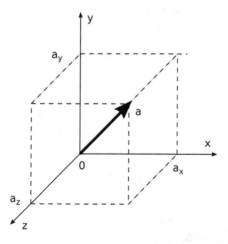

Figure 6.3   Vector in three-dimensional space

Multiplication of a sequence by a scalar is defined as follows:

$$c(a_1, a_2, \ldots, a_n) = (ca_1, ca_2, \ldots, ca_n).$$

The scalar multiplication is also defined for vectors:

$$a * b = (a_1, a_2, \ldots, a_n) * (b_1, b_2, \ldots, b_n) = a_1 b_1 + a_2 b_2 + \ldots + a_n b_n.$$

This product is called scalar because the result is an element of the Galois field, or scalar.

A subset of a vector space forms a subspace if it satisfies the same requirements. In practice, it is sufficient to check that the subset is closed under the operations of addition and multiplication by a scalar.

Here are some important statements that will be used later.

A linear combination of vectors $v_1, v_2, \ldots, v_k$ is a sum of the form:

$$u = a_1 v_1 + a_2 v_2 + \_ \ldots , + a_k v_k$$

where $a_1, a_2, \ldots, a_k$ are scalars, i.e. field elements.

The set of all linear combinations of a set of vectors from a vector space V is a subspace of V.

A set of vectors $v_1, v_2, \ldots, v_k$ is called linearly dependent if there exist scalars $c_1, c_2, \ldots, c_k$ (not all equal to zero) such that

$$c_1 v_1 + c_2 v_2 + \_ \ldots , + c_k v_k = 0.$$

A set of vectors is called linearly independent if it is not linearly dependent.

A set of linearly independent vectors generates a vector space if each vector of the vector space can be represented as a linear combination of the vectors of this set. This set is called the basis of the space. The number of linearly independent vectors generating a space is called the dimension of the space.

The sequence of symbols u_1, u_2 in the word $u = (u_1, u_2)$ can be considered as components of a vector in two-dimensional space, and the word itself as a vector in this space (Figure 6.4). There are four vectors in this space: $(0,0)$, $(0,1)$, $(1,0)$ and $(1,1)$. The four information words corresponding to these vectors were used in section 6.2.7. The sequence of symbols x_1, x_2, x_3 in the word $x = (x_1, x_2, x_3)$ can be considered as components of a vector in three-dimensional space, and the word itself as a vector

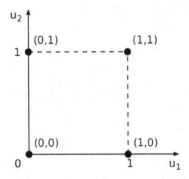

Figure 6.4   Information words as vectors in two-dimensional space

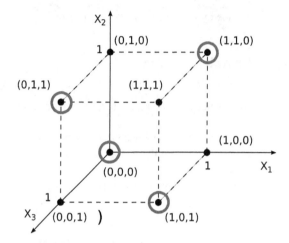

Figure 6.5   Subspace of code vectors in three-dimensional space of all vectors

in this space (Figure 6.5). There are eight vectors in the three-dimensional space. In the code of section 6.2.7 (code with one parity check), four vectors: (0,0,0), (0,1,1), (1,0,1) and (1,1,0) were chosen as code vectors. They are at a distance of 2 from each other and represent a subspace in the three-dimensional space of all vectors.

6.2.9.2   Linearity of codes

A code is called linear if the sum of any codewords gives a codeword. Addition of codewords means addition of vectors, which is performed as a bitwise addition of word symbols according to the rules of arithmetic of the field $GF(2)$. It is easy to verify that a code with one parity check belongs to the class of linear codes. For example, adding bitwise the second and third codewords of Table 6.4 (011 *and* 101), we get the fourth word 110: $(0 + 1) = 1$, $(1 + 0) = 1$, $(1 + 1) = 0$. It follows from the linearity property that the zero word (a word consisting of all zeros) is included in the set of code words of a linear code, since adding a word with itself, we get a zero word.

6.2.9.3   Minimum weight

The Hamming weight of a codeword is equal to the number of its non-zero symbols. Minimum code weight – the minimum number of non-zero symbols in a non-zero word. To find the minimum distance of a linear code, it is not necessary to compare all pairs of codewords. It is enough to find the words closest to the zero word. For a linear code, the minimum distance is equal to the minimum weight of a non-zero word. This means that you need to count the number of non-zero characters in the codewords closest to zero. It is easy to verify that the minimum weight of a code with one parity check is 2, therefore, the minimum distance $d* = 2$. With one parity-check code, you can detect one error in each codeword. As you can see, the estimation of the metric properties of linear codes is easier than non-linear ones.

6.2.9.4   Singleton bound

For linear codes, a simple inequality can be obtained that connects the parameters of the code. It is known as the Singleton bound. The minimum distance for any linear (n, k)-code satisfies the inequality: $d^* \leq n - k + 1$.

This inequality is easy to explain for a systematic code. There are systematic code words with one non-zero information symbol and $(n - k)$ check symbols. Even if all check symbols are equal to one, then the weight of such a codeword cannot be more than $(n - k + 1)$. Therefore, the minimum code weight cannot be more than $(n - k + 1)$. As noted above, the minimum distance of the line code is equal to the minimum weight. Therefore, the minimum code distance cannot be greater than $(n - k + 1)$. Each linear code is equivalent to a systematic linear code. Therefore, for any linear code, the minimum distance $d^* \leq n - k + 1$.

From the Singleton bound, it follows that to correct t errors, the code must include at least $2t$ check symbols. To correct one error, you must have at least two check symbols. The Singleton bound is an upper bound. Many codes that are considered good have a minimum distance much less than the bound gives. But there are codes whose parameters satisfy the Singleton boundary with equality. Such a code is called the maximum distance code. The maximum distance code provides exactly $2t$ check symbols to correct t errors. An example of the maximum distance codes are Reed-Solomon codes (RS codes).

6.2.9.5   Standard array

Let us describe the standard array using the example of a new code. The codes considered in sections 6.2.3 and 6.2.7 are in a sense antipodes. A repetition code, generally defined as an $(n, 1)$ code, has the minimum distance of n and excellent channel error correction capabilities, but its rate is minimum at $1/n$. A single parity-check code, generally defined as an $(n, n - 1)$ code, has a maximum rate of $n/(n - 1)$. But its minimum distance is 2 and it can only detect one error.

How to find a 'good' code that is somewhere in the middle, and has both good error correction capabilities and not very low speed? You can increase the number of check symbols by selecting them so that the code words differ from each other in the maximum number of positions. This approach is implemented in a systematic code of size $M = 4$ with parameters $(n, k) = (5,2)$, which is used to represent two-bit information words using five-bit codewords. The information word u is fed to the input of the channel encoder. The word at the output of the encoder x is formed by adding three check symbols (Table 6.7).

Let us find the minimum code distance by comparing pairs of codewords:

Table 6.7   Coding table for (5,2) code

Information words u	00	01	10	11
Code words x	00000	01011	10101	11110

$d(00000, 01011) = 3;$ $d(00000, 10101) = 3;$ $d(00000, 11110) = 4;$

$d(01011, 10101) = 4;$ $d(01011, 11110) = 3;$

$d(10101, 11110) = 3.$

As you can see, the minimum distance is $d^* = 3$ and the code is able to correct one error.

It would be possible to approach the determination of the minimum distance in a different way. Adding codewords, as described in section 6.2.9.2, we estimate the linearity of the code. For example, the sum of the codewords of the second and third columns of the coding table (Table 6.7) gives the codeword of the fourth column. The addition of other codewords gives similar results. Adding the codeword to itself gives the zero word, which is present in the coding table. After making sure of the linearity of the code, one could estimate the minimum weight of a non-zero codeword, which is also 3.

The input of the decoder will receive words with a length of five symbols. Among these words, there will be code words and other words that have been transformed from code words due to noise and distortion. It is necessary to describe the decisions that the decoder should make in response to each of the 32 words that may appear at the input of the decoder, i.e., create a decoding table. A reasonable way to compile the table is the standard array [10].

The first row of the standard array (Table 6.8) contains all codewords starting from zero (in the general case, these are q^k codewords of the (n, k) code, denoted as $(0, x_1, x_2, ..., x_{qk})$. From the remaining words, select any word (denote it by y_j) that has the unit weight and is at a distance of 1 from zero word, and write it down in the first column of the second row. In the remaining columns of the second row, we write the sums of the word y_1 and the codewords of each column. The following lines are constructed in the same way. At each step, select a word that is one of the nearest to zero and is absent in the previous rows. When after the next step there are no unwritten words left, the procedure will end. Each word is written to the standard array only once.

Table 6.8   Standard array for (5,2) code

Code words	00000	01011	10101	11110
	00001	01010	10100	11111
	00010	01001	10111	11100
	00100	01111	10001	11010
Cosets	01000	00011	11101	10110
	10000	11011	00101	01110
	11000	10011	01101	00110
	01100	00111	11001	10010

The set of the code words (the first line of Table 6.8, highlighted in red) can be considered as a subgroup of all words. Then the rest of the rows in the table are cosets for this subgroup. The words in the first column are called coset leaders.

If the decoder receives a word equal to the codeword (e.g., 00000), then in accordance with the maximum likelihood method, the decoder decides that this particular codeword was transmitted (in accordance with the coding table (Table 6.7), this means that at the input of the channel encoder source word 00 was given). This situation is described by the first row of the decoding table (Table 6.8), in which the symbols are marked in red.

In the next five rows of the table, marked in blue, the possible received words are shown, which differ from the code words in the first row of the corresponding column by the value of one symbol. For example, the second column of the second row contains the word 01010, which differs from the codeword in the second column of the first row 01011 by the value of the last symbol. The minimum code distance is 3, the word 01010 differs from all other codewords in the values of two symbols. The decoder decides that word 01011 was transmitted to the input of the channel, and information word 01 was transmitted to the input of the channel encoder. All five words resulting from an error in one symbol are located under the corresponding codeword. They are decoded into this codeword.

Earlier, a relationship was noted between the minimum distance d^* and the number of corrected errors t: $d^* \geq 2t + 1$. The set of words that are in the column of the standard array of the decoding table (Table 6.8) in the first six rows and marked in red and blue can be interpreted as a decoding sphere. The decoding sphere is a spherical region of radius t in the space of words centred at the point at which the codeword of this column is located. The distance in this space is the Hamming distance and is measured in the number of non-equal characters at the corresponding positions. A somewhat conditional geometric interpretation of the decoding spheres is shown in Figure 6.6 as a two-dimensional slice of the word space. Codewords are marked with red dots. The rest of the words are marked as blue dots. Decoding

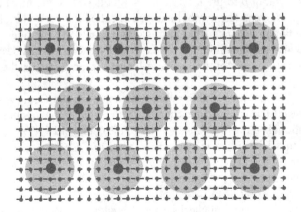

Figure 6.6   Decoding spheres (code words are marked by red colour)

spheres are highlighted in blue. They are translucent. Words that fall within the decoding spheres remain prominent.

If no more than t errors occurred in the channel, then the received word always lies inside some sphere and is decoded correctly. The set of all possible received words includes the decoding spheres of all codewords. But some received words containing more than t errors will not fall into any decoding sphere and will be in intermediate regions. In the standard array (Table 6.8), these words are located in the last two rows of the decoding table, marked in black. How to decode these received words? There are two possible answers to this question.

An incomplete decoder decodes only those words that are within one of the decoding spheres. The rest of the received words are not decoded and are considered unrecognized words containing errors. This result is also important. It means that the decoder corrects one error and detects some configurations of two or more errors. The decoder signals that an error has been detected in the received word, which can be corrected by other means.

The complete decoder decodes all words to the nearest codeword. If the words are in intermediate areas at equal distances from several decoding spheres, then one of these spheres is declared the nearest arbitrarily. In such a situation, when more than t errors occur, the decoder can sometimes decode correctly, sometimes not. The complete decoding mode is used when it is better to guess the message than to give no estimate.

6.2.10   Hamming code: encoding

The goal of developing systems that correct errors in communication channels is to create codes with special structural patterns. The presence of such structural patterns provides the possibility of practical implementation of encoding and decoding operations without compiling huge tables of encoding and decoding for long codes, when it is difficult even to list all the codewords.

Many good codes are based on parity-checking ideas. The code of section 6.2.7 uses one parity check. The word at the output of the encoder x is formed by adding one check symbol p to the information symbols so that the number of ones in each codeword is even. The symbol p added is called the parity bit. The resulting code has a minimum distance $d^* = 2$. It is only capable of detecting one error in a codeword. To increase the minimum distance, parity-check bits can be added, which must be calculated by performing some arithmetic procedures on the information word symbols.

Consider coding a four-bit binary information word $u = (u_1,u_2,u_3,u_4)$ using the Hamming code [10] by adding three check symbols: p_1,p_2,p_3. Let us consider the code to be systematic. Then the first four symbols of the codeword $x = (x_1,x_2,x_3,x_4,x_5,x_6,x_7)$ will be equal to the information symbols: $x_1 = u_1$, $x_2 = u_2$, $x_3 = u_3$, $x_4 = u_4$. The last three characters of the code word are check symbols: $x_5 = p_1$, $x_6 = p_2$, $x_7 = p_3$. The length of the codeword is $n = 7$, the size is $M = 2^k = 2^4 = 16$. This is a block code with parameters $(n,k) = (7,4)$.

Each parity-check symbol checks a set of predefined information symbols. Let us define parity-check symbols by equalities:

$$p_1 = u_1 + u_2 + u_3$$
$$p_2 = u_2 + u_3 + u_4 \qquad\qquad (6.7)$$
$$p_3 = u_1 + u_2 + u_4$$

The calculation results for 16 codewords are shown in Table 6.9. Determination of the most important parameter of the code – the minimum distance requires pairwise comparison of code words. Even for 16 words, this turns into a laborious procedure. But, as noted, the underlying mathematical structures of the code can help solve many problems.

The Hamming code belongs to the class of linear codes. Recall that this code is called linear if the sum of any codewords gives a codeword. It is easy to verify that the Hamming code is linear by adding bitwise the words of Table 6.9 according to the rules for adding in Table 6.5. For example, adding bitwise words from the second and third lines $(0 + 0) = 0$, $(0 + 0) = 0$, $(0 + 1) = 1$, $(1 + 0) = 1$, $(0 + 1) = 1$, $(1 + 1) = 0$, $(1 + 0) = 1$ we get the code word, which is in the fourth row of Table 6.9. The linearity property implies that the zero word (a word consisting of all zeros) is included in the number of code words, since adding a word with itself, we get a zero word.

To find the minimum distance of a linear code, it is not necessary to compare all pairs of codewords. It is enough to find the words closest to the zero word. For a linear

Table 6.9 Coding table for (7,4) Hamming code

Information words				Code words						
u_1	u_2	u_3	u_4	x_1	x_2	x_3	x_4	x_5	x_6	x_7
0	0	0	0	0	0	0	0	0	0	0
0	0	0	1	0	0	0	1	0	1	1
0	0	1	0	0	0	1	0	1	1	0
0	0	1	1	0	0	1	1	1	0	1
0	1	0	0	0	1	0	0	1	1	1
0	1	0	1	0	1	0	1	1	0	0
0	1	1	0	0	1	1	0	0	0	1
0	1	1	1	0	1	1	1	0	1	0
1	0	0	0	1	0	0	0	1	0	1
1	0	0	1	1	0	0	1	1	1	0
1	0	1	0	1	0	1	0	0	1	1
1	0	1	1	1	0	1	1	0	0	0
1	1	0	0	1	1	0	0	0	1	0
1	1	0	1	1	1	0	1	0	0	1
1	1	1	0	1	1	1	0	1	0	0
1	1	1	1	1	1	1	1	1	1	1

code, the minimum distance is equal to the minimum weight of a non-zero word. This means that you need to count the number of non-zero characters in the codewords closest to zero. It is easy to verify that the minimum weight of the (7,4)-Hamming code is 3, therefore, the minimum distance is $d* = 3$. With the (7,4)-Hamming code, one error in each codeword can be corrected.

There are 16 codewords in the coding table of the (7,4)-code. The decoding table for this code should contain 128 codewords, which makes it difficult to construct a decoder even for such a relatively simple code. But the linearity of the code allows you to find a relatively simple procedure if you introduce a matrix description of the code as a subspace in a vector space.

An n-dimensional vector space is a generalization of the three-dimensional space to the case of words with length n. The information word $u = (u_1, u_2, u_3, u_4)$ can be viewed as a vector in four-dimensional space. The codeword $x = (x_1, x_2, ..., x_7)$, obtained with the Hamming channel encoder by adding three check symbols, is a vector in the seven-dimensional space. The set of codewords of the (7,4)-Hamming code represents the code subspace in the space of all vectors of the seven-dimensional space.

6.2.11   *Matrix description of linear block codes*

6.2.11.1   Generator matrix

Let us return to the discussion of encoding a four-bit binary information word $u = (u_1, u_2, u_3, u_4)$ using the Hamming code $(n, k) = (7,4)$ by adding three parity symbols. The information word $u = (u_1, u_2, u_3, u_4)$ is a vector in four-dimensional space. The codeword $x = (x_1, x_2, ..., x_7)$ obtained with the channel encoder can be viewed as a vector in a seven-dimensional space.

Let us find a generalized description of the coding procedure. The last $(n - k) = 3$ symbols of the codeword are check symbols p_1, p_2, p_3, calculated using (6.7), which can be rewritten using the numbers p_{ij} in Table 6.10:

$$p_1 = u_1 + u_2 + u_3 = u_1 p_{11} + u_2 p_{21} + u_3 p_{31} + u_4 p_{41}$$
$$p_2 = u_2 + u_3 + u_4 = u_1 p_{12} + u_2 p_{22} + u_3 p_{32} + u_4 p_{42}$$
$$p_3 = u_1 + u_2 + u_4 = u_1 p_{13} + u_2 p_{23} + u_3 p_{33} + u_4 p_{43}$$

(6.8)

Table 6.10   Table of elements of parity matrix P for (7,4) Hamming code

		j		
p_{ij}		1	2	3
i	1	1	0	1
	2	1	1	1
	3	1	1	0
	4	0	1	1

In order for, for example, the first line (6.8) to coincide with the first line (6.7), it is necessary that $p_{11} = 1$, $p_{21} = 1$, $p_{31} = 1$, $p_{41} = 0$, which is written in Table 6.10. A longer notation of relations (6.8) made it possible to achieve uniformity in the forms of notation of all check symbols. As can be seen from (6.8), calculations are carried out according to similar formulas, the lines differ only in numbers p_{ij}, therefore (6.8) can be rewritten using one expression:

$$p_j = \Sigma_{i=1}^k u_i\, p_{ij}; \quad 1 \le j \le n - k. \tag{6.9}$$

Here it is necessary to digress and recall some information about matrices.

***

A $m \times n$ matrix is an ordered set of mn elements (in our case, the elements of the matrix are numbers from the Galois field) arranged in the form of a rectangular table that contains m rows and n columns:

$$\begin{bmatrix} a_{11} & a_{12} & \cdots & a_{1n} \\ a_{21} & a_{22} & \cdots & a_{2n} \\ \vdots & \vdots & \vdots & \vdots \\ a_{m1} & a_{m2} & \cdots & a_{mn} \end{bmatrix} = \lfloor a_{ij} \rfloor, 1 \le i \le m, 1 \le j \le n.$$

Matrices can be added element by element. They can be multiplied by a scalar by multiplying each element by it. Let us recall the rules for multiplying matrices as tables of numbers. You can multiply both square and rectangular matrices. It should be borne in mind that the product of matrices is not commutative, i.e., $A * B \ne B * A$. The product of matrices makes sense only if the number of columns in the first matrix is equal to the number of rows in the second matrix. Let us find the product $C = A * B$:

$$\left[c_{ij} \right] = \left[a_{ij} \right] \left[b_{ij} \right].$$

If matrix A has m rows and n columns, and matrix B has n rows and k columns, then matrix C will have m rows and k columns. The element of matrix C, which is at the intersection of row i and column j, is the scalar product of row i of matrix A and column j of matrix B:

$$c_{ij} = a_{i1}b_{1j} + a_{i2}b_{2j} + \cdots + a_{in}b_{nj} = \sum_{l=1}^{n} a_{il}\, b_{lj}; \ 1 \le i \le m, 1 \le j \le k.$$

The rows and columns of a matrix, which are strings of elements, can be thought of as vectors. A vector can be thought of as a one-row matrix. If the row vector A of n elements $\left[a_{1j} \right]$ is multiplied by the matrix B, then C is the row vector $\left[c_{1j} \right]$ of k elements:

$$c_{1j} = a_{11}b_{1j} + a_{12}b_{2j} + \cdots + a_{1n}b_{nj} = \sum_{l=1}^{n} a_{1l}\, b_{lj}; \quad 1 \le j \le k.$$

***

Let us return to the generalized description of the procedure for encoding the Hamming code. If we consider $u = (u_1, u_2, u_3, u_4)$ and $p = (p_1, p_2, p_3)$ as vectors, then expressions (6.8) and (6.9) represent the product of the vector u by the matrix P:

$$p = uP = [u_j][p_{ij}],$$ (6.10)

where

$$P = [p_{ij}] = \begin{bmatrix} p_{11} & p_{12} & p_{13} \\ p_{21} & p_{22} & p_{23} \\ p_{31} & p_{32} & p_{33} \\ p_{41} & p_{42} & p_{43} \end{bmatrix} = \begin{bmatrix} 1 & 0 & 1 \\ 1 & 1 & 1 \\ 1 & 1 & 0 \\ 0 & 1 & 1 \end{bmatrix}.$$ (6.11)

The matrix $P = [p_{ij}]$ is a matrix of four rows and three columns, the elements of which are taken from Table 6.10. It describes a linear transformation of vector u into vector p of parity symbols. The product of u and P is performed according to the matrix multiplication rules. But at the same time, u and p are considered as row vectors, or matrices from one row. These rules are actually described by (6.8) and (6.9). To obtain the jth symbol of the vector p, denoted as p_j, it is necessary to add the element-by-element products of the components of the row vector u and the jth column of the matrix P, as prescribed by (6.8) and (6.9).

Equation (6.10) specifies the formation of the three most significant bits of the code word, which contain check symbols. A similar expression must be obtained to form the entire codeword when encoding. As already noted, the code is systematic; therefore, the first k symbols ($k = 4$) of the codeword x_1, x_2, x_3, x_4 are equal to information symbols: $x_j = u_j$ for $j = 1,2,3,4$. Combining this equality with the rule (6.9) of the computation of check symbols, which are shifted to the right by k bits and located in the $(n - k)$ most significant bits of the codeword, we get:

$$x_j = u_j; \qquad 1 \le j \le k,$$ (6.12)

$$x_j = \Sigma_{i=1}^{k} u_i p_{i,j-k}; \qquad k+1 \le j \le n.$$ (6.13)

The rule for calculating the symbols of a code word can be written using one general expression instead of two expressions (6.12) and (6,13):

$$x_j = \Sigma_{i=1}^{k} u_i g_{i,j}; \qquad 1 \le j \le n,$$ (6.14)

where

$$\begin{aligned} g_{i,j} &= 1; & i = j, \ 1 \le j \le k, \\ g_{i,j} &= 0; & i \ne j, \ 1 \le j \le k, \\ g_{i,j} &= p_{i,j-k}; & k+1 \le j \le n. \end{aligned}$$

The numbers $g_{i,j}$ are given in Table 6.11 for the values of the indices $1 \le i \le k$ and $1 \le j \le n$. Index i indicates the number of the information symbol, which is included in the expression for calculating the code symbol; index j indicates the code symbol number.

Table 6.11   Table of elements of generator matrix G for (7,4) Hamming code

$g_{i,j}$		j						
		1	2	3	4	5	6	7
i	1	1	0	0	0	1	0	1
	2	0	1	0	0	1	1	1
	3	0	0	1	0	1	1	0
	4	0	0	0	1	0	1	1

Euqtaion (6.14) can be expressed in a more compact form if we consider the numbers in Table 6.11 as components of a matrix with dimensions of four rows and seven columns, and the information and code words as row vectors:

$$x = uG = \left[u_j \right] \left[g_{i,j} \right] \tag{6.15}$$

$$\text{where } G = \begin{bmatrix} g_{11} & g_{12} & \cdots & g_{17} \\ g_{21} & g_{22} & \cdots & g_{27} \\ \vdots & \vdots & \vdots & \vdots \\ g_{41} & g_{42} & \cdots & g_{47} \end{bmatrix} = \left[g_{ij} \right] \tag{6.16}$$

$$u = \left(u_1, u_2, u_3, u_4 \right) = \left[u_1 \ u_2 \ u_3 \ u_4 \right] \tag{6.17}$$

$$x = (x_1, x_2, ..., x_7) = \left[x_1 \ x_2 \ ... \ x_7 \right]. \tag{6.18}$$

Matrix G is called the generator matrix of the code. It defines a linear mapping of a set of 2^k information words into a set of 2^k codewords with block length n. A generator matrix is a compact description of the code, it replaces the coding table.

It was noted above that the Hamming code is a linear code. We confirm this using (6.15). Let the sum of two information words u_A and u_B gives an information word u_C. If the corresponding codewords are equal to $x_A = u_A G$ and $x_B = u_B G$, then

$$x_A + x_B = u_A G + u_B G = (u_A + u_B)G = u_C G = x_c. \tag{6.19}$$

From equality (6.19), it follows that the sum of two codewords is equal to another codeword.

From (6.14) and (6.15), it follows that if there is only one symbol equal to 1 in the information word, e.g., in position j, then the calculated code word is equal to the jth row of the generating matrix G (row j is denoted as g_j). Taking into account (6.19), we obtain that an arbitrary codeword can be represented as a linear combination of rows of the generating matrix G:

$$x = \sum_j u_j g_j. \tag{6.20}$$

Relation (6.20) means that the set of codewords is the row space of the generator matrix of the code G.

It can also be noted that the rows of the generator matrix are linearly independent. In general, the dimension of the entire word space is n. The number of rows of
the matrix k is equal to the dimension of the subspace of codewords. In total, there
are q^k codewords over the Galois field $GF(q)$. Thus, q^k different information sets of
length k can be mapped onto a set of codewords of length n.

Matrices can be split into blocks. The matrix P with dimensions $k = 4$ rows by
$(n - k) = 3$ columns was introduced earlier in (6.11). It can be selected as one block
of the generator matrix G. Then the second block will be the so-called identity matrix
with dimensions of four rows and four columns. The elements of the main diagonal of
the identity matrix are equal to 1. The remaining elements are equal to 0:

$$I_4 = \begin{bmatrix} 1 & 0 & 0 & 0 \\ 0 & 1 & 0 & 0 \\ 0 & 0 & 1 & 0 \\ 0 & 0 & 0 & 1 \end{bmatrix}. \tag{6.21}$$

Then the generator matrix G can be represented as a combination of the identity
matrix I_4 and the check matrix P:

$$G = \begin{bmatrix} I_4 \vdots P \end{bmatrix} = \begin{bmatrix} I_k \vdots P \end{bmatrix}. \tag{6.22}$$

It should be recalled that the generator matrix of the example under consideration was
found for a systematic code. The converse is also true. If the generator matrix consists
of two blocks of the form (6.22), then it maps information symbols into code words of
a systematic code.

6.2.11.2   Parity-check matrix

It was noted above that a generator matrix is a compact code description that replaces
a code coding table. Even more desirable is a compact code description replacing the
decoding table. To achieve this goal, a check matrix of the H code is introduced. It is
introduced using the example of the Hamming code under consideration.

To obtain a parity-check matrix, you must first find the matrix transposed to P.
The matrix transpose is performed by replacing the rows of the matrix with its columns. Replacing the rows of the matrix P in (6.11) with its columns, we get:

$$P^T = \begin{bmatrix} p_{11} & p_{21} & p_{31} & p_{41} \\ p_{12} & p_{22} & p_{32} & p_{42} \\ p_{13} & p_{23} & p_{33} & p_{43} \end{bmatrix} = \begin{bmatrix} 1 & 1 & 1 & 0 \\ 0 & 1 & 1 & 1 \\ 1 & 1 & 0 & 1 \end{bmatrix}. \tag{6.23}$$

The matrix P has four rows and three columns, while the transposed matrix P^T has
three rows and four columns. In accordance with (6.22), the generating matrix G contains the identity matrix I_4 in the first k columns, and the matrix P in the last $(n - k)$ columns. The check matrix H in the first $(n - k)$ columns contains the transposed matrix
P^T taken with the minus sign, and the last k columns contain the identity matrix I_3:

$$H = \left[-\boldsymbol{P}^T : \boldsymbol{I}_3 \right] \begin{bmatrix} p_{11} & p_{21} & p_{31} & p_{41} & 1 & 0 & 0 \\ p_{12} & p_{22} & p_{32} & p_{42} & 0 & 1 & 0 \\ p_{13} & p_{23} & p_{33} & p_{43} & 0 & 0 & 1 \end{bmatrix}$$

$$= \begin{bmatrix} 1 & 1 & 1 & 0 & 1 & 0 & 0 \\ 0 & 1 & 1 & 1 & 0 & 1 & 0 \\ 1 & 1 & 0 & 1 & 0 & 0 & 1 \end{bmatrix}. \tag{6.24}$$

For a binary Galois field, subtraction is performed according to the rules of addition, so the minus sign in expression (6.24) can be ignored. The transposed matrix $\boldsymbol{H}^T$ can be found by swapping rows and columns in expression (6.24):

$$\boldsymbol{H}^T = \begin{bmatrix} -\boldsymbol{P} \\ \cdots \\ \boldsymbol{I}_3 \end{bmatrix} = \begin{bmatrix} p_{11} & p_{12} & p_{13} \\ p_{21} & p_{22} & p_{23} \\ p_{31} & p_{32} & p_{33} \\ p_{41} & p_{42} & p_{43} \\ \cdots & \cdots & \cdots \\ 1 & 0 & 0 \\ 0 & 1 & 0 \\ 0 & 0 & 1 \end{bmatrix} = \begin{bmatrix} 1 & 0 & 1 \\ 1 & 1 & 1 \\ 1 & 1 & 0 \\ 0 & 1 & 1 \\ \cdots & \cdots & \cdots \\ 1 & 0 & 0 \\ 0 & 1 & 0 \\ 0 & 0 & 1 \end{bmatrix}. \tag{6.25}$$

The elements of the parity-check matrix $\boldsymbol{H}$ and the transposed parity-check matrix $\boldsymbol{H}^T$ for the (7,4) Hamming code are shown in Tables 6.12 and 6.13.

An important result allows one to obtain the multiplication of the generator matrix $\boldsymbol{G}$ and the transposed parity-check matrix $\boldsymbol{H}^T$, given for the systematic code by (6.22) and (6.25):

$$\boldsymbol{G}\boldsymbol{H}^T = \left[\boldsymbol{I}_4 : \boldsymbol{P} \right] \begin{bmatrix} -\boldsymbol{P} \\ \cdots \\ \boldsymbol{I}_3 \end{bmatrix} = -\boldsymbol{P} + \boldsymbol{P} = 0. \tag{6.26}$$

Table 6.12   Table of elements of parity-check matrix H *for (7,4) Hamming code*

h_{ij}		j						
		1	2	3	4	5	6	7
i	1	1	1	1	0	1	0	0
	2	0	1	1	1	0	1	0
	3	1	1	0	1	0	0	1

Table 6.13   Table of elements of transposed parity-check matrix H^T *for (7,4) Hamming code*

h^T_{ij}		j		
		1	2	3
i	1	1	0	1
	2	1	1	1
	3	1	1	0
	4	0	1	1
	5	1	0	0
	6	0	1	0
	7	0	0	1

Calculations in (6.26) were performed according to the rules of matrix multiplication, taking into account the fact that multiplication of some matrix by the identity matrix gives the original matrix $(I_4 * (-P) = (-P), P * I_3 = P)$.

As follows from (6.26), a remarkable property of the parity-check matrix is that a word x is a code word if and only if

$$x\,H^T = 0. \tag{6.27}$$

The product of the code vector and the transposed parity-check matrix is always zero. This property of the parity-check matrix can be used during decoding to check if the received word is a code word.

Several common properties of check matrices can be noted. Using the parity-check matrix H, it is possible to construct a linear code dual to the code built on the basis of the generator matrix G. For this, it is necessary to use the matrix H as a generator, i.e., multiply the information vectors by the matrix H. This will give the codewords of the dual code. The codewords of such a dual code will be linear combinations of rows of the matrix H. It is interesting that the matrix G will be a parity-check matrix for the dual code.

One way to design new code is to create a new generator matrix G. But you can use another way. To design a new code, you can create a new parity-check matrix H. This is not about creating a dual code, but about creating a new code, in which the starting point is to build a check matrix of this new code. This means that the first step is to develop a decoding procedure using the matrix H as a compact decoding description. To create a linear (n,k)-code correcting t errors, it is enough to find an $(n - k) \times n$ matrix H, in which $2t$ columns are linearly independent [10]. For the development of the encoder, at the second stage, a generator matrix G is created as a compact description of the encoding process. LDPC codes were invented based on the second approach, in which a code creation begins with building a parity-check matrix.

6.2.11.3 Syndrome

Decoding can be done using a decoding table. The systematized form of decoding is given by the standard array described in section 6.2.9.5. The use of a parity-check matrix allows you to formalize the construction of a standard array. The decoder multiplies the received word y by the transposed parity-check matrix H^T. The result of multiplication yH^T is equal to zero if and only if the received word is equal to the code word, as (6.27) sets. The result of the multiplication yH^T, which is not zero, can be used to find and correct an error. The product of the received word by the transposed parity-check matrix is called the syndrome:

$$s = y H^T. \tag{6.28}$$

The elements of the transposed check matrix h^T_{ij} are shown in Table 6.13. Multiplying the row vector y, which has n components (in the considered example, $n = 7$), and the H^T matrix, in which n rows and $(n - k)$ columns (in the considered example, this gives $n = 7$ rows and $(n - k) = 3$ columns), we get a row vector with $(n - k)$ components (for (7,4) code, these are three components):

$$s = (s_1, s_2, \ldots, s_{(n-k)}).$$

The components of the syndrome are calculated in accordance with the rules of matrix multiplication (in this case, it is the product of a vector, or a matrix with one row by a matrix with dimensions of seven rows and three columns):

$$s_j = \sum_{i=1}^{n} y_i h^T_{ij}; \quad 1 \leq j \leq n - k. \tag{6.29}$$

The block of symbols x transmitted over the communication channel may undergo distortion. The difference between the received block y and the sent block x can be interpreted as a noise block:

$$e = y - x. \tag{6.30}$$

Substituting into the expression for the syndrome (6.28) the representation of the received block as the sum of the code block sent to the communication channel and the noise block, taking into account (6.27), we obtain:

$$s = yH^T = (x + e) H^T = xH^T + eH^T = eH^T. \tag{6.31}$$

Thus, the syndrome does not depend on the sent code block and is determined only by the noise block. A syndrome in general is a group of symptoms that characterize a certain disease. When correcting errors, the syndrome row vector depends only on the error configuration, i.e. is a syndrome, or chracterization of errors. If the decoder can make a likely estimate of the noise block using the syndrome, then it can also obtain an estimate of the sent code block:

$$x' = y - e. \tag{6.32}$$

6.2.12   Hamming code: decoding and implementing an encoder and decoder

The results of calculating the syndrome for the Hamming code (coding was considered in section 6.2.10), using (6.31) for a zero noise block and noise blocks with single errors, are shown in Table 6.14. Recall that the minimum distance of the code under consideration is $d* = 3$; therefore, the code is only capable of correcting single errors. A symbol in a noise block in some bit equal to 1 means the defeat of the sent code block in this symbol. This means that the value of the symbol in this digit is reversed.

As follows from Table 6.14, in the absence of errors, the syndrome is equal to zero (all components of the syndrome as a row vector are equal to zero). Each error configuration (i.e., each error position) corresponds to a certain value of the syndrome.

Let us illustrate the work of the decoder. Let, e.g., the word y = 0000011 was received at the output of the communication channel. Multiplying y by H^T and calculating by (6.29) using the data in Table 6.13 give the following values of the syndrome components:

$$s_1 = 0 * 1 + 0 * 1 + 0 * 1 + 0 * 0 + 0 * 1 + 1 * 0 + 1 * 0 = 0$$
$$s_2 = 0 * 0 + 0 * 1 + 0 * 1 + 0 * 1 + 0 * 0 + 1 * 1 + 1 * 0 = 1 \qquad (6.33)$$
$$s_3 = 0 * 1 + 0 * 1 + 0 * 0 + 0 * 1 + 0 * 0 + 1 * 0 + 1 * 1 = 1$$

So the syndrome is s = 011. From Table 6.14, we find that the syndrome s = 011 corresponds to the error e = 0001000. Subtracting the error estimate from the received word, we obtain the estimate of the sent word x' = $y - e$ = 0000011 − 0001000 = 0001011. Since the code is systematic, the first four symbols indicate the estimate of the sent information word u' = 0001. Such a solution corresponds to the most likelihood decoding.

Table 6.14   *Syndrome components for noise blocks with single error*

Noise block e							Syndrome s		
e_1	e_2	e_3	e_4	e_5	e_6	e_7	s_1	s_2	s_3
0	0	0	0	0	0	0	0	0	0
1	0	0	0	0	0	0	1	0	1
0	1	0	0	0	0	0	1	1	1
0	0	1	0	0	0	0	1	1	0
0	0	0	1	0	0	0	0	1	1
0	0	0	0	1	0	0	1	0	0
0	0	0	0	0	1	0	0	1	0
0	0	0	0	0	0	1	0	0	1

Formula (6.29) and relation (6.33) for the considered code can be rewritten in a form that explicitly indicates the non-zero elements of the rows of the parity-check matrix (Table 6.12) and columns of the transposed parity-check matrix (Table 6.13) for the considered (7,4) Hamming code:

$$s_1 = y_1 + y_2 + y_3 + y_5$$
$$s_2 = y_2 + y_3 + y_4 + y_6 \qquad\qquad (6.34)$$
$$s_3 = y_1 + y_2 + y_4 + y_7$$

This form may turn out to be more convenient for the practical implementation of the decoder, just as (6.7) define a simple way of implementing the encoder.

The scheme of the encoder using the (7,4) Hamming code is shown in Figure 6.7. It is implemented using three modulo 2 adders, which form check symbols in accordance with (6.7). The scheme of the decoder of the (7,4) Hamming code is shown in Figure 6.8. It uses seven modulo 2 adders, which calculate the three components of the syndrome in accordance with (6.34). The error block estimator finds the elements of the error block, which are determined by the method of maximum likelihood in accordance with the data in Table 6.14. At the output of the adders modulo 2, estimates of the block of code symbols *x* sent to the communication channel are obtained. When using systematic coding, the first four output symbols $(x'_1,...,x'_4)$ represent the symbol estimates of the sent information block *u'*.

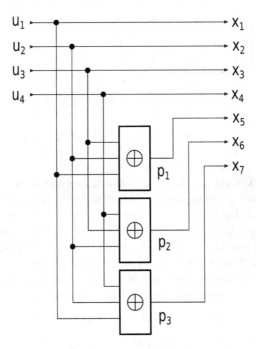

Figure 6.7   (7,4) Hamming code encoder

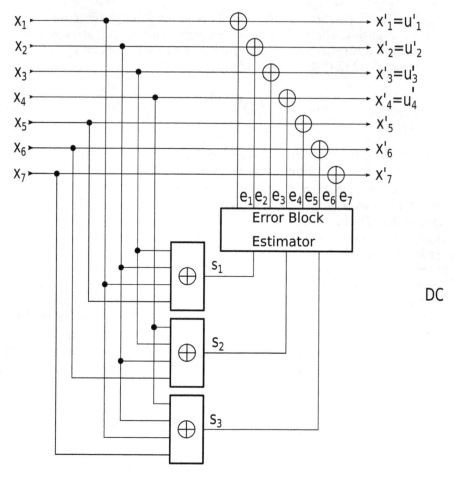

Figure 6.8   (7,4) Hamming code decoder

Using the data in Table 6.14, which is actually a compact form of the decoding table, the probability of correct decoding for a binary symmetric channel (Figure 6.2) can be calculated. Table 6.14 describes one error-free situation (first line) and seven cases of correct decoding with one error in the communication channel. Following the methodology given in the literature [16], and repeating the reasoning made when deriving (6.5), we obtain the probability of correct decoding P_{cor} and the probability of error P_{eer} when decoding by the maximum likelihood method for the (7,4) Hamming code in a binary symmetric channel without memory:

$$P_{cor} = (1 - \varepsilon)^7 + 7(1 - \varepsilon)^6 \varepsilon, \qquad P_{err} = 1 - (1 - \varepsilon)^7 - 7(1 - \varepsilon)^6 \varepsilon. \qquad (6.35)$$

If we take the error probability when transmitting one symbol in the channel equal to $\varepsilon = 0.01$, then the error probability when using the Hamming code will be $P_{err} = 0.002$.

6.3 About the strategy for designing good codes

A prominent scientist Claude Shannon in his work 'a mathematical theory of communication' [6] proved that with the help of channel coding it is possible to provide data transmission with an arbitrarily small error probability if the data transfer rate does not exceed the communication channel capacity. However, Shannon did not indicate how to find an algorithm for such an encoding. However, the history of error-correcting coding has convincingly proven that a long codeword is an inherent attribute of good codes. For example, Robert Gallagher, the inventor of LDPC codes, showed [16]. while studying block codes, that the probability P_{err} of an error decoding tends to zero with increasing block length n at any data rate R_c less than the throughput of channel C:

$$P_{err} \leq e^{-nEr(Rc)},$$

where $E_r(R_c)$ is a parameter of random coding, and $E_r(R_c) > 0$ for all $R_c < C$.

From the above relation, it follows that to achieve a data transmission rate approaching the channel capacity, it is necessary to increase the length of the code block n.

If the message is long and consists of a large number of symbols, then it is better to use one long code block than many short blocks [10]. This is due to the fact that the statistical structure of noise and interference causing errors in the transmission of message symbols in the communication channel is complex. Along with errors of a random nature and occurring from time to time with some probability, there are so-called burst errors, which are a series of errors that follow one another. The length of these error bursts is random, so from time to time a series of long errors may occur.

Let us compare the efficiency of codes with different codeword lengths in terms of correcting burst errors. Let there be a code with a codeword length, e.g., 2000 bits ($n = 2000$). Let each word be used to encode 1000 information symbols ($k = 1000$). The code is capable of correcting, e.g., 100 errors in each codeword ($t = 100$). To transmit 10,000 information symbols, ten code blocks must be used. The code is capable of correcting 1000 errors during the transmission of 10,000 information symbols using ten codewords with a total length of 20,000 code symbols. But in each word, the code is able to correct only 100 errors. In all, 1000 errors will be corrected if the errors are evenly distributed over ten codewords.

Let the second code be created, all parameters of which are increased ten times ($n = 20,000$, $k = 10,000$, $t = 1000$). Overall, 10,000 information symbols are transmitted using one codeword. The decoder of the second code is capable of correcting 1000 errors with any distribution. For example, a burst error of 1000 symbols can be corrected, which cannot be done using the first code. Codes with very long codewords are significantly more efficient at correcting errors with complex statistical structures.

It may seem that it is enough to set the requirements for the code, e.g., size, length of the codeword, the number of errors to be corrected and then find a good code by iterating over the set of all codes using machine search. Is it real? Each codeword is n symbols long. In total, $M = 2^k$ codewords should be found. A complete description of the code requires nM binary symbols to be generated. In total, you can count 2^{nM} options for choosing the symbols of this sequence. Therefore, there are $2^{n\cdot 2^k}$ options for encoding information words with length k using codewords with length n, i.e. $2^{n\cdot 2^k}$ different codes.

Using the data from the above example, we find the number of different codes: $2^{20000\cdot 2^{10000}}$. The number is so large that it is almost impossible to talk about finding good codes by sorting options. But even more difficult is the solution to the problem of creating practical algorithms for encoding and decoding. A code with a very long length could theoretically be described by encoding and decoding tables, but using such tables in encoders and decoders may be practically impossible.

Some coding theory must be developed to make good codes and find promising and practical coding and decoding algorithms. Such codes must have a special mathematical structure. The best results to date have been achieved in the field of codes that are algebraic in structure, and in the field of randomly selected codes. Their mathematical structure is used in order to achieve practical feasibility of encoding and decoding at large code word lengths.

6.4　Cyclic codes

6.4.1　Polynomials and Galois fields

6.4.1.1　Galois fields based on the ring of integers

The ideas of coding theory used in widely used cyclic codes are based on arithmetic systems of finite fields, or Galois fields. Recall that a field is a set of elements closed in two operations, called addition and multiplication. In section 6.9.2.1, a two-element field was introduced. Understanding cyclic codes requires the use of Galois fields with a large number of elements. The number of elements in a field is usually called the order of the field. It should be borne in mind that the order of the field cannot be arbitrary and Galois fields do not exist for all numbers of elements.

One way to construct a Galois field is based on the ring of integers. In the ring of integers, division is not always possible, but division with remainder is always possible. If the remainder of dividing the number a by q is equal to s, then this is written as $s = R_q[a]$. If q is a prime, i.e., it is divisible only by itself and by one, then the arithmetic of the field $GF(q)$ can be introduced as addition and multiplication modulo q. This is how the rules of arithmetic in the field of two elements $GF(2)$ were drawn up in Tables 6.5 and 6.6.

6.4.1.2　Galois fields based on the ring of polynomials

Another way of constructing Galois fields is based on polynomial rings. An expression of the form $f(z) = f_n z^n + f_{n-1} z^{n-1} + \ldots + f_0$ is called a polynomial over $GF(q)$ of degree n if the coefficients $f_n, f_{n-1}, \ldots, f_0$ are elements of the field $GF(q)$ and the first coefficient f_n is not zero. The symbol z in the polynomial cannot be interpreted as a variable or an unknown field element. It is an undefined symbol. In most cases, it is not $f(z)$ as a function that is of interest, but the sequence of elements $f_n, f_{n-1}, ... f_0$. Two polynomials are said to be equal if they correspond to the same sequence of coefficients. A polynomial is called normalized or monic if its leading coefficient is equal to 1. A polynomial that can be represented as a product of polynomials of lower degrees with coefficients from the field $GF(q)$ is called reducible, otherwise irreducible.

It is easy to define addition, subtraction and multiplication of polynomials according to the usual rules of addition and multiplication of polynomials, as was done in school, but actions with the coefficients of polynomials over the Galois field $GF(q)$ must be performed in accordance with the arithmetic of the field $GF(q)$. There is no operation inverse to multiplication, i.e., division of polynomials. The set of polynomials, together with the operations of addition, subtraction and multiplication, forms the ring of polynomials over the field $GF(q)$.

By analogy with arithmetic operations modulo a prime number, we can define the division of the polynomial $f(z)$ by the polynomial $p(z)$ as the remainder of dividing $f(z)$ by $p(z)$: $R_{p(z)}[f(z)]$. This remainder is called the residue of the polynomial $f(z)$ modulo the polynomial $p(z)$. One can also form the ring of polynomials modulo the monic polynomial $p(z)$. This will be the set of all polynomials over the field $GF(q)$, the degree of which is less than the degree of the polynomial $p(z)$, with operations of addition, subtraction and multiplication modulo the polynomial $p(z)$. Finally, you can enter the equivalent of a prime number. In the ring of polynomials, this will be a prime polynomial – an irreducible polynomial with a leading coefficient equal to one.

In the Galois field theory, it is proved that the polynomial ring modulo the monic polynomial $p(z)$ is a field if and only if $p(z)$ is a prime polynomial. This statement opens up the possibility of expanding the Galois field, i.e., of constructing a Galois field $GF(q^n)$ containing q^n elements if a prime polynomial of degree n is found over the Galois field $GF(q)$ containing q elements.

As an example, we construct the field $GF(4)$ by extending the field $GF(2)$ using the prime polynomial $p(z) = z^2 + z + 1$. There are four polynomials in total whose degree is less than the degree $p(z)$, i.e., does not exceed 1: 0, 1, z, $z + 1$. These polynomials are elements of the field $GF(4) = \{0, 1, z, z + 1\}$. Let us find the table of addition of the field elements: $0 + 0 = 0$, $0 + 1 = 1$, $1 + 1 = 0$, $z + 0 = z$, $z + 1 = z + 1$, $z + z = 0$, $z + 1 + 0 = z + 1$, $z + 1 + 1 = z$, $z + 1 + z = 1$, $z + 1 + z + 1 = 0$. You can also find a part of the multiplication table: $0 * 0 = 0$, $0 * 1 = 0$, $0 * z = 0$, $0 * (z + 1) = 0$, $1 * z = z$, $1 * (z + 1) = z + 1$. The degree of all the polynomials obtained as a result of the operations performed is less than the degree of $p(z)$, so finding the remainder from division by $p(z)$ was actually not required. But, e.g., the multiplication $z * z$ requires computation according to the complete rules of computation modulo $p(z)$. $z * z = z^2 = R_{p(z)}[z^2] = z + 1$, $z * (z + 1) = R_{p(z)}[z^2 + z] = 1$, $(z + 1) * (z + 1) = R_{p(z)}[z^2 + z] = z$. The results of the calculations performed are summarized in Tables 6.15 and 6.16.

After constructing the arithmetic tables of the $GF(4)$ field, you can add new representations of the field elements (Table 6.17). Binary representation is a set of coefficients of polynomials of the first degree, which were introduced as elements of the ring of polynomials modulo the monic polynomial $p(z)$. Integer representation can be introduced as decimal equivalents of binary sets.

Table 6.15   Addition table for GF(4)

+	0	1	z	z + 1
0	0	1	z	z + 1
1	1	0	z + 1	z
z	z	z + 1	0	1
z + 1	z + 1	z	1	0

Table 6.16   Multiplication table for GF(4)

×	0	1	z	z + 1
0	0	0	0	0
1	0	1	z	z + 1
z	0	z	z + 1	1
z + 1	0	z + 1	1	z + 1

Table 6.17   Elements of GF(4)

Polynomials representation	Binary representation	Integer representation	Power representation
0	00	0	0
1	01	1	α^0
z	10	2	α^1
z + 1	11	3	α^2

It is known from Galois field theory that the set of non-zero elements of the field $GF(q)$ forms a closed group by multiplication, or a multiplicative group. The number of elements in this group is equal to $(q - 1)$. The product of elements is always an element of the group because the group is closed. You can take some element of this group (e.g., an element β) and multiply it by itself. The powers of β will also belong to the group. The result is a series of numbers: β, $\beta\beta = \beta^2$, $\beta\beta^2 = \beta^3$, ... and so on. Since the group of non-zero field elements has a finite number of elements, a repetition will necessarily appear in the sequence of multiplication results. The number n of different elements that can be obtained by raising β to a power is called the order of β. Obviously, $\beta^n = 1$, since exactly the next multiplication by β will lead to repetition: $\beta\beta^n = \beta$.

There must be at least one primitive element among the field elements. A primitive element of the field α is an element such that all elements except zero can be represented as a power of the element α. This means that the order of the primitive element is $(q - 1)$. The nice thing about primitive elements is that you can easily build a multiplication table in a field if a primitive element is found.

When constructing a field extension in the form of a set of polynomials, it is convenient to choose a prime polynomial such that the polynomial z corresponds to a primitive element of the field α. Special prime polynomials that allow you to do this are called primitive. The prime polynomial $p(z) = z^2 + z + 1$ is primitive; therefore, the primitive element α of the field $GF(4)$ in Table 6.17 corresponds to the element z. Each element of the field, except for zero, corresponds to some degree of a primitive element. These degrees are shown as power representation for the elements of the field $GF(4)$ (Table 6.17).

The considered method of extending the Galois field as the ring of polynomials modulo the primitive polynomial is sufficient to obtain all finite fields. To summarize, there are several important properties of finite fields to note:

- The number of elements of any Galois field is equal to a prime power.
- For any prime p and any positive integer m, the smallest subfield of $GF(p^m)$ is $GF(p)$. The number p is called the characteristic of the field $GF(p^m)$.
- In a Galois field with characteristic $p = 2$, for each element β, the following equality holds: $-\beta = \beta$.

6.4.2 Principles of constructing cyclic codes

Cyclic codes are a subclass of linear codes that satisfy an additional structural requirement. The theory of Galois fields is used as a mathematical means for finding good cyclic codes. This theory leads to encoding and decoding algorithms that are efficient as computational procedures.

When considering the principles of constructing linear codes, the components of a sequence of n symbols (such sequences were called words) were considered as components of a vector in an n-dimensional space, and the sequence itself as a vector in this space. The set of codewords obtained with a channel encoder was considered as a subset with certain properties in the set of all vectors in n-dimensional space. It can also be argued that the set of codewords is a linear subspace in n-dimensional space. Each symbol is an element of the Galois field $GF(q)$ of q elements (in the examples considered above, $q = 2$, but all the results related to codes with parity checks can be generalized to non-binary symbols, i.e., to fields with a large number of symbols).

A cyclic code over the field $GF(q)$ is a linear code for which, for any cyclic shift of any codeword, another codeword is obtained. This means that if the sequence $x' = (x_1, x_2 ..., x_n)$ is a codeword, then $x'' = (x_2, x_3, ..., x_n, x_1)$ is also a codeword.

When describing cyclic codes, the designations are usually changed, numbering symbols not from the beginning, but from the end, i.e., from $(n - 1)$ to 0: $x = (x_{n-1}, x_{n-2}, ..., x_1, x_0)$. Each sequence of n symbols, or a vector, can be represented by a polynomial over the field $GF(q)$ in z with degree at most $(n - 1)$:

$$x(z) = x_{n-1}z^{n-1} + x_{n-2}z^{n-2} + ... + x_1z^1 + x_0. \tag{6.36}$$

The components of the vector are identified with the coefficients of the polynomial. The set of polynomials has the structure of the ring of polynomials modulo the polynomial $(z^n - 1)$ over the field $GF(q)$. This allows you to define a cyclic shift as a multiplication in that ring:

$$z * x(z) = R_{(z^n-1)}[x_{n-1}z^n + x_{n-2}z^{n-1} + \ldots + x_1z^2 + x_0z] =$$
$$= R_{(z^n-1)}[x_{n-1}(z^n-1) + x_{n-2}z^{n-1} + \ldots + x_1z^2 + x_0z + x_{n-1}] =$$
$$= x_{n-2}z^{n-1} + \ldots + x_1z^2 + x_0z + x_{n-1}.$$

Let us choose a non-zero monic code polynomial of the least degree in the subspace of code polynomials. Let us denote its degree as $(n - k)$. This polynomial is called the generator polynomial. Let us denote it as $g(z)$. In the theory of cyclic codes, it is proved that a cyclic code consists of all products of a generator polynomial by polynomials of degree at most $(k - 1)$. But a cyclic code exists only if $(z^n - 1)$ is divisible by $g(z)$ without a remainder. This means that there is a polynomial $h(z)$ such that $z^n - 1 = g(z) * h(z)$. Hence

$$R_{(z^n-1)}[g(z) * h(z)] = 0. \tag{6.37}$$

The polynomial $h(z)$ is called a parity-check polynomial. Each codeword satisfies the equality:

$$R_{(z^n-1)}[x(z) * h(z)] = 0. \tag{6.38}$$

The information sequence to be encoded can also be represented as a polynomial, the degree of which is $(k - 1)$:

$$u(z) = u_{k-1}z^{k-1} + u_{k-2}z^{k-2} + \ldots + u_1z^1 + u_0.$$

The set of information polynomials can be mapped to code polynomials in different ways. For example, you can multiply the information polynomials by the generator polynomial:

$$x(z) = u(z)g(z). \tag{6.39}$$

This coding is not systematic. In addition to coding according to the rule $x(z) = u(z) * g(z)$, there is a so-called systematic coding rule introduced above, in which the k highest coefficients of the codeword are set equal to the coefficients of the information polynomial: $x(z) = u_{k-1}z^{n-1} + u_{k-2}z^{n-2} + \ldots + u_0z^{n-k} + p_{n-k-1}z^{n-k-1} + \ldots + p_1z + p_0$. The $(n - k)$ least significant coefficients of the codeword $p = (p_{n-k-1}, p_{n-k-2}, \ldots, p_1, p_0)$, which are often called parity-check ones, are chosen such that the polynomial $x(z)$ is divisible by $g(z)$ without a remainder, i.e., $R_{g(z)}[x(z)] = 0$. This will be the case if the corresponding parity-check polynomial $p(z) = p_{n-k-1}z^{n-k-1} + \ldots + p_1z + p_0$ is calculated as $p(z) = -R_{g(z)}[z^{n-k}u(z)]$. The systematic coding rule gives codewords that are more convenient in practice, since information words are explicitly placed in the k most significant bits of code words.

All the code polynomials $x(z)$ are divisible by the generator polynomial $g(z)$ without remainder, i.e., $R_{g(z)}[x(z)] = 0$. This circumstance provides a key to decoding, which allows detecting and correcting errors that occur during data transmission. The words that are not evenly divisible by a generator polynomial are not code words and, therefore, contain errors.

A code block $x = (x_{n-p}, x_{n-2}, ..., x_p, x_0)$ transmitted over a communication channel or recorded on a medium may undergo distortion, e.g., due to noise. This can be described in general terms by adding to the code block the error set $e = (e_{n-p}, e_{n-2}, ..., e_p, e_0)$, which corresponds to the polynomial $e(z) = e_{n-1}z^{n-1} + e_{n-2}z^{n-2} + ... + e_1 z + e_0$. The accepted or reproduced set of symbols $y = (y_{n-p}, y_{n-2}, ..., y_p, y_0)$ corresponds to the polynomial $y(z) = x(z) + e(z)$. Having found the remainder of dividing the accepted polynomial $y(z)$ by the generator polynomial $g(z)$, one can understand whether there were actually errors. If $R_{g(z)}[y(z)] = 0$, then with a high probability we can assert that there were no errors and the received word is a code word, i.e., $y(z)$ is the transmitted word $x(z)$. If the remainder is not zero, then there were errors during transmission. The remainder of the division is a polynomial that depends only on the error polynomial:

$$R_{g(z)}[y(z)] = R_{g(z)}[x(z)] + R_{g(z)}[e(z)] = R_{g(z)}[e(z)] = s(z),$$

which is called a syndrome polynomial. The polynomial $s(z)$ depends only on the configuration of the errors, i.e., is a syndrome, or description of errors. If the number of errors does not exceed a certain limit, then there is a one-to-one correspondence between $e(z)$ and $s(z)$. This limit depends on the minimum code distance d^*. Each error polynomial with weight less than $d^*/2$ has the only syndrome polynomial.

Thus, to correct errors using a cyclic code, it is necessary to find the error polynomial $e(z)$ with the smallest number of non-zero coefficients that meets the condition:

$$s(z) = R_{g(z)}[e(z)].$$

This task can be solved, e.g., in a tabular way. For each error polynomial with weight less than $d^*/2$, a syndrome polynomial is calculated, the values of which are tabulated. The resulting table (Table 6.18) is called the table of syndrome values.

Table 6.18   Table of syndrome values for cyclic coding

Error polynomial e(z)	Syndrome polynomial s(z)
1	$R_{g(z)}[1]$
z	$R_{g(z)}[z]$
z^2	$R_{g(z)}[z^2]$
...	...
$1 + z$	$R_{g(z)}[1 + z]$
$1 + z^2$	$R_{g(z)}[1 + z^2]$
...	...

So, a number of polynomials were introduced, which are presented below in a systematized form, indicating their degrees:

information polynomial	$u(z)$, $deg[u(z)] = k - 1$,
code polynomial	$x(z)$, $deg[u(z)] = n - 1$,
generator polynomial	$g(z)$, $deg[u(z)] = n - k$,
received polynomial	$y(z)$, $deg[u(z)] = n - 1$,
parity-check polynomial	$h(z)$, $deg[u(z)] = k$,
error polynomial	$e(z)$, $deg[u(z)] = n - 1$,
syndrome polynomial	$s(z)$, $deg[u(z)] = n - k - 1$.

If a set of k symbols forms a block of information $u = (u_{k-p}u_{k-2}...,u_p u_0)$ encoded to detect and correct errors using a cyclic code, then encoding means the formation of a block $x = (x_{n-p}x_{n-2}..., x_p x_0) = (u_{k-p}u_{k-2}...,u_0 p_{n-k-p}...,p_p p_0)$, which has acquired some redundancy in the form of additional check symbols $p = (p_{n-k-p}...,p_p p_0)$. This redundancy has a strictly metered value in accordance with a given degree of noise immunity.

Decoding the received or reproduced symbol set $y = (y_{n-p}y_{n-2}...,y_p y_0)$ involves the following actions:

- finding the syndrome polynomial $s(z) = R_{g(z)}[y(z)]$;
- finding the error polynomial $e(z)$ in the table of syndrome values based on the calculated syndrome polynomial (there is no error if $s(z) = 0$);
- determination of the estimate of the transmitted code polynomial by calculating $x(z) = y(z) - e(z)$;
- determination of the transmitted block of information $u = (u_{k-1}u_{k-2}...,u_p u_0)$ using the k highest coefficients of the reconstructed code polynomial $x(z)$.

The described method determines the fundamental possibility of error correction based on channel coding using cyclic codes. A large number of efficient decoding schemes have been developed for many cyclic codes.

6.4.3   Cyclic codes: encoding and decoding

A cyclic code exists only if $(z^n - 1)$ is divisible by the generator polynomial $g(z)$ without a remainder, which follows from (6.38). Each polynomial $g(z)$ that divides the polynomial $(z^n - 1)$ generates a cyclic code. A natural approach to obtaining a generating polynomial is to factor the polynomial $(z^n - 1)$ into prime factors:

$$z^n - 1 = f_1(z) * f_2(z) * ...f_s(z),$$

where s is the number of prime factors.

The generator polynomial $g(z)$ can be obtained as the product of some subset of these prime factors. But how do you choose this subset? If all prime factors are different, then there are 2^s options for constructing the generator polynomial. If we exclude from these options the trivial cases $g(z) = 1$ (when no factor is included in $g(z)$) and $g(z) = (z^n - 1)$ (when all factors are included in the generator polynomial), then there are $(2^s - 2)$ options. It will be a large number for large values of n. What are the options for codes with a large minimum distance? To answer this question, the theory of cyclic codes traces the relationship of prime polynomials with their roots in the field extension. But first, we need to introduce the concept of a minimal polynomial.

Let $GF(q)$ be a field and $GF(Q)$ be an extension of this field. Let β be an element of the field $GF(Q)$. A prime polynomial $f(z)$ of the least degree over $GF(q)$ for which β is a root (this means that $f(\beta) = 0$) is called the minimal polynomial of β over the field $GF(q)$. The minimal polynomial always exists and is unique. If the minimal polynomial of β over the field $GF(q)$ is equal to $f(z)$ and β is a root of $g(z)$, then $f(z)$ divides $g(z)$.

Now the general approach to constructing a cyclic code can be formulated as follows. The cyclic code is constructed from the generator polynomial, which is given by its roots in the Galois field $GF(q^m)$. The Galois field $GF(q^m)$ is an extension of the $GF(q)$ field, over which the cyclic code is constructed.

Let $\beta_1, \beta_2, ..., \beta_r$ from the Galois field $GF(q^m)$ be the roots of the generator polynomial $g(z)$. Let $f_1(z), f_2(z), ..., f_r(z)$ denote the minimal polynomials of elements $\beta_1, \beta_2, ..., \beta_r$ from the Galois field $GF(q^m)$. The generator polynomial of a cyclic code is found as the least common multiple (LCM) of the product:

$$g(z) = LCM\left[f_1(z), f_2(z), ..., f_r(z)\right] \tag{6.40}$$

where $f_1(z), f_2(z), ..., f_r(z)$ are the minimal polynomials of the elements $\beta_1, \beta_2, ..., \beta_r$, i.e. roots of the generator polynomial $g(z)$.

The described general principle of constructing a cyclic code makes it possible to indicate a method for correcting errors using a field extension. Let $x(z)$ be the codeword obtained at the output of the channel encoder. If the codeword in the channel is affected by errors $e(z)$, then the polynomial describing the received word can be written as

$$y(z) = x(z) + e(z).$$

You can calculate the values of this polynomial on the elements of the extended Galois field $GF(q^m)$ at the points that are the roots of the generator polynomial $g(z)$, i.e., at points $\beta_1, \beta_2, ..., \beta_r$. This calculation gives the components of the syndrome

$$S_j = y(\beta_j), \quad j = 1, 2, ..., r.$$

The code polynomial at these points is equal to zero $x(\beta_1) = x(\beta_2) = ... x(\beta_r) = 0$, since it is the product of the information polynomial by the generator polynomial in accordance with (6.39). Therefore, the components of the syndrome depend only on the configuration of the errors:

$$S_j = y(\beta_j) = x(\beta_j) + e(\beta_j) = e(\beta_j), \quad j = 1, 2, ..., r.$$

Using the expression for the error polynomial in the form $e(z) = e_{n-1}z^{n-1} + e_{n-2} \cdot z^{n-2} + ... + e_1 z + e_0$, we find

$$S_j = e_{n-1}\beta_j^{n-1} + e_{n-2}\beta_j^{n-2} + ... + e_1\beta_j + e_0 = \sum_{i=0}^{n-1} e_i\beta_j^i, \quad j = 1, 2, ..., r.$$

The elements of S_j are not coefficients of the syndrome polynomial, but they provide equivalent information. So, a system of r equations has been obtained, which contains only the values determined by the errors. If these equations can be solved for the values of e_j, then it will be possible to calculate the error polynomial.

6.4.4   BCH codes

The Bose–Chaudhuri–Hocquenghem (BCH) codes are a subclass of cyclic codes. They are defined as follows. Let numbers q and m be given, let β be an element of the Galois field $GF(q^m)$, the order of which is n. For any positive number t and any integer j_0, the BCH code is the cyclic code with length n and a generator polynomial

$$g(z) = LCM\left[f_{j_0}(z), f_{j_0+1}, \ldots, f_{j_0+2t-1}\right]$$

(6.41)

where $f_{j_0}(z)$, $f_{j_0+1}(z)$, ..., $f_{j_0+2t-1}(z)$ are the minimal polynomials of $\beta^{j_0}, \beta^{j_0+1}, \ldots,$ β^{j_0+2t-1}.

Thus, $2t$ consecutive powers of an arbitrary element β of the extended Galois field $GF(q^m)$ are given as the roots of the generator polynomial. The length of the codeword over the Galois field $GF(q)$ is equal to the order of β, i.e., the smallest number n for which $\beta^n = 1$. The number j_0, equal to the initial value of the exponent of the element β, is often chosen equal to 1, which in many cases leads to the generator polynomial with the smallest degree. If a large code length is required, then the field element with the largest order is selected, i.e., primitive element α. The length of the code in this case will be $n = (q^m - 1)$. The number t defines the constructive number of corrected errors, i.e., the number of errors to be corrected, given when the code is constructed. It is related to the constructive minimum distance of the code d by the relation $d = 2t + 1$. The true minimum distance d^* may be greater than the constructive distance. The sequential powers of the primitive element α, the minimal elements of which are used to construct the generator polynomial, can be written as follows: $\alpha^1, \alpha^2, \ldots, \alpha^{2t} = \alpha^1, \alpha^2, \ldots, \alpha^{d-1}$.

Under the assumptions made, the algorithm for constructing the BCH code turns out to be as follows:

The numbers q and m are given.

The Galois field $GF(q^m)$ is constructed using a primitive polynomial of degree m.

The minimal polynomials $f(z)$, $j = 1, 2, \ldots, 2t$ are found for the powers of the primitive element $\alpha^1, \alpha^2, \ldots, \alpha^{2t}$, where t is the number of errors that need to be corrected.

The generator polynomial of the code is found and the length of the information word k is determined.

$$g(z) = LCM\left[f_1(z), f_2(z), \ldots f_{2t}(z)\right].$$

As an example, we construct the BCH code over the field $GF(2)$, correcting two errors, setting $q = 2$, $m = 4$, $t = 2$. The length of the code for the given parameters will be $n = (q^m - 1) = 15$.

The representation of the field $GF(2^4)$ constructed using the primitive polynomial $p(z) = z^4 + z + 1$ is given in Table 6.19.

Table 6.19 Elements of $GF(2^4)$ and minimal polynomials

#	Power representation	Polynomial representation	Binary representation	Integer representation	Minimal polynomials
1	0	0	0000	0	
2	α^0	1	0001	1	$z+1$
3	α^1	z	0010	2	z^4+z+1
4	α^2	z^2	0100	4	z^4+z+1
5	α^3	z^3	1000	8	$z^4+z^3+z^2+z+1$
6	α^4	$z+1$	0011	3	z^4+z+1
7	α^5	z^2+z	0110	6	z^2+z+1
8	α^6	z^3+z^2	1100	12	$z^4+z^3+z^2+z+1$
9	α^7	z^3+z+1	1011	11	z^4+z^3+1
10	α^8	z^2+1	0101	5	z^4+z+1
11	α^9	z^3+z	1010	10	$z^4+z^3+z^2+z+1$
12	α^{10}	z^2+z+1	0111	7	z^2+z+1
13	α^{11}	z^3+z^2+z	1110	14	z^4+z^3+1
14	α^{12}	z^3+z^2+z+1	1111	15	$z^4+z^3+z^2+z+1$
15	α^{13}	z^3+z^2+1	1101	13	z^4+z^3+1
16	α^{14}	z^3+1	1001	9	z^4+z^3+1

The minimal polynomials for the degrees of the primitive element α^1, α^2, α^3, α^4:

$$f_1(z) = z^4 + z + 1,$$
$$f_2(z) = z^4 + z + 1,$$
$$f_3(z) = z^4 + z^3 + z^2 + z + 1,$$
$$f_4(z) = z^4 + z + 1.$$

From the list of minimal polynomials in Table 6.19, it can be seen that the minimal polynomials for even degrees of the primitive element α are equal to polynomials for lower degrees. This makes it somewhat easier to find the generating polynomial of the code:

$$g(z) = LCM[(z^4 + z + 1), (z^4 + z + 1), (z^4 + z^3 + z^2 + z + 1), (z^4 + z + 1)] =$$
$$= (z^4 + z + 1) * (z^4 + z^3 + z^2 + z + 1) = z^8 + z^7 + z^6 + z^4 + 1.$$

The degree of the generator polynomial is 8. In general, this degree is written as $(n - k) = 8$, hence the length of the information word is $k = 7$. So, the $(15,7)$ BCH code is constructed, which is fifteen characters long and corrects two errors.

6.4.5 Decoding BCH codes: Peterson–Gorenstein–Zierler algorithm

BCH codes, being cyclic, can be decoded using shift register decoders created for cyclic codes. But for BCH codes, special algorithms have been developed that have better characteristics. The most important of them is the Peterson–Gorenstein–Zierler algorithm [9].

Based on the general approach described in section 6.4.3, we formulate the decoding problem using the field extension as applied to BCH codes. Let $x(z)$ be the codeword obtained at the output of the channel encoder. The polynomial describing the received word at the input of the decoder can be written as

$$y(z) = x(z) + e(z),$$

where $x(z)$ is the code polynomial corresponding to the codeword, $e(z)$ is the error polynomial.

We calculate the values of this polynomial on the elements of the extended Galois field $GF(q^m)$ at the points that are the roots of the generating polynomial $g(z)$, i.e., at the points $\alpha^1, \alpha^2, ..., \alpha^{2t}$. This calculation gives the components of the syndrome:

$$S_j = y(\alpha^j), \qquad j = 1, 2, ..., 2t.$$

The code polynomial at these points is equal to zero $x(\alpha^1) = x(\alpha^2) = ... \ x(\alpha^{2t}) = 0$, since the generator polynomial is equal to zero at these points. So

$$S_j = y(\alpha^j) = x(\alpha^j) + e(\alpha^j) = e(\alpha^j), \qquad j = 1, 2, ..., 2t.$$

Using the expression for the error polynomial in the form $e(z) = e_{n-1}z^{n-1} + e_{n-2}z^{n-2} + ... + e_1 z + e_0$, we find

$$S_j = e_{n-1}\alpha^{j(n-1)} + e_{n-2}\alpha^{j(n-2)} + ... + e_1\alpha^j + e_0 = \sum_{i=0}^{n-1} e_i \alpha^{ji}. \tag{6.42}$$

where $j = 1, 2, \ldots, 2t$.

As noted in section 6.4.3, the syndrome components S_j are not coefficients of the syndrome polynomial, but provide equivalent information. A system of $2t$ equations is obtained, which contains only the values determined by the errors. The error polynomial can be calculated by solving these equations for the values of e_i. The code is designed to correct no more than t errors, so we assume that no more than t coefficients in (6.42) are non-zero. Suppose that actually v errors occurred, with $0 \leq v \leq t$, and that these errors correspond to error positions $\{i_1, i_2, \ldots, i_v\}$ and error values $\{e_{i_1}, e_{i_2}, \ldots, e_{i_v}\}$.

The error polynomial can be rewritten as

$$e(z) = e_{i_1} z^{i_1} + e_{i_2} z^{i_2} + \cdots + e_{i_v} z^{i_v}.$$

The error positions $\{i_1, i_2, \ldots, i_v\}$ and the error values $\{e_{i_1}, e_{i_2}, \ldots, e_{i_v}\}$ are unknown. To correct errors, these numbers must be found using the components of the syndrome. The first component of the syndrome is the value of the accepted polynomial at the point α:

$$S_1 = y(\alpha) = x(\alpha) + e(\alpha) = e(\alpha) = e_{i_1} \alpha^{i_1} + e_{i_2} \alpha^{i_2} + \cdots + e_{i_v} \alpha^{i_v}.$$

It can be rewritten using a simplified notation:

$$S_1 = Y_1 X_1 + Y_2 X_2 + \cdots + Y_v X_v$$

where $Y_l = e_{i_l}$ is the value of the error at the i_l position, $X_l = \alpha^{i_l}$ is the field element that can be associated with the i_l error position and that can be called the error locator at the i_l position. Since the order of the element α is equal to n, then all the locators of the considered configuration of errors are different.

Calculating the values of the received polynomial for all powers of the primitive element α^j (here $j = 1, 2, \ldots, 2t$), we obtain $2t$ components of the syndrome and, accordingly, a system of 2t equations for unknown locators $X_1, X_2, \ldots, X_v$ and unknown values of errors $Y_1, Y_2, \ldots, Y_v$:

$$S_1 = Y_1 X_1 + Y_2 X_2 + \cdots + Y_v X_v$$
$$S_2 = Y_1 X_1^2 + Y_2 X_2^2 + \cdots + Y_v X_v^2$$
$$\cdots\cdots$$
$$S_{2t} = Y_1 X_1^{2t} + Y_2 X_2^{2t} + \cdots + Y_v X_v^{2t}.$$

This system of equations is effectively solved using the Peterson–Gorenstein–Zierler algorithm. A detailed description of the algorithm can be found in the literature [9]. In accordance with the algorithm, the actual number of errors is first determined, then the error locators and finally the error values. This allows you to find the error polynomial and evaluate the code polynomial transmitted over the communication channel.

6.4.6 RS codes

The Reed-Solomon (RS) codes are an important and widely used subset of BCH codes. The $GF(q)$ symbol field is the same as the $GF(q^m)$ error locator field, i.e., when constructing RS codes, the parameter $m = 1$. The length of the code in this case will be

$n = (q^n - 1) = (q - 1)$. Since the field of symbols and the field of locators are the same, then all minimal polynomials are linear. The minimal polynomial $f(\alpha)$ over the Galois field $GF(q)$ for the element α is $(z - \alpha)$. This allows you to write the generator polynomial of your code as:

$$g(z) = (z - \alpha)(z - \alpha^2)\ldots(z - \alpha^{2t}). \tag{6.43}$$

The power of the generator polynomial is always equal to $2t$; therefore, the parameters of the RS code are related by the relation $(n - k) = 2t$. This means that to correct one error, two parity-check symbols must be added. The minimum code distance is equal to the design distance: $d^* = d = n - k + 1$.

As an example, let us find a generator polynomial for the RS code over the Galois field $GF(q) = GF(16)$, which corrects two errors [10]. The representation of the elements of the Galois field $GF(16)$ is given in Table 6.19. In accordance with (6.44) and a given number of corrected errors, the generating polynomial must contain four factors:

$$g(z) = (z - \alpha)(z - \alpha^2)(z - \alpha^3)(z - \alpha^4) = z^4 + \alpha^{13}z^3 + \alpha^6z^4 + \alpha^3z + \alpha^{10}.$$

The length of the codeword is $q - 1 = 15$. The power of the generator polynomial is $n - k = 4$. Therefore, the length of the information word is $k = 11$. Each symbol of the Galois field $GF(16)$ is 4 bits long, so the information word is 44 bits long and the code symbol is 60 bits long.

The RS code was used as the outer code in the first generation of DVB television systems. This is a code over a Galois field of 256 elements $GF(2^8) = GF(256)$. A Galois field is an extension of a two-element Galois field using a primitive prime polynomial:

$$p(z) = z^8 + z^4 + z^3 + z^2 + 1.$$

The code was designed to correct $t = 8$ errors in the codeword, so the generator polynomial of the code consists of 16 factors:

$$g(z) = (z + \alpha^0)(z + \alpha^1)(z + \alpha^2)\ldots(z + \alpha^{15})$$

where α is a primitive element of the field $GF(256)$.

The length of the original systematic code is $n = (q - 1) = 255$. The number of parity-check symbols is $(n - k) = 2t = 16$. The length of the information word is $k = n - 2t = 239$. Thus, the original code is written as a (255, 239) RS code over the Galois field $GF(256)$. The number of elements in the field is 256, so each symbols of both the information and code word is 8 bits or 1 byte.

The code was applied to packets of the transport stream with a length of 188 bytes, which came to the input of the channel encoder. Before encoding, a word with a length of 188 bytes was supplemented with 51 zero bytes in the most significant digits to 239 bytes, which were encoded using the (255, 239) RS code. After encoding, 51 bytes in the most significant digits of the systematic codeword was removed. The communication channel received words with a length of $255 - 51 = 204$ bytes.

Therefore, the outer code was written in the specifications of the standard as a short-ened (204, 188) RS code.

6.5   LDPC codes

6.5.1   LDPC code concept

The main parameter of the codes for correcting errors in the communication chan-nel is the length of the codeword n. If the channel needs to be used efficiently, i.e., data transmission should be performed at the rate close to the channel capacity and with a low error probability, then the value of n should be large and very large. The main obstacle to achieving this goal is the complexity of the equipment and the long computational times required for encoding and decoding.

Low density parity-check (LDPC) codes are a special case of parity-check codes. Codewords in a parity-checked code are formed by combining blocks of binary infor-mation symbols with blocks of parity-check symbols. Each check symbol is the sum of some predefined set of information symbols. It is convenient to specify the rules for constructing such symbols using generator and parity-check matrices. Generator matrices are used for encoding; parity-check matrices are used for decoding.

Coding for parity-check codes is relatively straightforward. But the implemen-tation of decoding encounters much more difficulties, especially with large code-word lengths. Therefore, the goal of many studies was to find special classes of such parity-check codes for which there is an acceptable decoding method. Codes that are simpler in terms of decoding algorithm, memory size and number of operations are preferable even if they have a higher probability of error.

The biggest difference between LDPC codes and classic parity-check codes is the decoding method. Classic block codes are usually decoded using maximum likelihood algorithms. Maximum likelihood decoding has been described in the pre-vious sections. This method is convenient and minimizes the probability of error. However, the practical use of decoders using the maximum likelihood method and comparing the received word with all codewords presents great difficulties if the length of the codeword is large. This is because the size of the decoding table grows exponentially with increasing block length.

LDPC codes are decoded iteratively using a graphical representation of the parity-check matrix. Therefore, when creating codes with a low density of parity checks, the focus of attention is on the properties of the parity-check matrix.

LDPC codes, often called Gallager's codes after their inventor [11, 12], are based on an ingenious mathematical idea. The parity-check matrix of LDPC check codes consists mainly of zeros and contains a small number of ones. LDPC codes in the classical version proposed by Gallagher are a code whose parity-check matrix contains a small fixed number of ones c in each column and a small fixed number of ones r in each row. The size of the matrix is $m \times n$ (m rows and n columns). Such a LDPC code in which the block length is n is called an (n,c,r)-code.

The parity-check matrix of classical linear codes has linearly independent rows. This provision may not be completely true for LDPC codes. In this sense, the matrix of a LDPC code may not be a parity-check matrix in the classical sense. However, the simplicity of the structure of the parity-check matrix makes it possible to find simple decoding algorithms. These algorithms remain efficient at very large codeword lengths. Gallagher found that for a typical (n,c,r)-code with a low density of parity checks at $c \geq 3$, the minimum distance increases linearly with increasing block length n at constant c and r [12].

6.5.2 Decoding principles for LDPC Codes

The structure of the parity-check matrix proposed by Gallager is defined as follows [12]. The parity-check matrix is split vertically into c submatrices. Each column of each submatrix contains only one 1. In the first submatrix, all 1s are arranged in a stepwise order. Row i contains 1s in the columns numbered from $[(i − 1)r + 1]$th to irth. The rest of the submatrices are the permutations of the columns of the first one.

Table 6.20 shows the parity-check matrix for a LDPC code with parameters $n = 16$, $c = 3$, $r = 4$, i.e., for (16,3,4)-code [22]. The zeros are omitted in the table, which makes it easier to display matrix properties more clearly. In it, you can see three submatrices with sizes of 4 rows and 16 columns: from the 1st row to the 4th inclusive, from the 5th row to the 8th and from the 9th row to the 12th. In the first submatrix, groups of four 1s in each of the four rows are arranged in a stepwise order. The rest of the submatrices are permutations of the columns of the first one.

The parity-check matrix is a compact description of check equations similar to (6.35). It is essentially a system of linear homogeneous equations called parity-check equations. The set of solutions to this system of equations is a set of code words.

Gallagher proposed two methods for solving the system of equations and decoding [12]: hard and soft. In the first method, the decoder first decides on the value of each received symbol (zero or one). The parity checks are then computed. Part of the verification relationships can be fulfilled; part of the relationships can be unsatisfied. For each symbol, the number of occurrences in the parity checks that were not satisfied is determined. The values of the symbols that are contained in the largest number of unsatisfied check equations are changed/reversed. Then the parity checks are calculated again with the new changed symbol values and the symbols that are included in the largest number of unsatisfied check equations are found. The process is repeated until all the parity equations are satisfied. In this case, the decoding process ends successfully by decoding the received word. As follows from the above description, the method is iterative, the decoding process is iterative.

The second decoding method called soft is also iterative. But the input data at each stage are not the values of the symbols, but estimates of the probability that the values of the symbols are equal to one. Actually, these values represent a level of belief about the value of the symbols. The process is repeated until all check equations are satisfied. Then the decoding process ends.

It is important to note that the number of operations per symbols for each iteration does not depend on the length of the code.

Table 6.20   *Parity-check matrix for (16,3,4) low-density parity-check code*

h_{ij}	j															
i	1	2	3	4	5	6	7	8	9	10	11	12	13	14	15	16
1	1	1	1	1												
2					1	1	1	1								
3									1	1	1	1				
4													1	1	1	1
5	1				1				1				1			
6		1				1				1				1		
7			1				1				1				1	
8				1				1				1				1
9	1					1					1					1
10		1					1					1	1			
11			1					1	1					1		
12				1	1					1					1	

6.5.3   The Tanner graph

The decoding of LDPC codes is often represented graphically using the Tanner graph. Gallagher's methods using the Tanner graph are applicable to decoding other parity-check codes. To demonstrate the method using a simpler example, let us first consider decoding the (7,4) Hamming code, considered in sections 6.2.10 and 6.2.12. The parity-check matrix of the (7,4) Hamming code is presented in Table 6.12. The parity checks used in calculating the syndrome components by formulas (6.34) can be written in the form of a system of three linear homogeneous test equations (the equations are assigned numbers that will be used later):

(1) $y_1 + y_2 + y_3 + y_5 = 0$

(2) $y_2 + y_3 + y_4 + y_6 = 0$ (6.44)

(3) $y_1 + y_2 + y_4 + y_7 = 0$

LDPC codes are often represented graphically using the Tanner graph. There are two types of vertices in a Tanner graph: vertices for codeword symbols (bit vertices) and vertices for test equations (check vertices). Figure 6.9 shows fragments of the Tanner graph for the (7,4) Hamming code. The length of the codeword is $n = 7$, so there are 7 bit vertices on the graph. The numbering of the bit vertices is the same as the numbering of the bits in the codeword. There are three check vertices on the graph, which corresponds to three check equations of the code $((n - k) = 3)$. The numbering of the check vertices is the same as the numbering of the check equations in (6.44). Bit vertices are represented by round nodes. Check vertices are represented by square nodes. The edges of the graph illustrate the interaction of bit and check nodes.

The Tanner graph is two-sided. The round nodes of the graph in Figure 6.9a show the bits that go into the check equations. The edges of the graph in Figure 6.9a show

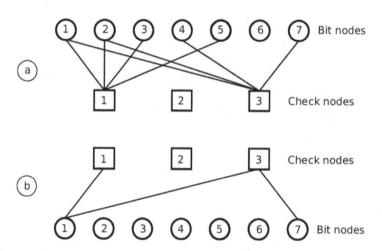

Figure 6.9   Fragments of the Tanner graph for Hamming (7,4) code (a – bit nodes for equations 1 and 3 in formulas (6.45)); b – check nodes (equations in formulas (6.45)) with bits 1 and 7)

that bits 1–3, 5 form the check set of the first check equation. Bits 1, 2, 4, 7 form the check set of the third check equation. Figure 6.9a actually illustrates the first and third rows of the check matrix of the (7,4) code (Table 6.12). The edges of the graph in Figure 6.9b show that bit 1 is included in the first and third test equations. It also illustrates that bit 7 is included in the third check equation. Figure 6.9b actually displays the first and seventh columns of the parity-check matrix of the (7,4) code (Table 6.12). The rest of the edges in Figure 6.9 are not shown so as not to clutter up the drawing and to show the main features of the Tanner graph.

Figure 6.10 illustrates the decoding of the Hamming code using the iterative method. As the received word to be decoded, the same word $y = 0000011$ that was

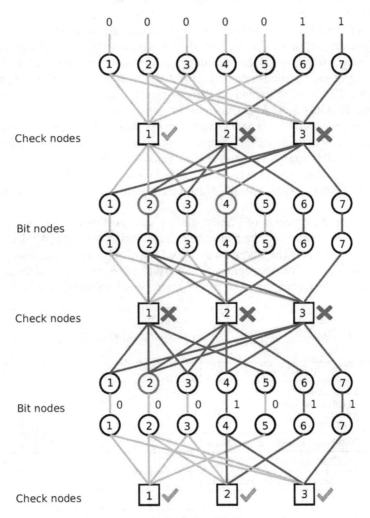

Figure 6.10   *Iterative decoding of (7,4) Hamming code word (0000011 – received word, 0001011 – decoded word)*

decoded using the most likelihood method in section 6.2.12 is used. For clarity, bit lines with zero values are shown in green, bit lines with ones are shown in blue. Inputting the received word into the decoder can be considered as an initialization of the process, after which the steps are repeated until all the equations are satisfied.

Check equations are calculated first. On the graph, this can be done by calculating the sum of the number of blue edges coming in the check nodes. If the number of incoming blue edges is even, the equation is satisfied and the parity is satisfied (this corresponds to the arithmetic of the two-element Galois field). If the number is odd, then the equation is not satisfied. Equations 6.2 and 6.3 are not satisfied (they are marked with red crosses).

The edges of the graph connecting the check nodes of the satisfied equations with the bit nodes are shown in green. The edges of the graph connecting the check nodes of the unsatisfied equations with the bit nodes are shown in blue. The number of occurrences in the check equations that were not satisfied is determined for each symbol. Bit 5 is not included in the unsatisfied equations. Bits 1, 3, 6, 7 come in the unsatisfied equations one time. Bits 2 and 4 enter the unsatisfied equations two times. This is the maximum number, so the values of bits 2 and 4 are changed/reversed. Bit nodes 2 and 4 are shown in red in this step.

The second check of the parity checks shows that not all check equations are not satisfied. Bit 2 has the maximum number of occurrences of unfulfilled relationships (bit node 2 at this stage is shown in red), its value is reversed.

The third check shows that all test equations are now satisfied. Decoding is complete. Thus, the word $x' = 0001011$ is taken as the estimation of the transmitted word.

The described process of iterative decoding, consisting of three iterations, is also shown in Table 6.21. The first stage is initialization, the last is the declaration of the decoded word. The algorithm described is a member of the message-passing algorithm class because its operation can be explained by the passing of the decoded words (messages) along the edges of the Tanner graph back and forward between

Table 6.21   Iterative decoding of (7,4) Hamming code word

Stage	Bit node number	1	2	3	4	5	6	7
1	Initialization (received word)	0	0	0	0	0	1	1
2	Checked word	0	0	0	0	0	1	1
	Number of unsatisfied equations	1	2	1	2	0	1	1
3	Checked word	0	1	0	1	0	1	1
	Number of unsatisfied equations	2	3	2	2	1	1	1
4	Checked word	0	0	0	1	0	1	1
	Number of unsatisfied equations	0	0	0	0	0	0	0
5	Decoded word	0	0	0	1	0	1	1

the bit and check nodes iteratively. This algorithm is also called bit-flipping decoding because the messages passing are binary.

The graph of Figure 6.10 and Table 6.21. should be considered only as an illustration of decoding using the iterative method using a simple example. The (7,4) code is not a LDPC code. There are cycles in the graph that do not allow decoding of all possible received words. The algorithm 'loops' if an error occurs in one of the bits of the word carrying check symbols ($y5$, $y6$, $y7$). A modification of the coding algorithm would have to be performed to eliminate cycles.

6.5.4 *Iterative decoding algorithm for LDPC codes*

Consider the decoding process using the hard iterative method using the example of a (16,3,4) LDPC code, the parity-check matrix of which is given in Table 6.20. The parity-check equations can be written in the form of a system of 12 linear homogeneous equations:

$$
\begin{aligned}
&(1) && y_1 + y_2 + y_3 + y_4 = 0 \\
&(2) && y_5 + y_6 + y_7 + y_8 = 0 \\
&(3) && y_9 + y_{10} + y_{11} + y_{12} = 0 \\
&(4) && y_{13} + y_{14} + y_{15} + y_{16} = 0 \\
&(5) && y_1 + y_5 + y_9 + y_{13} = 0 \\
&(6) && y_2 + y_6 + y_{10} + y_{14} = 0 \\
&(7) && y_3 + y_7 + y_{11} + y_{15} = 0 \\
&(8) && y_4 + y_8 + y_{12} + y_{16} = 0 \\
&(9) && y_1 + y_6 + y_{11} + y_{16} = 0 \\
&(10) && y_2 + y_7 + y_{12} + y_{13} = 0 \\
&(11) && y_3 + y_8 + y_9 + y_{14} = 0 \\
&(12) && y_4 + y_5 + y_{10} + y_{15} = 0
\end{aligned}
$$

$$(6.45)$$

Figure 6.11 shows fragments of the Tanner graph that can be used to decode the LDPC code. Each check vertex is connected by edges with four bit nodes, since there are four ones in each row of the parity-check matrix. Figure 6.11a displays the 1st and 12th check equations, which correspond to the 1st and 12th rows of the parity-check matrix (Table 6.20). The edges of the graph connect check vertices 1 and 12 with bit nodes that are included in the 1st and 12th equations. A fragment of the graph in Figure 6.11b shows which check equations include the 1st and 16th bits. Each bit node has three edges, since there are three ones in each column of the matrix.

The complete graph for decoding the codewords is shown in Figure 6.12. Figure 6.12a illustrates all the check equations (from 1 to 12) that are displayed on the graph of the rows of the check matrix. The edges of the graph connect check vertices with bit nodes that are included in the check equations. The graph in Figure 6.12b shows in which parity checks the bits are included in. Figure 6.12b

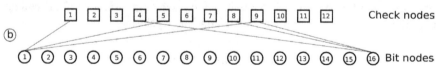

Figure 6.11   *Fragments of the Tanner graph for (16,3,4) LDPC code (a – bit nodes for equations 1 and 12 in formulas (6.46)); b – check nodes (equations in formulas (6.46)) with bits 1 and 16)*

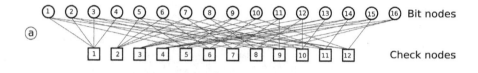

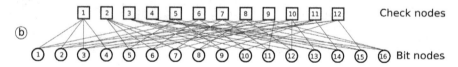

Figure 6.12   *Tanner graph for (16,3,4) LDPC code (a – bit nodes for all the equations in formulas (6.46) (from 1 to 12); b – check nodes (equations in formulas (6.46)) with bits 1 through 16)*

is a mirror image of Figure 6.12a. They both display the same check matrix. But the graph in Figure 6.12a is convenient for the stage at which the rows of the check matrix are used and the check equations are calculated. The graph in Figure 6.12b is convenient for the stage at which the columns of the check matrix are used and the number of unsatisfied check equations is calculated for each bit nodes.

The decoding algorithm is implemented on the basis of the graph in Figure 6.12 for two examples of the received word. But the execution of the process would be difficult to trace in a form similar to the scheme in Figure 6.10, with the number of bit vertices equal to 16 and the number of check vertices equal to 12. The stages of implementation are shown in Tables 6.22 and 6.23.

The calculation results of example 1 at each stage of the algorithm implementation are presented in Table 6.22 for the received word $y = 1000010110100100$.

Table 6.22 *Iterative decoding of (16,3,4) LDPC code word 1*

Stage	Bit node number	1	2	3	4	5	6	7	8	9	10	11	12	13	14	15	16
1	Initialization (received word)	1	0	0	0	0	1	0	1	1	0	1	0	0	1	0	0
2	Checked word	1	0	0	0	0	1	0	1	1	0	1	0	0	1	0	0
2	Number of unsatisfied equations	2	1	3	2	0	1	1	2	1	0	2	1	1	2	2	3
3	Checked word	1	0	1	0	0	1	0	1	1	0	1	0	0	1	0	1
3	Number of unsatisfied equations	0	0	0	0	0	0	0	0	0	0	0	0	0	0	0	0
4	Decoded word	1	0	1	0	0	1	0	1	1	0	1	0	0	1	0	1

Table 6.23   *Iterative decoding of (16,3,4) LDPC code word 2*

Stage	Bit node number	1	2	3	4	5	6	7	8	9	10	11	12	13	14	15	16
1	Initialization (received word)	1	0	0	0	0	1	0	1	1	0	1	0	0	0	0	1
2	Checked word	1	0	0	0	0	1	0	1	1	0	1	0	0	0	0	1
	Number of unsatisfied equations	1	2	2	1	0	1	1	0	0	1	1	0	1	2	2	1
3	Checked word	1	1	1	0	0	1	0	1	1	0	1	0	0	1	1	1
	Number of unsatisfied equations	1	3	2	2	1	1	2	0	0	2	1	1	2	2	3	1
4	Checked word	1	0	1	0	0	1	0	1	1	0	1	0	0	1	0	1
	Number of unsatisfied equations	0	0	0	0	0	0	0	0	0	0	0	0	0	0	0	0
5	Decoded word	1	0	1	0	0	1	0	1	1	0	1	0	0	1	0	1

Calculation of the parity checks at stage 2 after initialization shows that all bits except the fifth and tenth are included in the unsatisfied check equations. Bits 3 and 16 enter such unsatisfied check equations the maximum number of times – three each (these vertices are highlighted in blue at this stage, and the changed bit values are highlighted in red). Their values are flipped. After the next stage, all equations are satisfied, therefore, $x' = 1010010110100101$ is taken as an estimate of the sent codeword.

The calculation results of example 2 at each stage of the algorithm implementation are presented in Table 6.23 for the received word $y = 1000010110100001$. Calculation of the parity checks at the first stage shows that bits 2, 3, 14, 15 enter the unsatisfied check equations the largest number of times – two times. Their values are changed. After the second calculation of the parity checks, it turns out that bits 2 and 15 enter the unsatisfied equations the maximum number of times – three times. Their values are flipped. After the third stage, all equations are satisfied. As an estimate of the sent codeword, $x' = 1010010110100101$ is taken.

Above was illustrated a hard method of decoding. It shows well the principle of the iterative decoding approach. A feature of the iterative approach is the ability to perform parallel operations with the symbols of the received block, which allows you to achieve more in comparison with other methods of performance. It should also be noted that the number of operations per symbol in the iterative decoding of LDPC codes grows at most logarithmically with the block length.

6.6 Error correction in digital television systems

The most advanced error correction coding systems in use today are the second-generation DVB systems (DVB-T2, DVB-S2, DVB-C2) and the third-generation ATSC system ATSC 3.0. These coding systems are called forward error correction (FEC) systems in the standards. There are a lot of similarities between the coding algorithms in these standards. Next, there will be details for the second generation of DVB systems.

Coding for error correction in all digital television systems is performed by using two coding systems: outer and inner coding. For outer coding, these systems use BCH codes. For inner coding, LDPC codes are used. Both outer and inner coding is systematic. A block of data called Base Band Frame (BBFrame) is encoded for error correction in digital television systems. The BBFrame data block is processed to make outer BCH encoding first. The BCH code block is then encoded by inner LDPC coding system.

The sizes of information and code blocks for outer and inner coding are given in Table 6.24 for the normal FEC output block size of 64,800 bits (normal FECFRAME). Normal mode provides the highest possible transmission quality. A variant of the so-called short frame (short FECFRAME) is also provided, which can be used in cases where a low latency is required even at the cost of degrading the transmission quality. Table 6.24 data are given for different modes, the main parameter of which is the rate of the inner code.

Table 6.24   Coding parameters for DVB-T2/S2/C2 (normal FECFRAME n_{ldpc} = 64,800)

LDPC code R_{ldpc}	BCH information block k_{bch}	BCH code block n_{bch} LDPC information block k_{ldpc}	BCH t-error correction	$n_{bch} - k_{bch}$	LDPC code block n_{ldpc}
1/2	32,208	32,400	12	192	64,800
3/5	36,688	38,880	12	192	64,800
2/3	43,040	43,200	10	160	64,800
3/4	48,408	48,600	12	192	64,800
4/5	51,648	51,840	12	192	64,800
5/6	53,840	54,000	10	160	64,800

Table 6.25   BCH polynomials (for normal FECFRAME n_{ldpc} = 64,800)

$g_1(z)$	$1 + z^2 + z^3 + z^5 + z^{16}$
$g_2(z)$	$1 + z + z^4 + z^5 + z^6 + z^8 + z^{16}$
$g_3(z)$	$1 + z^2 + z^3 + z^4 + z^5 + z^7 + z^8 + z^9 + z^{10} + z^{11} + z^{16}$
$g_4(z)$	$1 + z^2 + z^4 + z^6 + z^9 + z^{11} + z^{12} + z^{14} + z^{16}$
$g_5(z)$	$1 + z + z^2 + z^3 + z^5 + z^8 + z^9 + z^{10} + z^{11} + z^{12} + z^{16}$
$g_6(z)$	$1 + z^2 + z^4 + z^5 + z^7 + z^8 + z^9 + z^{10} + z^{12} + z^{13} + z^{14} + z^{15} + z^{16}$
$g_7(z)$	$1 + z^2 + z^5 + z^6 + z^8 + z^9 + z^{10} + z^{11} + z^{13} + z^{15} + z^{16}$
$g_8(z)$	$1 + z + z^2 + z^5 + z^6 + z^8 + z^9 + z^{12} + z^{13} + z^{14} + z^{16}$
$g_9(z)$	$1 + z^5 + z^7 + z^9 + z^{10} + z^{11} + z^{16}$
$g_{10}(z)$	$1 + z + z^2 + z^5 + z^7 + z^8 + z^{10} + z^{12} + z^{13} + z^{14} + z^{16}$
$g_{11}(z)$	$1 + z^2 + z^3 + x^5 + z^9 + z^{11} + z^{12} + z^{13} + z^{16}$
$g_{12}(z)$	$1 + z + z^5 + z^6 + z^7 + z^9 + z^{11} + z^{12} + z^{16}$

The outer BCH code is capable of correcting 12 or 10 errors in the BCH code-word. The length of the parity-check data block is 192 or 160 bits, respectively. The size of the information block for outer coding k_{bch} varies widely depending on the mode (from 32,208 to 53,840 bits).

The generator polynomial of the BCH code is obtained by multiplying the first t polynomials of Table 6.25 corresponding to the normal size of the FEC output block (t is the number of the errors corrected by this BCH code).

A code block of the BCH code of length n_{bch} is then encoded by the LDPC code at different rates. R_{ldpc} ranges from ½ to 5/6. The code block size of the

LDPC code in different modes is one and the same and is equal to $n_{ldpc} = 64,800$ bits for the normal size of the FEC system output block (normal FECFRAME).

LDPC codes are specified using a parity-check matrix. But to achieve high efficiency, the length of the codeword must reach tens of thousands bits. With such a length, the problem arises of describing the code and performing the encoding and decoding operation. Therefore, a number of restrictions are imposed on the structure of the parity-check matrix.

For encoding, you need to have a generator matrix of the code. For any linear code, the generator matrix can be obtained from the parity-check matrix using matrix transformations. But the generator matrix obtained in this way no longer has a low density of ones. This leads to difficulties in the coding process. Therefore, additional restrictions are imposed on the parity-check matrix of the code. The parity-check matrix with the of size $(n - k)$ rows and n columns must consist of two blocks, called matrix A with the size of $(n - k)$ rows and k columns, and a square matrix B, having $(n - k)$ rows and $(n - k)$ columns:

$$H_{(n-k) \times n} = \left[A_{(n-k) \times k} \vdots B_{(n-k) \times (n-k)} \right].$$

Matrix B is a bit like the identity matrix, but has a special staircase lower triangular structure (Table 6.26). The matrix has ones along the central diagonal and zeros in the regions above and below [7].

The block of information $\boldsymbol{u} = (u_1, u_2, ..., u_k)$, coded for error detection and correction using the LDPC code, is converted into a code block using systematic coding: $x = (x_1, x_2, ..., x_n) = (u_1, u_2, ..., u_k, p_1, p_2, ..., p_{n-k})$.

The product of the code word by the transposed parity-check matrix must be equal to zero. Therefore, the check bits are found from the equation:

$$xH^T = 0.$$

The parity-check bits are found using a recursive algorithm to solve the above equation for the parity-check bits:

$$a_{11}u_1 + a_{12}u_2 + \cdots + a_{1k}u_k + p_1 = 0 \quad \dashrightarrow \quad p_1.$$

Table 6.26   *Structure of parity-check matrix B of DVB-T2/S2/C2 LDPC code*

1						
1	1					
	1	1		0		
		1	.			
		.	.			
		.	.	1		
0				1	1	
					1	1

$$a_{21}u_1 + a_{22}u_2 + \cdots + a_{2k}u_k + p_1 + p_2 = 0 \quad \text{-->} \quad p_2$$

.....

$$a_{(n-k)1}u_1 + a_{(n-k)2}u_2 + \cdots + a_{(n-k)k}u_k + p_{n-k-1} + p_{n-k} = 0 \quad \text{-->} \quad p_{n-k}.$$

The constraints imposed on the parity-check matrix H lead to a very small increase in the threshold signal-to-noise ratio (by 0.1 dB). The parity-check matrix of the code is not directly written in the standard. It is defined by special formulas that allow calculating the numbers of bit nodes and numbers of check nodes that are connected by the edges of the Tanner graph.

6.7   Conclusions

The channel coding systems of DVB-2 and ATSC 3.0 made it possible to achieve a Near Shannon limit performance and provide quasi-error-free delivery of television programs. These indicators are the facts of today. Time will pass, new coding systems will be proposed that will allow to surpass these indicators. The facts of today will be forgotten, just as today we are beginning to forget the channel coding performance of first-generation DVB and ATSC television. And that's okay. Claude Adrien Helvétius said: 'Knowledge of certain principles easily supplies the knowledge of certain facts'. Knowledge of what principle must remain with us in order to understand the facts of the next generation of digital television broadcasting systems? This principle is the theory of channel coding, which has become a tool for designing digital systems and a component of scientific and technical culture.

References

[1]   ETSI EN 302 755: "Digital Video Broadcasting (DVB); Frame structure channel coding and modulation for a second generation digital terrestrial television broadcasting system (DVB-T2)".

[2]   ETSI EN 302 307-1: "Digital Video Broadcasting (DVB); Second generation framing structure, channel coding and modulation systems for Broadcasting, Interactive Services, News Gathering and other broadband satellite applications; Part 1: DVB-S2".

[3]   ETSI EN 302 769: "Digital Video Broadcasting (DVB); Frame structure channel coding and modulation for a second generation digital transmission system for cable systems (DVB-C2)".

[4]   ATSC: "ATSC standard: ATSC 3.0 system," Doc. A/300:2022-04, Advanced Television Systems Committee, Washington, DC. 2022.

[5]   DVB Fact Sheet. Second Generation Terrestrial – The World's Digital Terrestrial TV Standard. Produced by the DVB project office – dvb@dvb.org. 2016

[6]   Shannon C.E. 'A mathematical theory of communication'. *Bell System Technical Journal*. 1948, vol. 27, pp. 379–423, 623–656.

[7]   Eroz M., Sun F.-W., Lee L.-N. 'DVB-S2 low-density parity-check codes with near shannon limit performance'. *International Journal of Satellite Communications and Networking.* 2004, vol. 22(3), pp. 269–79.

[8]   DVB Fact Sheet. 2nd Generation Cable – The World's Most Advanced Digital Cable TV System. Produced by the DVB project office – dvb@dvb.org. 2012.

[9]   Peterson W.W., Weldon E.J. *Error-correcting codes.* Second edition. Cambridge, Massachusetts, and London, England: The MIT Press; 1972.

[10] Blahut R.E. *Theory and practice of error control codes.* Addison-Wesley Publishing Company, Inc; 1983.

[11] Gallager R.G. 'Low-density parity-check codes'. *IEEE Transactions on Information Theory.* 1962, vol. 8(1), pp. 21–28.

[12] Gallager R.G. *Low-density parity-check codes.* Cambridge, Massachusetts: M.I.T. Press; 1963. Available from https://direct.mit.edu/books/book/3867/Low-Density-Parity-Check-Codes

[13] MacKay D.J.C., Neal R.M. 'Near shannon limit performance of low-density parity-check codes'. *Electronics Letters.* 1997, vol. 33(6), pp. 457–58.

[14] ATSC: "ATSC standard: physical layer protocol," Doc. A/322:2022, Advanced Television Systems Committee, Washington, DC. 2022.

[15] ATSC: "ATSC recommended practice: guidelines for the physical layer protocol," Doc. A/327:2022, Advanced Television Systems Committee, Washington, DC. 2022.

[16] Gallager R.G. *Information theory and reliable communication.* John Wiley and sons; 1968.

[17] Huffman W.C., Pless V. *Fundamentals of error-correcting codes.* Cambridge University Press; 2003.

Chapter 7

Compression

Konstantin Glasman[1]

7.1 Data compression

7.1.1 Data compression basics

Digital communication systems use discrete signals for transmitting messages that take a finite set of values. The need to use discrete signals is determined by the fact that all digital systems have a limited speed and can only work with finite amounts of data presented with finite accuracy. Discrete representation is natural when displaying messages in the form of sequences of discrete characters selected from a finite alphabet. Examples of such discrete messages, called data, are telegrams, letters, number sequences, or machine codes.

Primary ideas and concepts of data compression are easy to explain using text messages. If you enter a space character to indicate the space between words in addition to the 26 letters of the English alphabet, you get an alphabet of 27 characters (we denote the number of characters of the alphabet as M). From the standpoint of information theory, the text is generated by a message source having M states with the numbers $m = 1, 2, ..., M$. A binary code of K binary symbols can be assigned to each source symbol by writing the state number m in the binary representation. The integer K is found from the condition: $M \leq 2^K$. Each letter of the text message alphabet, including the space character, can be encoded with a combination of $K = 5$ binary characters. A message of N letters will require $K * N = 5 * N$ binary characters. But whether this message will bring $5 * N$ binary units of information, or bits (the term "bit" is multivalued—"bit" means both a binary symbol and a binary unit of measure for the amount of information).

The average information per message symbol is called entropy H in the information theory [1]. If the characters in the messages are independent, the entropy is calculated as $H = \Sigma(-p_m * \log_2 p_m)$, where p_m is the probability that the letter m appears in the message, and summation is performed over all the characters of the alphabet. As you can see, the entropy is determined by the statistical properties of the message source. In the case of equiprobability and independence of letters, the entropy $H_0 = \log_2 M = \log_2 27 = 4.75$ bits/symbol.

But in messages in English, as well as in any other, letters appear with a different probability. The most probable letters are e, t, and a, and the least probable are x, q, and z. Entropy, calculated taking into account the different probability of

[1] Television Department, St. Petersburg Electrotechnical University (LETI), Russia

the appearance of letters, gives the value of $H_1 = 4.1$ bits/symbol. This means that the description of each letter of five binary digits is redundant. The message of N symbols actually contains $4.1 * N$ bits of information with entropy $H_1 = 4.1$ bits/symbol. To reduce the volume of a message (to compress the message), one can use short code words for the most probable letters and long code words for the least probable letters. Such coding, which takes into account the statistical properties of the symbols of the alphabet and allows for the presentation of a message with a lower consumption of binary symbols, is often called entropy coding.

The statistical structure of the language is complex. In the texts, one can notice certain frequently occurring multi-letter combinations reflecting correlations between the letters of the texts. For example, the letter T is most likely followed by the letter H, and the letter Q is often followed by U. If we take into account the probabilities of two-letter combinations, then the entropy becomes $H_2 = 3.5$; taking into account the three-letter combinations, the entropy decreases to $H_3 = 3.0$ bits/symbol. If we confine ourselves to eight-letter combinations, then we can estimate the entropy of meaningful text as $H = H_8 = 2.3$ bits/symbol.

The coefficient $R_S = (H_0 - H)/H_0 = 1 - (H/H_0)$ can characterize the redundancy of the message source. It shows how effectively alphabet characters are used by this source. The coefficient R takes values in the range from 0 to 1. The smaller R, the greater the amount of information generated by the source per letter. It is equal to 0 (there is no redundancy) if $H = H_0$. This may be the case if all the characters of the alphabet are independent and equiprobable. The entropy H at the same time takes the maximum possible value H_0. Each message symbol carries H_0 bits of information. The coefficient R tends to 1 (or 100%) if $H \ll H_0$, which means low information content and a high degree of predictability of the text of this message. This may be the case if the symbols of the alphabet are characterized by significantly different probabilities of the appearance in the text messages, e.g., when one symbol of the alphabet is much more probable than others. This can also be with close probabilities of the appearance of individual symbols, if in the texts of messages one can find strong correlation links, manifested in the frequent repetition of certain multi-letter combinations. This circumstance allows for the possibility of predicting the part of the message text that has not yet been received (or at least its fragment) on the basis of the part of the message already received. As can be seen, the coefficient R shows by what percentage the length of the message can be reduced if economical coding is applied.

The coefficient $R_C = (K - H)/H = (K/H) - 1$ can characterize the redundancy of the code representation of messages. The parameter K defines the bit length per symbol, i.e., the number of bits in the symbol code of the alphabet. In the above example, the primary uniform binary coding assumes the parameter $K = 5$ for all symbols of the alphabet. If variable-length coding is used, then to estimate the redundancy of the code representation, it is necessary to find the average value of the bit length per symbol.

A message of N characters contains $H * N$ information. With uniform binary encoding, its length will be $K * N$ binary characters. Economical coding, taking into account the probability of occurrence of symbols and combinations, allows, in principle, to reduce the length of the message to $H * N$ binary digits. The value $C_R = (K * N)/(H * N) = K/H$ can be called the maximum possible compression ratio.

It should be borne in mind that, determining the amount of information in messages and evaluating their redundancy, we consider only the statistics of the appearance of individual symbols of the alphabet and their combinations and do not affect the semantic and pragmatic aspects of information.

As follows from the above reasoning, the redundancy of the English text exceeds 50%. The redundancy of other common European languages is also great. However, this is often helpful. It is the redundancy of the language that makes it possible to understand the speech of a foreigner, speak with accent and errors, edit the text if there are typos, and also restore the original message in case of missing letters and even whole words.

7.1.2 Huffman coding

In communication technology, one of the widely used methods of economical representation is the Huffman code [2]. When encoding message symbols with variable length combinations, there is also the problem of separating one combination from another. The Huffman code is prefix, i.e., none of its code combinations are the beginning of another combination, which makes it possible to dispense with delimiters between the combinations in the text of a coded message. The use of a Huffman code reduces the length of a message from $K * N$ binary digits (with uniform binary encoding) to almost $H_1 * N = 4.1 * N$ binary digits. To encode a single character, an average of 4.1 binary digits will be spent, and the message length will be reduced.

We illustrate this with a somewhat humorous example of encoding text messages that use only eight letters and the space character: E, C, A, H, I, M, S, T, and BL (BLANK). One of these messages is "MICE EAT CHEESE." To evenly encode these letters and the space character, you must use a combination of four binary digits. In this case, to encode a phrase of 15 characters (letters and spaces), 60 binary digits are required. But spending four binary digits per letter does not mean that each binary combination carries 4 bits of information. If we take the phrase "MICE EAT CHEESE" as a typical representative of texts, then it is possible to estimate the probability of occurrence of each letter (Figure 7.1). Entropy calculation gives a value of

		Probabilities							
		Reduced sets							
	S_0	S_1	S_2	S_3	S_4	S_5	S_6	S_7	S_8
E	0.32	0.32	0.32	0.32	0.32	0.32	0.40	0.60	1.00
BL	0.13	0.14	0.14	0.14	0.26	0.28	0.32	0.40	
C	0.13	0.13	0.14	0.14	0.14	0.26	0.28		
A	0.07	0.13	0.13	0.14	0.14	0.14			
H	0.07	0.07	0.13	0.13	0.14				
I	0.07	0.07	0.07	0.13					
M	0.07	0.07	0.07						
S	0.07	0.07							
T	0.07								

Figure 7.1   Coding table of the Huffman code

2.9 bits/symbol. It should be borne in mind that the entropy value of this example does not apply to texts composed of all 27 characters.

The Huffman code is formed in the process of successive compression and splitting of the set of characters. An example of the formation of the Huffman code is shown in Figure 7.1. The set of characters (E, C, A, H, I, M, S, T, and BL) is in column S_0. All symbols are arranged in descending order of their probabilities. Each of the sets S_1–S_8 is obtained by compressing the previous set. The two lower ones, i.e., the least probable symbols are combined into one, to which the total probability is attributed. Digit 0 is assigned (as the symbol of the code word) to the upper symbol of this association, and digit 1 is assigned to the lower one. The union occupies the position corresponding to its general probability in the new grouping of symbols. The two least probable characters are again combined according to the same rules. This process is repeated until the merger results in a total probability equal to one (S_8 set).

To compile the code of each source symbol, it is necessary to trace the path to this symbol from the last grouping. This tracking is conveniently performed using a code tree (Figure 7.2). From the point corresponding to probability 1, two branches are sent, and the branches with the greater probability are to be assigned the value 0, and the branches with the lesser probability are to be assigned the value 1. Such splitting (or branching) continues until the symbol of the original set of symbols is reached. Moving through the code tree from the bottom up, you can write for each symbol the corresponding code combination.

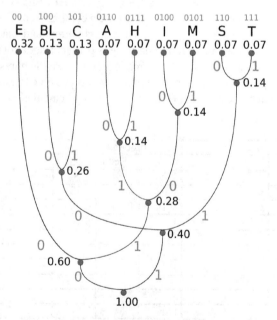

Figure 7.2   *Code tree of the Huffman code*

The resulting code is optimal. To send the message "MICE EAT CHEESE," consisting of 15 letters, using the contained code combinations, 44 binary digits are required. A ratio of 44 binary digits to 15 letters of the text gives a value that is 2.93 bits/symbol. This value is only slightly larger than the entropy H_1, equal to 2.9 bits/symbol. The redundancy of the code representation is equal to only 1%: $R_C = (K - H)/H = (K/H) - 1 = 0.01$.

7.1.3   Run length coding

The Huffman coding described above is ineffective if one of the symbols generated by the message source has a probability close to one. A typical example of such messages is, e.g., the video signal of contour images in the form of white lines on a black background. In this case, an easy-to-implement method for run length encoding (RLE) is often used [3, 4]. Consider the basic principles of encoding RLE using the example of a binary message source, the alphabet of which consists of two characters: B and W, and the probability of the symbol W is much less than the probability of the symbol B. An example of such a coding is given in Table 7.1.

The length of the codeword determines the maximum length of the message block. When the length of the code word is 3, the message block has a length not exceeding seven characters. There is at most one W character in each block. As you can see, the code word is numerically equal to the length of the B character series in the block. With the help of such blocks, you can describe any message of infinite length. If the message length is finite, the last character of the message must be the W character [or after the last W character indicating the contour of the image to be encoded, some special character must be inserted—the end of the block (EOB), which appears only once in the message code].

Consider the message of a binary source of the form: BBBBWBBBBBBB BBBBBBBBW, whose length is equal to 20 characters. Assuming that the probability of the appearance of the symbol B is 0.9 and the probability of the symbol W is 0.1, we find the source entropy: $H = \Sigma(-p_m * \log_2 p_m) = -(0.9 * \log_2 0.9 + 0.1 * \log_2 0.1) = 0.47$.

Table 7.1   *Source coding by RLE code*

Block of symbols	Length of block	Code word
W	1	000
BW	2	001
BBW	3	010
BBBW	4	011
BBBBW	5	100
BBBBBW	6	101
BBBBBBW	7	110
BBBBBBB	7	111

The use of Huffman coding in the version described above would give a trivial, but the only possible code, in which one character of the alphabet corresponds to the bit 0 of a single-bit code word, and the second one—bit 1. The length of the coded message will be equal to 20 bits, which corresponds to the bit length per symbol $K = 1$ bit/symbol, and the redundancy ratio of the code representation of the message is equal to $R_c = 1.13 = 113\%$.

Using the RLE encoding rules described above, we write the message as a sequence of blocks of variable length: BBBBW BBBBBBB BBBBBBB W. Representing each block as fixed-length code words in accordance with Table 7.1, we get the message encoded with the RLE code: 100 111 111 000. Now the average number of code bits per message symbol is $K = 12$ bits/20 symbols = 0.61 bit/symbol, which gives the redundancy ratio of the code representation of the message $R_c = 0.29 = 29\%$.

There are a large number of options for the practical implementation of the method of RLE. For example, if a text message is transmitted: AAAAAAABBBBB BBBBBCCCCCCCC, in which there are three series of identical characters: seven characters A, nine characters B, eight characters C, the message size is 24 bytes when encoding individual letters using ASCII code. The use of run length coding allows to reduce the amount of data by encoding each run in pairs (run length, character code): (7, A), (9, B), (8, C). The coded message is recorded as a sequence of 6 bytes: 07 65 09 66 08 67 (instead of the characters A, B, and C, their ASCII codes are used). The use of the RLE code has reduced the message length by four times.

If the message text contains not only long series of identical symbols but also groups of different symbols, then the application of the RLE code in the described version may not be effective. For example, applying coding of lengths of series to a message of 24 bytes AAAAAAAAADBDBDBCDCDCCCCC in length and recording each series in pairs (run length, character code), we get (9, A), (1, D), (1, B), (1, D), (1, B), (1, D), (1, B), (1, C), (1, D), (1, C), (1, D), (5, C). Writing down the character codes, we get a coded message with a length of 24 bytes: 09 65 01 68 01 66 01 68 01 66 01 68 01 66 01 67 01 68 01 67 01 68 05 67. If there are many groups of different characters, the coding of the lengths of the runs can lead to the so-called negative compression, when the length of the coded message is more than the original text. A possible solution that improves coding efficiency in this case is to use a special byte, or a flag (this can be, e.g., byte 255), which signals the start of a series of duplicate characters and the use of run length coding for this series. In this case, the encoding allows you to achieve data compression and bring the length of the encoded message: 255 09 65 68 66 68 66 66 67 68 67 68 255 05 67 to 16 bytes. In any case, the effectiveness of the method of run length coding is higher, if the proportion of long series of identical characters in the text of the message being encoded is greater.

7.1.4 Arithmetic coding

In the Huffman method, each symbol of the encoded sequence is assigned to its own codeword, which is determined from the coding table. Therefore, the bit length per symbol cannot be less than $K = 1$ bit/symbol. This means that the redundancy of the code will be great if the entropy of the source of the message is less than one binary digit of information per symbol. To improve efficiency, you can combine source symbols into blocks, thereby increasing the number of symbols. For example, the symbols B and W, generated by the message source, can be combined in pairs, moving to a new alphabet of four symbols: (B, B), (B, W), (W, B), and (W, W) and creating a Huffman code for it. The greater the number of symbols in a block, the higher the coding efficiency will be; however, the complexity of the encoding increases exponentially.

An effective alternative to the Huffman code is arithmetic coding [3–5]. The essence of arithmetic coding is that a sequence of symbols created by the source of messages is put into correspondence with a number that uniquely identifies this sequence of symbols. The whole coded sequence consumes, of course, an integer number of bits; however, the average bit length per source symbol can be less than $K = 1$ bit/symbol.

Consider the principle of arithmetic coding by example. The alphabet of the message source consists of three symbols: A, B, and C, and the probability of occurrence of symbols is equal to 0.4, 0.2, and 0.4, respectively. On a known probability distribution, the probability intervals for each symbol are found. The interval $[0,1)$ is divided into three non-intersecting intervals whose lengths are equal to the probabilities of the symbols Table 7.2.

Let the "CABAC" sequence of symbols be encoded. The arithmetic coding procedure begins with the establishment of the initial current interval $[0.0; 1.0)$ at stage 0 (Table 7.3). Obtaining the first symbol of the encoded sequence of symbols, the encoder splits the current interval $[0.0; 1.0)$ for segments proportional to the

Table 7.2   Probability distribution and probability intervals for source symbols (arithmetic coding)

Symbol	Probability	Interval
A	0.4	$[0.0; 0.4)$
B	0.2	$[0.4; 0.6)$
C	0.4	$[0.6; 1.0)$

Table 7.3   Arithmetic coding of sequence "CABAC"

Stage	Symbol	Start of interval	Length of interval	End of interval
0	–	0.0	1.0	1.0
1	C	0.6	0.4	1.0
2	A	0.60	0.16	0.76
3	B	0.664	0.032	0.696
4	A	0.6640	0.0128	0.6768
5	C	0.67168	0.00512	0.67680

probabilities of the symbols and selects the interval [0.6; 1.0), corresponding to the first symbol of the message "C." This procedure is reflected as step 1 in Tables 7.3. The width of the current interval in step 1 is 0.4.

At stage 2, the encoder splits the interval [0.6; 1.0) for segments that are proportional to the probabilities of the symbols and selects the interval [0.60; 0.76), corresponding to the second symbol of message "A." The encoder sequentially performs these actions for each symbol of the message being encoded.

The coding process is illustrated in Table 7.3 and Figure 7.3. As can be seen, each next current interval is nested in the previous interval. After encoding all symbols of the message, the final interval is obtained [0.67168; 0.67680). As a result of arithmetic coding, you can take from this interval any number containing the least number of digits, e.g., $A_C = 0.672$. Since digital systems use binary representation of numbers, the result of arithmetic coding should be a number containing the smallest number of binary digits: $A_C = 0,a_1a_2a_3...a_i$, where $a_1a_2a_3...a_i$ can be 0 or 1. The number of A_C must satisfy the system of their two inequalities: $A_C \geq 0.67168$ and $A_C < 0.67680$.

To recover the encoded sequence of symbols on the side of the decoder, the probability distribution and probability intervals of the source symbols must be known. Initially, the current interval [0; 1) is selected. This current interval is divided in proportion to the probabilities of the symbols. Next, it should be defined, in the interval of which symbol is the number A_C. This gives the first symbol of the sequence. Then the process is repeated cyclically. The next decoded symbol is located, and

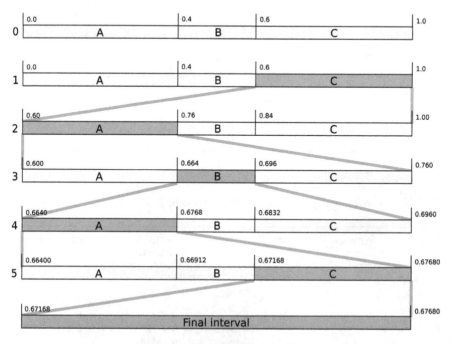

Figure 7.3   Arithmetic coding

its probability interval is selected as the next current interval, etc. The length of the sequence of encoded symbols must be known in advance to the decoder. An alternative is to use a special symbol at the end of the coded sequence.

The above reasoning reveals the general ideas of the method of arithmetic coding and involves the use of a hypothetical computing device that operates with an infinite digit bit depth of the representation of numbers. The implementation of arithmetic coding on real computing devices has the specifics associated with the finite bit depth of the representation of numbers.

In most cases, the use of arithmetic coding leads to more efficient data compression compared with the above methods of Shannon-Fano, Huffman, coding of run lengths (RLE). However, this has to be paid for by the significantly greater computational complexity of the implementation of the arithmetic coding method. Nevertheless, in modern video compression systems, context-adaptive binary arithmetic coding (CABAC—context-adaptive binary arithmetic coding) is used.

7.2   Video compression

7.2.1   Video compression basics

The use of digital technologies allows us to significantly increase the accuracy of representation of such continual processes as an image. The parameters characterizing the quality of the reproduced image in digital systems for recording and transmitting signals can significantly exceed those values that were typical for analog systems. However, the introduction of digital technology creates new problems. The frequency band of digital signals is much larger than the band of their analog "predecessors." The use of broadband channels with the necessary bandwidth may be technically impossible. This may also be economically disadvantageous since the cost of the communication channel increases with increasing capacity. An effective way to solve this problem is encoding, which has the goal of a compact representation of a digital signal by compression. The development of digital television broadcasting is inextricably linked with the improvement of methods and technical means of video compression.

In modern television, the image is represented by a sequence of frames following each other with a strict periodicity determined by the frame rate. A frame is described by an array of image elements, or pixels, located at the nodes of a rectangular grid, the horizontal lines of which form TV lines (Figure 7.4). Each pixel of a color image has three components, which correspond to the luminance signal Y and two color difference signals C_R and C_B. Another possible set of color image components is the signals of the red R, green G, and blue B components of the image. Thus, from an information point of view, each image frame consists of three rectangular matrixes of image samples. The sampling structure, in which all three components of an image are sampled in the same way, is denoted as 4:4:4. The number of pixels in a line and, accordingly, the image definition are determined by the video signal sampling frequency. The sampling frequency is a harmonic of the horizontal frequency, which provides a fixed orthogonal structure of the TV image samples. In the digital representation, the value of the video signal in each pixel is expressed by

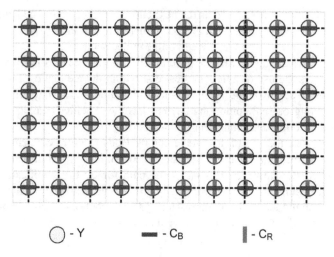

○ - Y   ▬ - C_B   ▌ - C_R

Figure 7.4   Bitmap representation of the image (4:4:4 sampling structure)

a binary code number, the number of bits of which determines the number of quantization levels and, accordingly, the accuracy of the digital representation.

The capacity of the memory required to store the data of one component of one full frame of an image can be found by multiplying: the number of bits of the code word describing the component of the image and the number of pixels in the frame. Interfaces for transmitting a digital signal involve the sequential transmission of code words that carry information about image samples [6–9]. The bit rate of digital video data passing through the interface and carrying information about a single component of a TV image can be found by multiplying the amount of data per frame and frame rate. The bit rate of the full stream of video data carrying an image component can also be found by multiplying the length of the code word describing one image sample and the sampling rate. The examples below show the requirements for the interfaces used to transmit a stream of digital video data with a sampling structure and quantization at a bit depth of 10 bits per sample, which allows up to 1024 quantization levels.

To store one full frame of standard definition digital TV image (576 active lines, 720 active elements per line, frame rate 25 Hz) with a 4:3 aspect ratio, a memory of 16.2 Mbits is necessary. This corresponds to a bit rate of 405 Mbits/s at a sampling frequency of 13.5 MHz. One full-frame digital TV image of standard definition (576 active lines, 960 active elements per line, frame rate 25 Hz) with an aspect ratio of 16:9 has a volume of 21.6 Mbits. The bit rate is 540 Mbits/s at a sampling rate of 18 MHz.

The capacity of the memory required to store the data of the active part of one frame of the image of high definition 2K (1 920 × 1 080) is 62.208 Mbits or 7.776 MB. This value increases to 31.104 MB and 124.416 MB in ultra-high-definition formats 4K (3 840 × 2 160) and 8K (7 680 × 4 320), respectively. The bit rate of video data carrying the active part of the frame in the 8K format is 49 766.4 Mbits/s = 49.7664 Gbits/s with a frame rate of 50 Hz and 119.43936 Gbits/s with a frame rate of 120 Hz.

Above were given estimates of the memory capacity and the bit rate of video data transmission in digital TV. But do the given values of the memory capacity required to store a frame correspond to the amount of information in the images? As will be shown below, the presentation of images as a sequence of frames, each of which is described by an array of pixels, is redundant for images that are typical for TV broadcasting. Redundancy means that the bit rate of video data can be reduced without compromising the accuracy of the description of the transmitted image.

In general, structural, statistical, and psychophysical redundancies are distinguished. Structural redundancy is due to the presence of blanking intervals on the line and frame. To eliminate structural redundancy, you can, e.g., reduce the clock frequency by about 25% and fill the blanking intervals with data from the active part of the lines and frame. This transformation of the data stream is a form of reducing the bit rate of video data or video compression. The blanking intervals can also be used to transmit additional information, such as audio signals, teletext, etc.

The statistical redundancy of the raster representation is associated with the properties of typical television subjects. Most of the image of a single frame usually falls on fields that have a constant or slightly varying brightness in space, and sharp light transitions and small details occupy a small fraction of the image area. This will be especially noticeable if you build an oscillogram of the interelement difference, which for the majority of the line interval is very small and only occasionally deviates significantly from zero (Figure 7.5). The correlation coefficient of neighboring image elements, describing the statistical relationship between the brightness of these elements, is close to 1. Images of neighboring frames in television are usually very similar to each other, even when shooting moving objects. Transitions from plot to plot are rare. The interframe difference over a large part of the image area is usually close to zero. The correlation coefficient of adjacent frames is also close to 1.

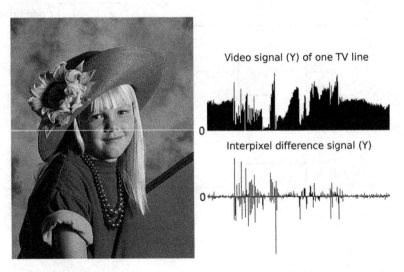

Figure 7.5   Single-line image signal and interpixel difference signal

High correlation means that the brightness and chromaticity of an element of a typical TV image can be predicted with high probability if you know the values of the pixels surrounding the predicted in space and in time. By transmitting the value of one pixel, you can refuse to transfer the next one, since the value of the transferred pixel is a good prediction for the next pixel located near the transmitted one in space or in time. It can be limited to the transmission of a prediction error (prediction residual), which is the difference between the actual value and the predicted value (Figure 7.6).

For typical images, the prediction error/residual is usually small; therefore, a smaller number of codeword bits are required for its transmission, which can reduce the bit rate of the video data transmitted. The possibility of reducing the bit rate is a consequence of the redundancy of the raster representation of images, which is manifested in connection with the statistics of typical television subjects and can be called statistical redundancy. It can be divided into spatial redundancy, which is due to correlations between the elements of one image frame, and temporal redundancy, which appears due to the high correlation of neighboring frames.

The described method of compression, or coding, in the process of which you can achieve redundancy in the representation of the image and reduce the amount of transmitted data, is known as differential pulse code modulation (DPCM) [3, 4]. The procedure for reducing the redundancy of the presentation using the DPCM is reversible. When decoding, to restore the image element, you need to add the prediction residual transmitted to the prediction—the value of the first pixel (Figure 7.6). During the encoding process, no distortions are introduced, and during decoding, the original image is completely restored. Such coding, in which the bit rate of video data is reduced without image distortion, is called lossless compression, or mathematically accurate compression.

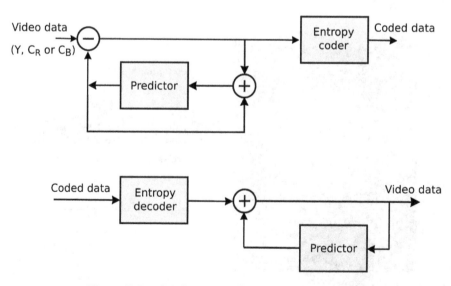

Figure 7.6   DPCM as a video compression method

An important advantage of lossless compression is that re-encoding the decoded image to eliminate redundancy does not also cause distortion. Repeated application of lossless compression does not lead to the accumulation of distortions. However, the degree of mathematically precise compression that is currently achieved, showing the proportions of the original and coded data volumes, is small; usually, it does not exceed 2:1.

Significant reserves for reducing the bit rate of the digital stream are the use of the properties of vision. The degree of compression can be increased if during the coding process, which reduces the bit rate of the data stream, some distortions are allowed, which, however, are not noticed by observers or viewers. For example, quantization noises are well distinguished by the eye in the form of false contours in areas of the image with a constant or smoothly varying brightness. However, they are hardly noticeable on sharp drops in brightness and on areas of the image with small details. Image distortions are not noticeable to the eye for a few tenths of a second after a sudden change of plot.

Introducing the quantization of the difference signal (Figure 7.7) and assuming the appearance of artifacts and distortions unnoticeable to the viewer in the vicinity of steep changes in the brightness of the image, it is possible to achieve in the coding process more significant compression of video data. The possibility

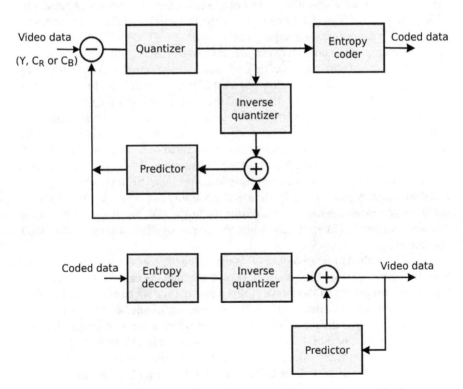

Figure 7.7   DPCM with quantization of the prediction residual

of such compression can be interpreted as the presence of psychophysical redundancy in the representation of the image, since the original and decoded images are evaluated by the viewer as equivalent, which can be regarded as psychophysically accurate transmission of video data. The degree of compression, during which psychophysical redundancy is eliminated and which can be called compression without visual loss, strongly depends on the characteristics of the image. Repeated application of such compression can lead to the accumulation of distortions.

The degree of compression can be increased to large values by introducing a coarser quantization of the difference signal, but this will lead to distortions and artifacts that are visible to the viewer. This encoding is called visually lossy compression. The degree of video compression with visual losses also depends on the characteristics of the image; it can reach tens and hundreds. Repeated application of this compression leads to accumulation of distortions.

7.2.2 Video compression algorithms

7.2.2.1 Intraframe coding

7.2.2.1.1 Discrete cosine transform

The purpose of intraframe coding is to reduce spatial redundancy within the frame (or field) of the television image. As noted above, this redundancy is caused by strong correlations between the elements of the image. One of the means of reducing the coded data stream rate using correlation links is DPCM. The difference signal at the output of the predictive encoder is characterized by smaller correlations. However, there are other means of de-correlating image samples. An effective way is to decompose the image function into a set of orthonormal basis functions using unitary transformations [3, 4]. If we find the corresponding unitary transformation, then we can convert the array of image samples into a matrix of coefficients that will be correlated with each other to a much lesser extent. As a result of the conversion, most of the image energy is concentrated in a small number of coefficients, which makes it possible to reduce the amount of data that must be used when transmitted. The complete decorrelation is provided by the Karhunen–Loève transformation. A good approximation of the Karhunen–Loève transform for typical television images is the discrete cosine transform (DCT), which allows you to transform an array of image samples into the frequency domain as an array of DCT coefficients [3, 4].

DCT is applied to small blocks of image elements, e.g., 8 × 8 (Figures 7.8 and 7.9). In Figure 7.9, four blocks are marked, for which the calculation of the DCT brightness component of the image is illustrated below. As a result of calculating the DCT, a matrix of coefficients is found—the amplitudes of the basis cosine functions (waves) of different frequencies, from which a block of image elements can be combined. For most blocks of typical images, only a small part of the coefficients has a significant value. If within the block, the brightness of the image changes little, which will occur quite often due to the high correlation of close

Figure 7.8   Image separation into blocks

Block 1 of picture samples

Figure 7.9   Blocks for which the DCT calculation is illustrated

image elements, then only the constant component and several low-frequency basis functions have significant values {the constant component [DC (direct current) component] is located in the block of DCT coefficients in the upper left corner, with black corresponding to the maximum value}. An example is block number 1 (Figure 7.10).

An image block in which the brightness changes, e.g., vertically, can be composed mainly of basic cosine waves "running" in the vertical direction. The amplitudes of such waves are given in the left column of the DCT matrix, with the amplitude of the highest-frequency wave located in the lower left corner. This case illustrates the calculation of the DCT for block 2 (Figure 7.11). An image block in which the brightness varies horizontally can be composed mainly of basic cosine waves "running" in the horizontal direction. The amplitudes of such waves are given in the upper row of the DCT matrix, with the amplitude of the highest-frequency

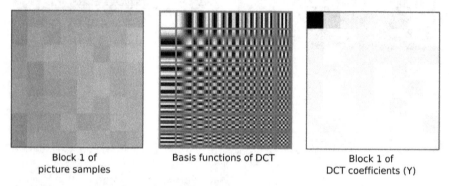

Block 1 of   Basis functions of DCT   Block 1 of
picture samples   DCT coefficients (Y)

Figure 7.10   *DCT of block 1 (component Y, black color in the block of DCT coefficients corresponds to the maximum value)*

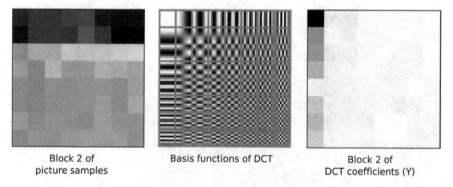

Block 2 of   Basis functions of DCT   Block 2 of
picture samples   DCT coefficients (Y)

Figure 7.11   *DCT of block 2 (component Y, black color in the block of DCT coefficients corresponds to the maximum value)*

wave located in the upper right-hand corner. This case illustrates the calculation of DCT for block 3 (Figure 7.12).

Block 4, in which there is a bright flare point on the bead (Figure 7.13), consists of cosine waves of all directions: horizontal, vertical, and diagonal (the amplitude of the highest-frequency diagonal wave is indicated in the lower right corner of the DCT matrix).

The amplitudes of the high-frequency components for many image blocks are very small or equal to zero. If the DCT coefficients are transmitted instead of the brightness values of the image elements, then the reduction in the data transfer rate can be achieved because zero coefficients can be simply excluded. At the receiving side, this reduced amount of data is sufficient to completely restore the original image block based on the transmitted transform coefficients. This method of reducing the redundancy of the video signal is based on the use of the

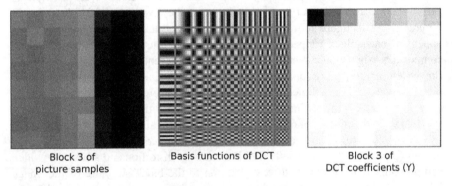

| Block 3 of picture samples | Basis functions of DCT | Block 3 of DCT coefficients (Y) |

Figure 7.12   *DCT of block 3 (component Y, black color in the block of DCT coefficients corresponds to the maximum value)*

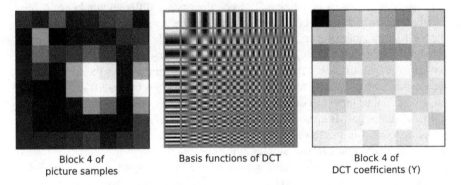

| Block 4 of picture samples | Basis functions of DCT | Block 4 of DCT coefficients (Y) |

Figure 7.13   *DCT of block 4 (component Y, black color in the block of DCT coefficients corresponds to the maximum value)*

statistical properties of the image, and there is no distortion, except for rounding errors when calculating the DCT. To neglect these errors, DCT coefficients are calculated with greater accuracy than image brightness values. If, e.g., image elements are specified in 8-bit words, then DCT coefficients should be calculated as 11-bit numbers.

7.2.2.1.2   Quantization of DCT coefficients

Further reduction of the data transfer rate can be achieved by eliminating psychophysical redundancy. As noted, visual perception allows for a greater level of noise and quantization errors in areas of the image with a significant level of high-frequency components. This means that you can either completely discard the coefficients of the high-frequency components of the DCT matrix with small amplitudes or quantize them into a small number of levels, thereby achieving a reduction in the number of binary digits needed to transmit a block of DCT coefficients.

The exclusion of high-frequency components of the matrix of DCT coefficients before the reconstruction of image blocks using the inverse DCT is illustrated in Figure 7.14. This figure shows the image recovery results when one constant component of each block (upper left picture) is used, four lowest-frequency cosine components (right upper picture) are used, 16 low-frequency cosine components (lower left picture) are used, and 36 low-frequency cosine components (lower right picture) block DCT coefficients (total in block 64 coefficient) are used. As can be seen, even direct compression with an immediate 4:1 ratio due to the simple rejection of 48 coefficients out of 64 (the upper right picture) allows us to obtain the quality of the reconstructed image that is suitable in a number of applications. The same images (especially the upper left) well illustrate the distortion and artifacts of video compression. It is necessary to pay attention to the block structure of the image inherent in this method with significant degrees of compression (when viewing and evaluating images, one should keep in mind possible distortions in the process of printing a book).

Higher quality of recoverable images in combination with significant compression can be achieved by quantizing DCT coefficients at a different number of levels. For the constant component (DC component) and the lowest-frequency coefficients, the DCT quantization is performed in small steps, which makes it possible to transmit these coefficients with high accuracy. High-frequency coefficients are quantized with a large quantization step, i.e., transmitted less accurately. As the frequency of the cosine components increases (with distance from the upper left corner of the DCT block), the number of quantization levels decreases, reaching values of several units for the largest frequencies. For example, the highest-frequency component, which is located in the lower right corner, can only be quantized into two levels. This means that you can use single-digit binary numbers to transfer it. But this, of course, also means high quantization noises when transmitting high-frequency components, sharp edges, and image contours. If these distortions are practically not noticeable to the viewer, then video compression, by reducing psychophysical redundancy, reaches its goal.

*Figure 7.14   Rejection of high-frequency components of matrix of DCT
coefficients before restoration of image blocks using reverse DCT*

Practically, quantization is performed by element-by-element division of an array of DCT coefficients by a quantization matrix, the values of which elements increase with distance from the upper left corner and closer to the lower right corner. When decoding at the receiving side, the DCT array coefficients are multiplied by the elements of the quantization matrix, which restores the correct values of the coefficients, but with a rounding error, which is small for low-frequency components of the image sample block, but large for high-frequency ones.

The distribution of the number of quantization levels, determined by the quantization matrix, is subordinate to the goals of reducing the visual visibility of the resulting quantization noise. Rough quantization of high-frequency DCT coefficients with relatively accurate low-frequency quantization means that the majority of quantization noise is concentrated in the high-frequency components of the image, i.e., in areas with a fine-grained structure and in the neighborhood of sharp edges. But the human eye sees quantization noise well in the image areas where the brightness changes smoothly (quantization noise appears as false contours). In the vicinity of sharp edges and against the background of small details, the quantization noise is much less noticeable. Therefore, quantization of DCT transform coefficients can be interpreted as a means of compression, achieved by eliminating the psychophysical redundancy of the image description. Details that the eye does not see are eliminated from the image in the coding process, and distortions are introduced that are almost imperceptible to the eye.

If quantization is significant, which can be the case for a high degree of compression, the quantization noise becomes noticeable. Examples of the distortions that occur during quantization are illustrated by the images in Figures 7.15–7.18. These figures show the original image blocks, blocks of DCT coefficients, blocks of quantized DCT coefficients, and reconstructed image blocks. During quantization, high-frequency components with small amplitudes are lost. In this case, as can be seen, the images of some blocks become "flat" (block 1, Figure 7.15), and in the images of others, small details and contours are distorted (this is especially noticeable on block 4, Figure 7.18). Quantization distortions are irreversible, but for the eye, they may be hardly noticeable if the compression is not excessively large.

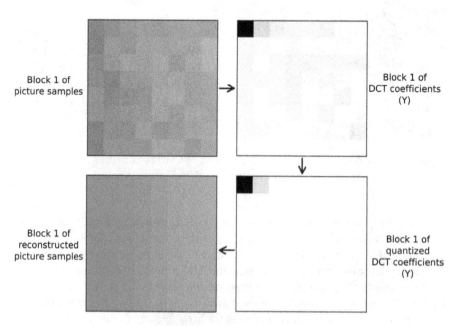

Figure 7.15   Quantization of DCT coefficients (block 1)

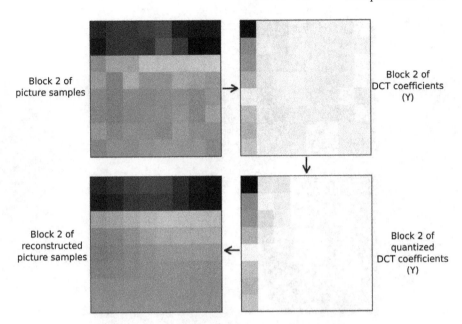

Figure 7.16   Quantization of DCT coefficients (block 2)

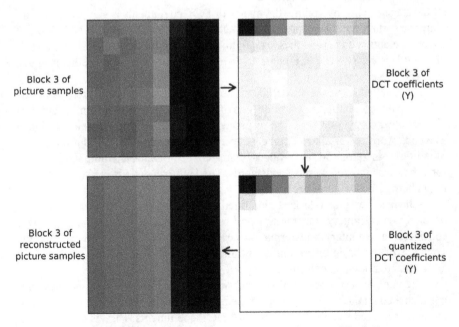

Figure 7.17   Quantization of DCT coefficients (block 3)

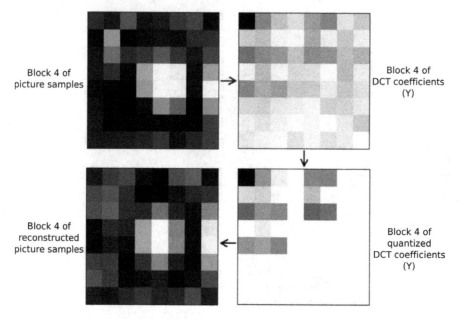

Block 4 of
picture samples

Block 4 of
DCT coefficients
(Y)

Block 4 of
reconstructed
picture samples

Block 4 of
quantized
DCT coefficients
(Y)

Figure 7.18   Quantization of DCT coefficients (block 4)

7.2.2.1.3   Generic codec scheme

At the stage of entropy coding, the matrix of DCT coefficients is subjected to scanning of a certain type (e.g., zigzag), at which it is possible to group together most of the coefficients with zero values, usually located in the right lower part of the DCT block (this is clearly seen in Figures 7.15–7.18). Instead of all zero coefficients, you can send one character—the EOB.

The scanned sequence of coefficients undergoes recoding. From a series of coefficients, pairs of numbers are formed, one of which is equal to a non-zero coefficient and the other to the number of zeros preceding this element. Each pair is assigned as a code word of variable length, e.g., according to the rules of the Huffman code. Small series of zeros and small values of nonzero coefficients are more likely; therefore, they are assigned as short codewords. The Huffman code is prefix, so no separators between codewords are needed.

Schemes that illustrate in a simplified form intraframe coding and decoding on the basis of unitary transformations (in the previous arguments, the DCT was considered as a unitary transformation) are shown in Figure 7.19. They show the sequence of the basic operations of intraframe coding during video compression: DCT, quantization and entropy coding.

The encoder can operate in two possible modes. The first one involves coding with a constant level of quality of the restored image, which is possible due to the use of a fixed quantization matrix of DCT coefficients. But, as follows from the above reasoning, this leads to a variable bit rate of data following the output

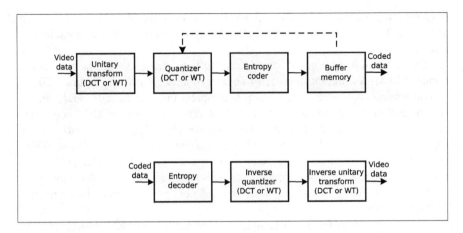

Figure 7.19   Intraframe coding: coder and decoder schemes

of an entropy coder, or a coder with a variable word length. The data bit rate will increase with increasing level of high-frequency components of the image (this case illustrates Figures 7.13 and 7.18) and decrease when coding images with insignificant small detail [an example is block 1, shown in Figures 7.10 and 7.15, for which the block end symbol (EOB) will replace the overwhelming part of DCT coefficients that become zero after quantization]. This mode can be used in professional applications that require further image processing while maintaining high quality and not associated with restrictions in data transfer rate and memory capacities.

When transmitting and distributing television programs, fluctuations in the data transmission rate may be unacceptable. In this case, the second mode of operation of the encoder is used, which ensures a fixed speed of data rate. This can be done by changing the quantization matrix, which, accordingly, affects the quality of the restored image. Buffer memory is included in the coder. Data are stored in memory at variable rates and read from a constant rate. To prevent buffer overflow or its full release, which can lead to disruptions in the compression system, adaptive quantization is used. Information about the degree of filling of the buffer memory serves as a control signal (dashed line in Figure 7.19), which regulates the quantization scale. If, e.g., the encoded image is characterized by high-frequency detail, then the number of nonzero elements of the matrix of DCT coefficients increases. The volume of transmitted data also increases, so the buffer is filled at an increased rate. Thanks to the feedback, quantization becomes more coarse and the rate at which data enter the buffer is reduced, but due to an increase in the quantization noise and image degradation. If a simple "flat" image without small details is encoded, the number of zero elements of the matrix of DCT coefficients increases, and the rate of data entry into the buffer memory is reduced compared with the average value. Then quantization becomes less coarse (an increasing number of DCT coefficients are quantized to the maximum number of levels). Thanks to feedback, the buffer filling rate is on average maintained at a constant level.

Examples of images encoded and decoded in accordance with the schemes of Figure 7.19 with different degrees of compression are shown in Figure 7.20. At a compression ratio of 2:1 (left image), compression is achieved mainly due to entropy coding and elimination of an array of zeros at the end of DCT coefficient blocks (this array is replaced by the EOB symbol). The influence of quantization of DCT coefficients is small, and irreversible image distortion is almost absent. With a large degree of 35:1 (right image), compression is achieved through coarse quantization of high-frequency DCT coefficients. This leads to significant distortion of the image (when viewing, one should keep in mind possible distortions in the process of printing a book).

The nature of artifacts and compression distortions is well manifested in an artificial image (Figure 7.21, upper picture). This image contains many "flat" areas with constant brightness and is well subjected to compression; therefore, to obtain noticeable artifacts and distortions in the reconstructed image, we had to use a very large degree of compression equal to 54:1 (lower picture). The block structure becomes noticeable (as in Figure 7.14), sharp luminance and color transitions are distorted, and sharpness and clarity decrease. Noise pulses appear in the vicinity of the contours, similar to mosquitoes (this is sometimes called mosquito noise). It seems that dirty glass has been applied to the image.

(a)   (b)

Figure 7.20   Images coded at different compression ratio: 2:1 (a) and 35:1 (b)

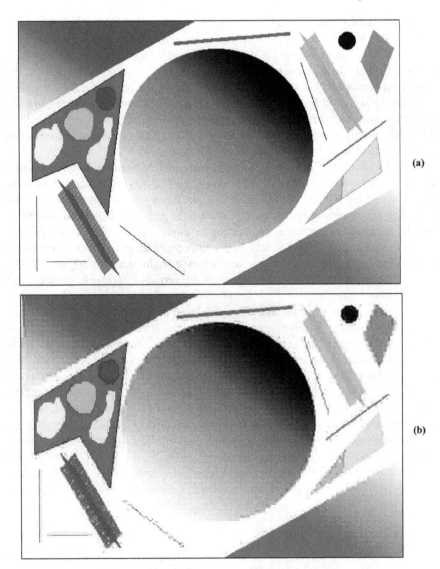

(a)

(b)

*Figure 7.21   Compression distortions and artifacts [intraframe coding with a
compression ratio of 54:1 (b) and original (a)]*

7.2.2.1.4   Wavelet transform

Wavelet transform (WT) [3, 4] can be used as a unitary transformation in video
compression systems. WT can be considered as a representation of a signal in the
form of a superposition of some basic functions—wave packets. A feature of these
wave packets is that they are all obtained from a single prototype wave by stretching

or dilating and displacement. The prototype wave can be considered as the impulse response of the base filter. Then the WT is reduced to a set of filtering and decimation processes (Figure 7.22). The signal is filtered using low-pass and high-pass filters that divide the frequency band of the original signal into two halves. Both the low-frequency and high-frequency components of the signal, obtained by filtering, have a twice narrower band of frequency components. Therefore, they can be sampled at a frequency equal to half the sampling rate of the original signal. After low-frequency and high-frequency filtering using digital filters, every second sample can be simply eliminated, which means decimation.

The application of one-dimensional filtering and decimation in the horizontal direction to the television image is shown in Figure 7.23. On the left there is the original image. In the left part of the right image, there is a filtered and decimated low-frequency component of the image and in the right part, the filtered and decimated high-frequency one. Since after decimating the number of samples in each component is halved in each television line, both components are placed on the area of the original image. Application to the resulting image (right picture in Figure 7.23) of one-dimensional filtration and decimation in the vertical direction is shown in the left picture of Figure 7.24.

As follows from the scheme of Figure 7.22, at the second stage of the conversion, the low-frequency component is again divided into low-frequency and high-frequency components. After decimation, the low-frequency component may again undergo a separation into low-frequency and high-frequency components in the next stage of the conversion. The results of the three stages of two-dimensional filtering and decimation are shown in the right-hand picture of Figure 7.24. After each of the three stages, the image is divided into one low-frequency and three high-frequency components. At each stage, the image is first filtered and decimated horizontally, and the resulting two components are filtered using filters of the lower and upper vertical spatial frequencies and decimated vertically.

Of the three high-frequency components, one displays the high-frequency component of the original image obtained by sequentially turning on the

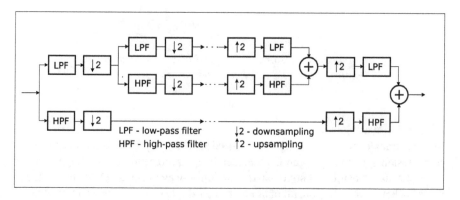

Figure 7.22   Wavelet transform

Figure 7.23   *WT: original (a) and the result of one-dimensional horizontal filtering and decimation (b)*

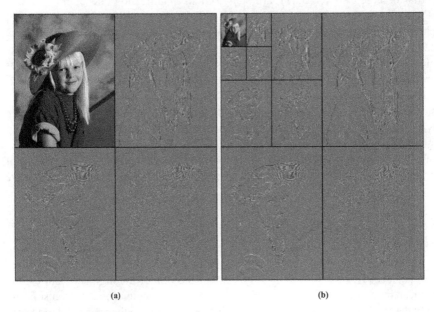

Figure 7.24   *WT: single (a) and threefold two-dimensional horizontal and vertical filtering and decimation (b)*

horizontal spatial frequency high-pass filter and the vertical spatial frequency low-pass filter. This component is located in the upper right corner of the combined image (see the left image of Figure 7.24), it has noticeable vertical boundaries of brightness transitions and vertical lines. The second high-frequency component displays the vertical high spatial frequency components (horizontal brightness transitions and horizontal lines). It is located in the lower left corner of the left image of Figure 7.24. In the lower right corner of the left image in Figure 7.24, the third high-frequency component is located, which is associated with diagonal spatial frequencies and displays diagonal brightness transitions in the original image.

After three stages, the image was divided into one low-frequency component and nine high-frequency components with different spatial frequencies and different frequency bands. It should be noted that after the third stage, the definition of the low-frequency component located in the upper left corner of the right picture in Figure 7.24 is eight times less than the definition of the original image (the frequency band of the component obtained in the third stage is 1/8 of the band of the original signal). At the stage of the inverse WT, each component of the transformed signal is first stretched twice, i.e., after each sample, an additional zero sample is inserted (Figure 7.22). The stretched component undergoes filtering, as a result of which interpolated values are placed in place of the zero samples.

Actually, video compression based on the WT is performed in the same way as compression based on cosine transform. The components of the video signal obtained after the WT are subjected to quantization and entropy coding (Figure 7.19). The principal difference from DCT compression is in the method of obtaining frequency components of the image. DCT allows us to obtain frequency components that occupy equal bands at all average frequencies (e.g., 1/8 of the maximum frequency of the signal). WT gives components whose frequency bands are halved as the average frequency decreases (e.g., 1/2, 1/4, and 1/8 of the maximum signal frequency, etc.). The WT does not require the formation of blocks, so the distortion of the decoded image is more "natural," i.e., looks less alien on typical images than, e.g., artifacts in the form of a block structure, manifested in systems based on DCT. However, with low degrees of compression, the benefits of compression based on the WT are not so noticeable regarding the appearance of artifacts. It should also be noted that the compression codecs based on the WT are much more complex.

7.2.2.2   Interframe coding

The purpose of interframe coding is to reduce the temporal redundancy of video data caused by strong correlations between adjacent frames. Since the images of two adjacent frames are usually very similar, the rate of the digital stream can be reduced by predicting the current frame based on the previous one and transmitting only the prediction error (residual)—the difference between the actual and predicted image of the current frame. The interframe and intraframe coding combination leads to a scheme in which the coding consists of the following main stages: DPCM, unitary transform-based decorrelation, quantization of the transform coefficients, and entropy coding (Figure 7.25).

The encoder calculates the prediction error, i.e., the difference between the actual and predicted image blocks to reduce temporal redundancy. The prediction error is subject to DCT, quantized, and encoded in a variable-length entropy encoder. Such a process reduces both temporal, and spatial, and psychophysical redundancy. As the simplest predictor, you can use the memory per frame. But the prediction is not performed by delaying the frame of the input image signal received directly from the source. This is due to the fact that the prediction must be also performed in the decoder, which cannot use source images. The images restored in the decoder underwent a quantization procedure in the process of intraframe coding and, therefore, underwent some irreversible transformations. Therefore, in the encoder, inverse quantization and inverse DCT are performed, thanks to which the image is reconstructed in the same way as in the decoder.

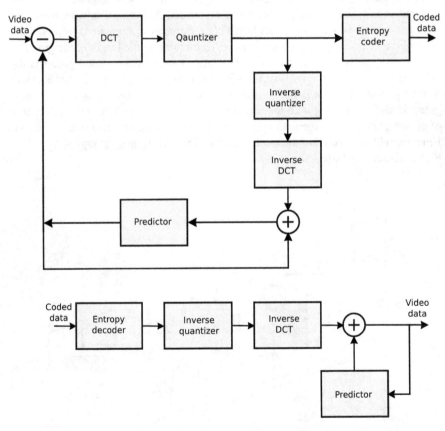

Figure 7.25   Interframe coding: encoder and decoder schemes

The decoder performs inverse quantization, inverse DCT, as a result of which a prediction error is generated. The prediction error is added to the decoded image of the previous frame, forming a decoded image of the current frame.

In accordance with the possible methods of prediction in compression systems with a reduction of temporal redundancy, three types of images are used. Type I (intracoded) images are encoded independently of other images. They only reduce spatial redundancy. Images of type P (predictive-coded) are encoded with a prediction based on the previous image of type I. They themselves can also be used as the reference for further prediction. They reduce both spatial and temporal redundancy. Type B images (bidirectionally predicted-coded) are encoded using bidirectional prediction based on previous and subsequent images. Both spatial and temporal redundancies are reduced in them, and the highest degree of compression is achieved, since the bidirectional prediction is the most accurate. At a fixed quality level, a P-image is less than an I-image by an average of three times, and a B-image is about four times less.

Images of various types are combined into repeated series, called groups of images (Figure 7.26). A group begins with a type I image, forming a reference signal for prediction when encoding other types of images. A group of images should be large enough if it is necessary to achieve a high degree of compression. However, the dynamics of the television program imposes restrictions on the length of the group.

Groups can be closed and open. The last element of the closed group is the P-image predicted on the basis of the I-image of the same group. The last element of the open group is the B-image, for prediction of which the I-image of the next group is also used along with the preceding P-image (Figure 7.26). The advantage of closed groups is that they allow direct switching to be performed before the next I-picture without decoding the digital stream. The use of open groups allows for a higher degree of compression.

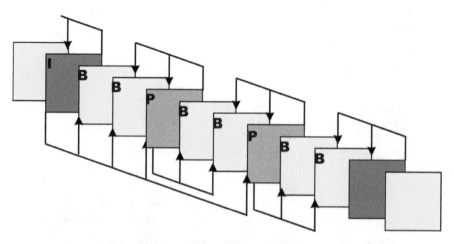

Figure 7.26   Video sequence and group of pictures

Until now, it was assumed that when coding a certain block of the image of the current frame, blocks of the previous and subsequent frames with the same spatial coordinates are used for prediction. But the faster the moving objects move, the more the encoded and predicted blocks differ and the more data about the prediction error must be transmitted, and this reduces the efficiency of compression. Prediction accuracy in the encoding of images of moving objects can be increased by motion compensation. The principle of DPCM with motion compensation is illustrated in Figure 7.27.

The motion vector estimator estimates the speed of motion of the image objects. The motion vector is called the difference in spatial coordinates between the corresponding points of the object in adjacent frames. Motion is compensated by prediction. The more accurately the motion vector is estimated, the better and more accurate the prediction and the smaller the amount of data to be transmitted. Not only the prediction error is sent to the receiver (the difference between the actual image block and the predicted one) but also the motion vector. The motion vector is also encoded with variable length words. The coded prediction error is combined with the motion vector codes, after which the encoded digital stream is generated. The decoder scheme is similar to the lower scheme of Figure 7.25; only it should use a predictor with motion compensation, operating on the basis of the motion vector data transmitted from the encoder.

One of the simplest methods for estimating the motion vector is block matching. The current block is projected onto the previous reference image and compared with all blocks of the previous frame within a certain search area. The block, which

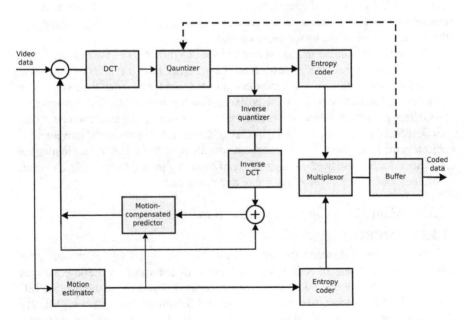

Figure 7.27   Interframe coding: DPCM encoder with motion compensation

is the least different from the current one (e.g., by the standard deviation), is taken as the reference block in the formation of the prediction. The two-coordinate offset between the projected and the reference blocks determines the motion vector.

Predictors with motion compensation in modern video compression systems can use a variety of methods. For example, a block can be predicted on the basis of the previous image, on the basis of the subsequent image, and also on the basis of both the previous and the subsequent, as described above. In interlaced systems, the fields of one frame can be predicted separately using different motion vectors or together using a common vector. There is also the possibility of zero prediction (if no suitable reference block is found). In this case, the current block itself will be encoded instead of a prediction error, which means the rejection of interframe coding and the transition to intraframe coding. For each current block, the encoder selects a prediction method that provides the highest quality of the decoded image, taking into account restrictions on the rate of data transfer. Information about the prediction method is included in the general stream and transmitted to the decoder for correct reconstruction of the image.

The determination of the magnitude and direction of displacement of moving objects from frame to frame, called the motion vector, is performed at the macroblock level. The macroblock is made up of smaller blocks (e.g., with dimensions of 8×8 pixels, as described above), on which DCT is performed. Each macroblock contains a group of four blocks with luminance samples (from an image area with dimensions of 16×16 pixels) and a group of blocks with samples of color difference signals taken from the same image area as the samples of brightness blocks. The number of blocks with samples of color difference signals depends on the sampling format: one C_B block and one C_R block in 4:2:0 format, two C_B blocks and two C_R blocks in 4:2:2 format, and four C_B blocks and four C_R blocks in 4:4:4 format. At the macroblock level, the encoder also decides the type of prediction in accordance with the characteristics of the image being encoded.

The estimation of the motion vector and the determination of the best prediction strategy require the use of complex computational procedures that must be carried out in real time. The encoder is much more complicated than the decoder; the compression system based on DPCM with motion compensation is asymmetric. Motion compensation during prediction reduces the prediction error. However, the prediction error cannot be reduced to zero, especially in the case of the complex nature of the movement of objects in the image field. This is hampered not only by inaccuracies in determining the motion vector but also by changing the size of a moving object, its turns, simultaneous movements of several objects in different directions, etc.

7.2.3   Main video compression standards

7.2.3.1   MPEG-1

Video compression became the real driving force of digital television only after its international standardization. The first result of the work on the international standardization of video compression systems for TV was the standard ISO/IEC 11172—1.5 Mbit/s (encoding image and sound at compressed data up to 1.5 Mbit/s). This standard, published in 1993, became known as Moving Picture Experts Group (MPEG)-1 [10].

The MPEG-1 encoder eliminates spatial, temporal, and psychophysical redundancy of images. Key components of the spatial redundancy reduction system are as follows: DCT, quantization, and entropy coding with variable word length. The reduction of the temporal redundancy is performed using DPCM, supplemented by motion vector estimation and motion compensation in prediction. MPEG-1 codecs are designed to compress video streaming in multimedia systems. In a typical case, they use progressively scanned images of the CIF format and compress video data up to 1.2 Mbit/s while maintaining quality at the level of home video (VHS). The role of this standard was great; it made interactive video on CDs possible.

7.2.3.2 MPEG-2

The ISO/IEC 13818 standard—Information Technology—Generic Coding of Moving Pictures and Associated Audio Information (generic image and audio coding has become known worldwide as MPEG-2 [11–13]). It uses the same methods for reducing redundancy as in the MPEG-1 standard. The most significant improvement in the MPEG-2 standard is supporting interlaced images, expanding the possibilities for estimating the motion vector and prediction modes. MPEG-2 (video) coding is generic, or typical, in the sense that it involves a large range of the rates of the encoded data streams. MPEG-2 includes the production of television programs, distribution programs using terrestrial and satellite communication lines, and multimedia systems.

Summarizing the requirements of typical and most important applications determined the syntax and semantics of the video stream. For the most effective application in practice and to ensure a high degree of interoperability of devices operating within the MPEG-2 standard, but developed and manufactured by various manufacturers, several subsets, called profiles, are defined within the MPEG-2 syntax. But even within the syntactic boundaries of each profile, there can be a huge number of combinations of parameters of a digital stream. Therefore, there are several levels in each profile, determined by a set of restrictions imposed on the parameters of the digital stream within the profile syntax. In other words, a profile is a subset of the standard for specialized applications that defines algorithms and compression tools. The levels inside each profile are mainly related to the parameters of the compressible image.

Main profiles of MPEG-2:

- Simple—simple profile. It uses only I and P images, only unidirectional prediction is possible in it.
- Main—main profile. It supports I, P, and B images.
- High—high profile. The compression ratio is minimal, and the image quality is the highest.
- 422—Studio profile. It provides full resolution compliant with ITU-R 601 recommendation.

MPEG-2 also supports two scalable profiles that imply the separation of a digital stream into layers during encoding. The base layer can be decoded by itself, regardless of the higher layers. Base layer decoding provides acceptable image quality. Decoding of all layers allows us to reduce quantization noise and improve spatial resolution. Scalable profiles are convenient when transmitting video data in packet form over data networks.

Levels of MPEG-2:

- Low—low level.
- Main—main level. It supports images with a resolution of 720 × 576.
- High-1440—high level-1440. Supports high-definition television images with resolutions up to 1 440 × 1 152.
- High—high level. Supports widescreen high-definition television images with resolutions up to 1 920 × 1 152.

Images can be interlaced or non-interlaced. The number of elements in a frame can vary from 176 × 144 (144 active lines with 176 elements per line) for the low level of the main profile (MP @ LL) to 1 920 × 1 152 for the high level of the high profile (HP @ HL). The maximum compressed data rate can be in the range from 4 Mbits/s (MP @ LL) to 100 Mbits/s (HP @ HL). Some of the possible combinations of profiles and levels are quite developed and accepted as standard.

7.2.3.3 DV

Video compression is the most important and most complex component of a signal processing system, defined by the DV recording standard (IEC 61834—helical-scan digital video cassette recording system using 6.5-mm magnetic tape for consumer use) [14]. It eliminates the spatial redundancy inherent in the frame of a typical television image, bringing the rate of the compressed video data stream to 25 Mbits/s. The main elements of DV video compression: DCT, quantization of DCT coefficients, and entropy coding of a sequence of quantized DCT coefficients. DV video compression is performed by intraframe coding, but it is an adaptive system that adapts to the movement of the imaged objects. Therefore, as part of a compression scheme, there is a device that performs motion estimation, on the basis of which a decision is made on the DCT mode and the features of quantization of DCT coefficients.

The object of the main operations of video compression is a small part of the frame of the television image, called the video segment. The amount of data space occupied by a compressed video segment is defined by the standard and cannot be exceeded. A feature of the DV compression scheme is the absence of feedback in the form of information about the degree of filling of the space reserved for the compressed video segment. The quantization control of the DCT coefficients is performed based on the entropy estimate of the input data.

From the heading of the IEC 61834 standard, it follows that the DV format regulates the household video recording system. However, the potential of the format turned out to be so significant that the equipment of the DV format found application not only in home appliances but also in the fields of applied and broadcast television. The processing and compression algorithm for the recorded data, defined by the

DV format, is very effective. It served as the basis for the creation of a number of new video formats for applied and broadcasting applications.

7.2.3.4 JPEG2000

The enhanced version of the JPEG compression system, known as the ISO/IEC 15444 standard—Information Technology—JPEG2000 image coding system, does not contain a DCT. Instead of DCT, the WT described above is used as a means of de-correlating an array of image samples [15]. JPEG2000 supports both lossless compression and lossy compression; 8-, 10-, and 12-bit quantization is used. One of the main applications is the digital intermediate process for digital cinema, television applications of the highest quality level. Using image decomposition into sub-bands makes it easy to implement scalability. Only intraframe coding is used, which does not significantly increase the efficiency of compression. It is only 20–40% higher than in the JPEG system.

7.2.3.5 MPEG-4

The ISO/IEC 14496 Standard—Information Technologies—Generic Coding of Audio-Visual Objects is known by the name MPEG-4 [16]. The main difference between MPEG-4 and all other standards is the object-oriented presentation of audiovisual scenes, which are formed using separate objects located in a certain way in space and time. The coded representation of video objects defines part 2 of the standard (Part 2: Visual). Under part 2 of the standard, 21 profiles have been developed. The most common are ASP and Simple Profile, which is a subset of the ASP. The MPEG-4 Part 2 document was originally created for multimedia applications designed for low-rate digital streams of compressed video data but was subsequently extended to the field of television broadcasting. The MPEG-4 standard as a whole was the beginning of a new approach based on the coding of audiovisual objects. Separate coding of objects allows for more efficient compression, but with respect to natural television images, the gain in comparison with MPEG-2 was not too great (comparative efficiency means the difference in the bit rates of the compressed video data of two systems with the same image quality levels).

The advanced video compression system has been standardized as ISO/IEC 14496 Part 10 (part 10 of the MPEG-4 standard) called Advanced Video Coding (AVC), and also as H.264 (ITU-T Recommendation H.264), which gives grounds to call this system video compression AVC/H.264 [17]. The AVC system preserved the generic scheme based on the DCT and DPCM, but each stage was significantly improved. At the stage of DPCM, the prediction accuracy is improved. In AVC, a prediction is generated based on an analysis of five frames of an image (in MPEG-2, only two frames). Spatial prediction is more widely used. Due to these factors, a gain in coding efficiency of 5–10% is achieved. The accuracy of estimating the motion vector is increased to ¼ pixel (in MPEG-2, the motion vector is estimated with an accuracy of ½ pixel). This allows you to increase the efficiency by about 20%. Conversion to the frequency domain is performed on blocks of different sizes. The minimum block size is smaller than MPEG-2. This allows you to divide the image into blocks whose dimensions adapt to the content of the image. Instead of DCT, an integer transform with similar properties is used, which makes it possible to avoid round-off errors when calculating the transform coefficients at the encoder and the image samples when decoding at the receiver. Advanced quantization matrices are used, which more closely correspond to the peculiarities of visual perception of quantization noises. This allows you to increase the coding efficiency by another 15–20%. More complex but more efficient entropy coding algorithms are used—CABAC, which increases the coding efficiency by 10–15%. To reduce the visual visibility of the most unpleasant of video compression artifacts—image blockiness, an adaptive filter is used, smoothing the block structure, but not affecting the transfer of fine image details inside the blocks. This allows you to increase the coding efficiency by another 5–10%.

So, there is no one element of the AVC standard, which in itself has provided high coding efficiency. There were a lot of relatively small improvements, which provided a significant gain (about two times) in comparison with MPEG-2.

AVC tools are divided into subsets—profiles:

- baseline profile—for applications requiring low end-to-end latency, e.g., video conferencing;
- extended profile—for mobile apps;
- main profile—for broadcast applications of standard definition;
- high profiles—the high profile was originally designed for use in HD-DVD systems in accordance with the specifications of the DVD forum, BD-ROM in accordance with the specifications of the Blu-Ray Disc Association, as well as in DVB broadcast systems. In order to expand the scope of the standard and ensure its applicability in the field of studio editing, program layout, a family of four high profiles was developed.

  - High profile supports video encoding with 4:2:0 sampling structure and 8 bits per sample quantization.
  - High 10 profile supports video encoding with 4:2:0 sampling structure and 10-bit quantization per sample.
  - High 4:2:2 profile supports video encoding with a 4:2:2 sampling structure and 10 bits per sample quantization.
  - High 4:4:4 profile (H444P) supports video encoding with 4:4:4 sampling structure and quantization with 12 bits per sample, as well as RGB encoding.

The system parameters are set in accordance with the 16 levels: 1, 1b, 1.1, 1.2, 1.3, 2, 2.1, 2.2, 3, 3.1, 3.2, 4, 4.1, 4.2, 5, and 5.1. Levels define the upper bounds of a number of system parameters, such as image size, maximum frame rate, data buffer sizes, compressed data rate, etc. AVC supports the widest range of TV image formats, among which are SQCIF format (128 × 96 elements), QCIF format (176 × 144), CIF format (352 × 288), the format of VGA (640 × 480), 525SD format (720 × 480), 625SD format (720 × 576), 720HD format (1 280 × 720), 1080HD size (1 920 × 1 088),

4K × 2K size (4 096 × 2 048), and the format of 4 096 × 2 304 (4 096 × 2 304 elements). The number of samples of the brightness signal in the frame is in the range from 122 888 (SQCIF format) to 9 437 184 (4 096 × 2 304 format). The maximum sampling rate can be in the range from 380 160 Hz (for the baseline profile and level 1) to 251 658 240 Hz (for the H444P profile and level 5.1). The maximum bit rate of the coded video data ranges from 64 kbits/s (for the baseline profile and level 1) to 960 Mbits/s (for the H444P profile and level 5.1). The maximum frame rate can be 7.6, 12.5, 15, 25, 30, 60, and 172 Hz.

7.2.3.6 HEVC/H.265

High Efficiency Video Coding (HEVC)—the last implemented project of two research groups from standardization organizations—the ITU-T Video Coding Experts Group and the ISO/IEC MPEG, united to work within the Joint Collaborative Team on Video Coding. This video compression system was standardized in the form of an ISO/IEC 23008-2 standard (MPEG-H Part 2) called HEVC and also in the form of ITU-T Recommendation H.265, which gives grounds to call this video compression system HEVC /H.265 [18]. The first version of the standard was published in 2013.

The HEVC system retained the general scheme using DPCM, DCT, motion compensation, and entropy coding, which is implemented in MPEG-2 and AVC/H.264 systems, but each stage has been significantly improved. The AVC/H.264 standard has grown from the MPEG-2 standard, providing a gain of about two times in comparison with MPEG-2 due to a large number of relatively minor improvements. Similarly, the HEVC/H.265 standard provides the same significant gain in comparison with AVC/H.264 due to various improvements: interframe prediction with more accurate motion vector estimation and improved motion compensation, wider use of intraframe prediction with more modes, flexible adaptive structure of the areas of motion vector estimation, intraframe prediction and unitary coding (integer transform, approximating DCT), improved entropy coding—CABAC, and adaptive filters smoothing block structure (adaptive deblocking filter and sample adaptive offset filter).

The design of the HEVC allows improving performance thanks to parallel processing. The HEVC standard supports tiles that allow dividing a picture into a grid of rectangular regions when both encoding and decoding. These regions can be independently encoded and decoded simultaneously. Parallel calculations can be used in the in-loop deblocking filter and waveform processor as well.

The HEVC standard supports the following main profiles: Main profile, Main 10 profile, Main Still Picture profile, and Main 10 Still Picture profiles. Already in the first version, the main profile supported the length of the code word (bit depth) of 8 bits for the luminance signal and the color difference signals, the Main 10 profile—8 or 10 bits. The Main Still Picture Profiles provided single still picture coding with the same characteristics as the Main profiles. The scope of the standard is significantly expanded due to extensions:

Format range extensions profiles (the Monochrome, Monochrome 10, Monochrome 12, and Monochrome 16 profiles; the Main 12 profile; the Main 4:2:2 10 and Main 4:2:2 12 profiles; the Main 4:4:4, Main 4:4:4 10, and Main 4:4:4 12 profiles; the Main Intra, Main 10 Intra, Main 12 Intra, Main 4:2:2 10 Intra, Main 4:2:2 12 Intra, Main 4:4:4 Intra, Main 4:4:4 10 Intra, Main 4:4:4 12 Intra, and Main 4:4:4 16 Intra profiles; the Main 4:4:4 Still Picture and Main 4:4:4 16 Still Picture profiles)

High throughput profiles (the High Throughput 4:4:4, High Throughput 4:4:4 10, and High Throughput 4:4:4 14 profiles; the High Throughput 4:4:4 16 Intra profile)

Screen content coding extensions profiles (the Screen-Extended Main and Screen-Extended Main 10 profiles; the Screen-Extended Main 4:4:4 and Screen-Extended Main 4:4:4 10 profiles; the Screen-Extended High Throughput 4:4:4, Screen-Extended High Throughput 4:4:4 10, and Screen-Extended High Throughput 4:4:4 14 profiles)

The standard defines 13 levels combined into two tiers: Main (1-3.1) and High (4-6.2). The HEVC family of levels is similar to the AVC system levels. The main difference is levels 6–6.2, which regulate support for 8K ultra-high-definition video. The maximum number of samples of the brightness signal in the frame reaches 33 554 432 for the 8 192 × 4 096 format and 35 651 584 for the 8 192 ×4 320 format.

High-quality parameters of the HEVC system should significantly affect the broadcast industry. Its use will reduce the data rate (or file size) by about two times compared with the AVC system. And if you keep the flow rate (or file size) at the level of AVC requirements, then you can significantly improve the image quality. The HEVC system will be able to promote the spread of 4K and 8K ultra-high-definition television. It can lead to an increase in the number of channels in satellite, cable, and IP television networks. In the mobile communications market, using HEVC can lead to lower costs for content delivery. The use of HEVC in OTT services can lead to an improvement in the quality of QoE and the achievement of the standard broadcasting level.

7.2.3.7   AVS

The Audio Video Coding (AVS) standard—a national standard developed in China called "Information Technology, Advanced Audio Video Coding, Part 2: Video" (AVS1 for short, Standard No.: GB/T 20090.2-2006)—was approved in 2006 [19]. Part 2 of the standard regulates the encoding of the video within two profiles:

- Jizhun profile—describes the means of delivery of materials and programs of terrestrial, satellite, and cable television broadcasting.
- Zengqiang profile—describes the recording of audio-visual materials on DVDs, HD-DVDs, and Blu-ray disks.

Part 7 specifies video coding for mobile broadcasting with reduced clarity within the Jiben Profile.

Ten years later, in 2016, the standard "Information Technology, Advanced Audio Video Coding Part 16: Radio Television Video" (AVS + for short, Standard No.: GB/T 20090.16-2016) [20] was adopted. MPEG-2 can be considered its predecessor, it competes with AVC. The AVS + compression system achieves AVC performance, but, according to the developers, with lower processing power requirements for data processing devices in codecs.

The primary application target for the second generation AVS standard, referred to as AVS2, is ultra HD (4K above) and high dynamic range videos. AVS2 should also support 3D video, multi-view, and virtual reality video. The General Administration of Quality Supervision, Inspection and Quarantine of the People's Republic of China and Standardization Administration of People's Republic of China issued AVS2 as the national standard "Information Technology—High Efficient Media Coding—Part 2: Video" (Standard No.: GB/T 33475.2-2016) in 2016. The tests made by the Radio and Television Metering Center of State Administration of Radio, Film, and Television show that AVS2 is twice as efficient as the previous generation standards AVS+ and H.264/AVC [21]. It means that AVS2 is a competitor with the HEVC/H.265 standard.

7.3 Audio compression

7.3.1 Basics of audio compression

The basis of the compression is the quantization of the components of the audio signal. However, the resulting quantization noise degrades the quality of the sound reproduced during decoding and limits the degree of compression and packing density. The goal of compression algorithms can be to minimize the level of introduced quantization noise. The compression algorithms used today in television broadcasting suggest the introduction of the maximum level of quantization noise, but on the condition that they remain inaudible. This approach, based on the use of a psychoacoustic model of hearing and applied in the development of MPEG audio compression standards, provides a combination of efficiency and quality. This approach is used in the development of all new advanced audio data compression systems, which can therefore be called psychoacoustic algorithms.

The evolution of sensory systems possessed by living beings followed a simple but dramatic way: "To distinguish in order to survive." The ear as a sensory analyzer must distinguish sounds according to their frequency composition. However, the response to the sound stimulus must be fast, which means that the signal processing in the ear and the nervous system must be done in a short time. The requirements of a high-frequency and temporal discriminative ability of the analyzer are contradictory. The result of evolution is the optimal combination of these indicators. So, the human auditory system has a finite resolution both in the time and in frequency domains. This leads to the fact that sounds that have a small level and are in the time and frequency "neighborhood" with a stronger sound may be inaudible. We can say that weak sounds are masked by strong ones.

Coders that are equipped with a hearing model extract everything that the ear cannot hear from the audio signal and quantize the signal so that the quantization noise is inaudible. The psychoacoustic model of hearing provides a quantitative description of the effects that limit the distinguishing capabilities of the auditory system. A generalized scheme for the implementation of psychoacoustic algorithms is shown in Figure 7.28. The input audio signal is divided into frequency components

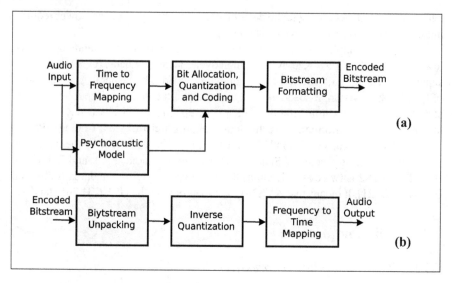

Figure 7.28   A generalized scheme for the implementation of psychoacoustic algorithms (a—encoder and b—decoder)

(time-to-frequency mapping block), e.g., using a filter bank. The signal is also used to calculate the signal-to-mask ratio (SMR) for each frequency component (psycho-acoustic model block). The value found (SMR) is used in the bit allocation block to decide how many bits must be used to quantize each frequency component of the signal in order to make the quantization noise inaudible. The last coder block (bitstream formatting) formats the quantized frequency components and packs them into a stream of coded compressed audio data. The decoder performs the reverse operations. It unpacks the data stream (bitstream unpacking), restores the values of the frequency components of the signal using quantized values (inverse quantization), and restores the audio signal from the frequency components (frequency-to-time mapping).

Thus, the key to understanding the principles of the work of the psychoacoustic algorithm is understanding the mechanisms of hearing and studying the properties of the human auditory system.

7.3.2   Psychoacoustic model of hearing

7.3.2.1   Human auditory system

In the human auditory system, there are two main stages in the processing of sound waves [22–25]. At the first stage, sound waves are converted into electrical impulses that transfer sound information to the brain via the auditory nerve. The processing of nerve impulses in the brain, performed in the second stage, leads to the appearance of auditory sensations.

The processing of the first stage is performed using the ear, in which it is customary to isolate the outer ear, middle ear, and inner ear. The outer ear includes the pinna—the visible part of the ear and the auditory canal. The auditory canal is a tube about 27 mm long and 7 mm in diameter, open on one side and closed on the other, where the eardrum is located—a tympanic membrane. The sound pressure wave is directed by the pinna into the auditory canal and acts on the eardrum, which is the interface between the outer and middle ear. In a column of air enclosed in the auditory canal, standing waves arise, and the auditory canal acts as a resonator. The first resonance at a frequency of approximately 3 kHz is in the frequency range of human speech sounds, which leads to amplification of oscillations in the frequency range 2–5 kHz, and the amplification is 5–10 dB. Thus, the external ear provides increased sensitivity of the auditory system to sound pressure waves corresponding to human speech.

The middle ear begins with the eardrum, which oscillates under the action of sound pressure waves propagating in the air medium of the auditory canal. The tympanic membrane is connected with three small ossicles (bones) in the air cavity of the middle ear. They are called malleus, incus, and stapes. These three bones are also commonly referred to as hammer, anvil, and stirrup. A system of three bones transmits mechanical vibrations of the eardrum to the inner ear and excites a pressure wave in the fluid of the inner ear. The elastic properties of the inner ear fluid and the air are different. The middle ear can be regarded as an impedance transducer, matching the external ear's air environment with the internal ear's fluid. The best impedance matching is performed at the frequency of 1 kHz, at which the pressure rises by almost 20 dB.

The inner ear includes the cochlea, from which mechanical vibrations emanating from the oval window are transformed into electrical impulses. The cochlea or ear labyrinth is a spiral-shaped tube of about 35 mm length, coiled into approximately 2.5 turns. An incompressible lymphatic fluid is in a rigid tube. The middle of the cochlea is divided into partitions by a flexible plate–the basilar membrane that passes through the cochlea labyrinth from the base to the apex. A traveling acoustic wave is created in the fluid of the cochlea. This wave is excited in the upper part through the membrane of the oval window, which oscillates under the action of the stirrup. The pressure wave propagates in the cochlea and, gradually fading away, finally leaves the cochlea through a round window located under the oval window in the lower part. The wave in the cochlea fluid causes mechanical oscillations of the cochlea's basilar membrane. Mechanical vibrations of the membrane excite sensitive cells of the organ of Corti [22, 26], which is located on the membrane along its entire length. Electrochemical reactions in the hair cells of the organ of Corti are sources of electrical nerve signals transmitted through the fibers of the auditory nerve to the brain. The processing of nerve impulses in the central nervous system leads to the appearance of auditory sensations.

7.3.2.2   Audible sounds

Sound waves—longitudinal mechanical vibrations—are emitted by a sound source—an oscillating body—and are distributed in solids, liquids, and gases in the form of pressure fluctuations. There are certain connections between the pressure oscillations that generate the sound wave in the air and the sound sensation. When clarifying these connections and studying the properties of the auditory system, models of sounds that a human hears in real life are usually used.

A wave in the form of a sinusoidal oscillation gives a pure musical tone. The amplitude of this wave determines the volume of the sound sensation, and the frequency determines the pitch. A sound wave of several pure tones gives a consonance, or musical sound, which remains a periodic signal, but no longer a sinusoidal oscillation. The tone of the lowest frequency determines the overall pitch, and the remaining tones determine the timbre or color of the sound.

A mixture of multiple oscillations, the frequencies of which fill a certain part of the frequency spectrum, is noise. If this frequency band is wide and all spectral components contribute equally, then the resulting sound is white noise. Narrowband noise can be obtained if white noise is subjected to frequency filtering using a band-pass filter.

A short-term sound effect in the form of a pressure surge is a sound pulse, which, depending on the shape and size, can be felt as a click or an explosion. If we select a short time interval from a pure tone, then we get a tonal impulse. The selection of a time interval from the noise allows you to get a noise pulse.

The human auditory system is able to sense sound waves in the form of pure musical tones, the frequencies of which lie in the band from 20 Hz to 20 kHz. The effective value of audible tone level is in the range of about 10 micro pascals (uPa), which corresponds to the absolute threshold of hearing in the central part of the perceived frequency range of musical tones, up to 100 Pa, which corresponds to a pain threshold. To compare sound waves in such a wide amplitude range, a logarithmic scale is used—the sound pressure level (SPL) in decibels, or the multiplied by 20 logarithm of the ratio of effective sound pressure values of two waves: $L = 20*\log(p_2/p_1)$ dB.

Decibel is a dimensionless quantity, but it can be used as a unit for measuring the sound level, if the SPL is always calculated in relation to the same reference level, the value $p_0 = 20.4$ uPa is taken as SPL $= 20*\log(p/p_0)$ dB. When using this unit, the level of thunder rolls is estimated at about 120 dB, aircraft noise or music at a rock festival corresponds to 110 dB, train passing noise is 100 dB, and noisy street sounds are 80 dB. A conversation in the room corresponds to a sound level of about 50–60 dB and a whisper—20–30 dB.

7.3.2.3   Frequency response: audibility of pure tones

With one and the same SPL, the sensation of loudness of pure tones of different frequencies turns out to be different. It turns out to be different and the minimum sound pressure at which there is still an auditory sensation or the threshold of hearing of pure tones of different frequencies. The threshold of hearing depends on the conditions of experience. The minimum SPL at which a harmonic sound wave is detected in the absence of other sounds is called an absolute threshold of hearing, or a threshold of hearing in silence [25]. The averaged curve of the threshold audibility of pure tones is shown in Figure 7.29.

The shape of the curve of the threshold audibility of pure tones is largely determined by the frequency filtering carried out in the outer and middle ear. The ear

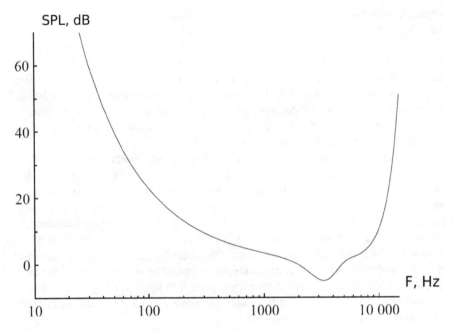

Figure 7.29   Threshold audibility of pure tones

device is best adapted for perceiving fluctuations in air pressure in the range in which the frequency spectrum of human speech sounds is located. This is reflected in the form of the frequency response of the threshold audibility. The value of the threshold pressure is also associated with the level of physiological noise of the human body, which is the result of, e.g., muscle contraction, limb movement, respiration, and blood flow in the vessels. There are also internal noises of the inner ear. The threshold level should be slightly above the level of physiological noise in order for these noises to be inaudible. The frequency distribution of physiological noise in combination with the frequency characteristics of the ear determines the frequency dependence of the hearing threshold.

7.3.2.4   Biomechanical spectrum analyzer

As already noted, a wave in the fluid of the inner ear causes mechanical oscillations of the basilar membrane of the cochlea. The profile and elasticity of the basilar membrane vary considerably along its length. Therefore, the spatial distribution of membrane oscillations depends on the frequency properties of the sound wave.

Each point along the basilar membrane can be associated with a so-called characteristic frequency—the frequency of the sound that causes the maximum response at that point of the basilar membrane. In the frequency range of sound up to 500 Hz, the distance from the apex to the point with a certain characteristic frequency is approximately proportional to the frequency; at frequencies above 500 Hz, this distance is proportional to the logarithm of the characteristic frequency.

Vibrations of the basilar membrane in response to a certain tone cannot be localized at a point, i.e., infinitely small area of the membrane. This is not surprising, the membrane is a rather elastic film, but one cannot make one point of this film oscillate, keeping the others still. Pure tone excites a rather wide area of the basilar membrane, and the selected point responds to close frequencies. Thus, each point of the main membrane can be associated with a bandpass acoustic filter with a central frequency equal to the characteristic one [27–30].

The band of the auditory filter, representing an equivalent description of the excitation of the membrane point, is called critical. The critical band in the frequency range below 500 Hz is almost constant and equal to 100 Hz. In the frequency range exceeding 500 Hz, the critical band grows in proportion to the frequency, reaching 4 kHz as the frequency approaches 20 kHz. In practical applications, the ear is modeled by a finite set of filters with non-overlapping bands equal to the critical bands at the center frequencies of the filters [31, 32]. The upper bound of the passband of one filter is equal to the lower bound of the passband of the next. There are only 25 such filters or frequency groups in the frequency range of audible sounds. Interestingly, the critical band corresponds to a displacement of about 1.5 mm along the basilar membrane.

Based on the results of measurement of critical bands as a function of frequency, it is possible to construct a scale of subjective frequency, or pitch, at which the width of all critical bands is taken to be equal to one unit of the subjective frequency (the Bark scale). This unit is called the bark. This means that the distance along the frequency axis, equal to the critical band, is always equal to one bark.

The described mechanism for the transformation of vibrations in the inner ear allows us to formulate a statement about the frequency-selective nature of the response of the auditory system to sound. A characteristic frequency is associated with each point of the basilar membrane. The flow of nerve impulses transmitted to the brain from this point of the membrane corresponds to the response of the auditory filter, the central frequency of which is equal to the characteristic frequency and the bandwidth to the critical band. Thus, the main membrane of the inner ear acts as a kind of biomechanical spectrum analyzer. Frequency selection in this analyzer determines the ability of the human auditory system to distinguish different sounds.

7.3.2.5   Frequency masking

The frequency resolution of the spectrum analyzer described in the previous section is limited by the bandwidth of the membrane point as an auditory filter, i.e., critical band. The spectral components that fall into the passband of the filter interact with each other in this auditory filter. When evaluating the results of the interaction of sound components, one must bear in mind that the inner ear is a nonlinear system. As it is known from the theory of systems, in nonlinear resonant systems, there are such effects as the suppression of a weak signal by a strong signal. Numerous psychoacoustic experiments have shown that a similar effect, called masking, exists in the human auditory system.

The effect of masking can be explained as follows. As noted above, even a pure tone excites a fairly wide area of the basilar membrane. Suppose that a second sound appears—a pure tone with a smaller amplitude and a frequency slightly different from the frequency of the first tone. The second tone should vibrate the area of the membrane, which is already oscillating under the action of the first tone. If the second tone were quiet, then it would have excited the membrane in the corresponding region and would have been heard. But the membrane in this area is already oscillating, so the second tone may be inaudible against the background of the first tone.

If the ear hears a tone with an SPL, e.g., 40 dB, and then a near frequency tone is added with a level of 20 dB falling in the same critical band as the first tone, then the ear integrates sound intensity within the critical band. Adding a second tone increases the level of the resultant sound only by 0.04 dB (it must be borne in mind that the intensities of the sounds, and not their effective pressure, are added) [32]. If such a small additive is below the threshold for distinguishing the organ of Corti, the second sound will not be audible. For the second tone to become audible, its level must be equal to or greater than the level of the first tone. But then the first tone may already be inaudible against the background of a more intense second tone. This effect is called frequency or spectral masking.

Masking is described quantitatively by changing the threshold of hearing in the presence of a masker – a masking sound of a large level (Figure 7.30). The dotted line of the diagram shows the threshold of audibility of a pure tone during masking. A sound of the level that is higher than the absolute threshold, but lower than the threshold of hearing during masking (maskee), is inaudible.

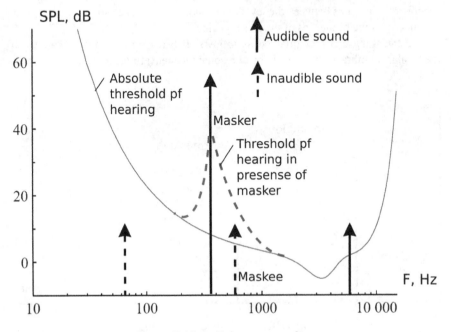

Figure 7.30   Frequency masking

So, two sounds with close frequencies that are at a frequency distance less than the width of the critical band can mask each other, because they excite almost the same area of the basilar membrane and interact in a nonlinear manner. Training or the desire to hear cannot help to detect the masked tone. The processes of higher nervous activity cannot eliminate masking, since information about the masked tone is not formed in the inner ear and is not directed along the auditory nerve to the brain.

7.3.2.6   Temporal masking

Frequency masking is always manifested in the conditions of the simultaneous existence of a masking and masked sound. General considerations about the limited frequency and time resolution of the auditory system allow us to conclude that one pulse sound is able to mask another sound. Due to the finite value of the temporal resolution of the auditory system, masking is manifested in an increase in the threshold of hearing of a weak sound within a certain time interval exceeding the interval of exposure to the masking sound. Indeed, there is the effect of masking sounds that follow each other with a time delay.

In Figure 7.31, it is shown the line of the hearing threshold of a short tonal impulse with a duration of several milliseconds, which comes with a delay after the masking sound—a long tonal impulse with a duration of several hundred milliseconds. The dashed line of the timing diagram shows the threshold of hearing of a short tonal impulse during masking. Tonal impulses with a level below the line of the graph are inaudible. Since the masked sound follows the masking sound, this masking is called post-masking. The duration of post-masking depends on the duration of the masking sound and can reach several hundred milliseconds [25,32]. The specific relationships between the levels of masking and masked sounds are strongly dependent on the properties of the two signals.

Postmasking can be explained as follows. It takes time for the vibrations of the basilar membrane (as a mechanical resonant system) to reach a stationary level

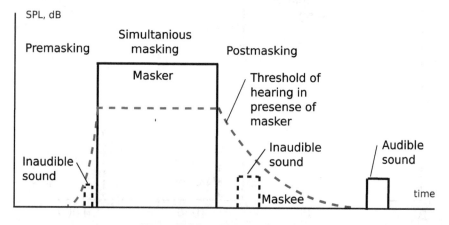

Figure 7.31   Temporal masking

when excited. The vibrations of the basilar membrane do not disappear instantly after the cessation of sound, they fade out gradually. As a result, the masking effect is manifested not only with the simultaneous existence of two sounds but also when the masking signal of high intensity disappears before the appearance of a weak masked signal.

Also important is the nature of the response of the internal cells of the organ of Corti to tonal impulses [27]. The internal cells respond to the front of the pulse with a large burst of the signal that enters the brain through the auditory nerve. The intensity of nerve impulses resembles the response of a high-pass filter, e.g., a differentiating circuit. If the tonal sound pulse is long enough, the signal level comes to a stationary value that is significantly less than the initial burst. In response to the end of the sound pulse, a negative surge appears, which means the loss of sensitivity of the internal cells. It takes time to restore the full sensitivity, and this time is proportional to the duration of the sound pulse. If a second sound pulse appears during the sensitivity recovery interval, the internal cells will not be able to respond to it, and it will be inaudible.

It is only necessary to bear in mind that the effect of postmasking will appear only if the first and second sound pulses have common frequency components. If the frequency components of the second pulse are significantly different from the frequency components of the first pulse, then cells that are in a different area of the basilar membrane and have not lost their sensitivity will react to the second pulse.

The engineer's view should have long ago focused on the left side of the diagram in Figure 7.31. It seems that the diagram contradicts the fundamental law of physics—the principle of causality. It shows that the masking sound changes the threshold of hearing a weak sound before its appearance. However, such advanced masking, or pre-masking, really exists [25, 27, 33]. An increase in the threshold of hearing of a pulsed sound preceding a masking sound pulse occurs in a relatively short interval, the duration of which can be as long as several tens of milliseconds. The explanation is that it takes some time for the auditory system to form a feeling out of sound. The stronger the sound, the sooner the auditory system responds to it. Forming a feeling of a weak signal requires more time, which is spent on processing in the central nervous system. A strong masking sound is already heard at the time of the formation of the sensation of a weak test sound. This explains the pre-masking effect.

Temporal masking that exists in the auditory system is one of the important effects taken into account when creating a psychoacoustic model of hearing and designing audio compression codecs.

7.3.2.7 Hearing limitations and concept of audio data compression

If the critical bands of the auditory system as a biomechanical spectrum analyzer were very small and the equivalent filters were extremely narrowband, then the effect of frequency masking would practically disappear. However, in this case, the temporal resolution of the auditory system would be very bad (when the bandwidth of the bandpass filter tends to zero, parameters such as the duration of the impulse response and, accordingly, the rise time of the output signal in response to the sound impulse tend to infinity). In reality, the critical band (the bandwidth of the equivalent filters of the auditory analyzer) depends on the center frequency and the widest band at high frequencies. This gives the worst frequency resolution, but the best temporal resolution. At the lower frequencies of the audio range, the bandwidth is minimal, which gives the best frequency discrimination, but the worst time resolution.

So, the auditory system as a discriminating organ has a finite resolution in the time and in the frequency domain. Approaching the problem of temporal and frequency resolution of hearing from the standpoint of system theory, we conclude that sounds that are relatively small and that are in frequency or in temporal proximity with a stronger sound can be masked. The sensation of one sound may weaken or disappear completely in the presence of another sound. In quantitative terms, masking is manifested in an increase in the threshold of hearing of one sound in the presence of a second, stronger one. It is characterized by the amount by which the threshold of hearing of the masked signal increases (relative to the threshold of hearing in silence) in the presence of a masking signal. Due to the finite frequency and time resolution of the auditory system, masking can be in the frequency and time domains.

The hearing properties described above make it possible to formulate a general principle of data transmission in the auditory system, evaluated from the standpoint of systems theory and communication technology. A broadband signal with a large dynamic range, such as sound, is transformed into a set of narrowband signals, each of which is transmitted in a small dynamic range and with a limited time resolution. Information that is lost in the process of such a transformation (losses are described by a masking process) is not transmitted to the brain and cannot be audible. The inner ear acts in a sense as a lossy compression coder. The overwhelming part of the information that the human ear does not hear is lost in this coder—the inner ear. The principles described underlie the operation of efficient digital audio data compression systems based on psychoacoustic algorithms [28, 29, 31–34].

7.3.3   Main algorithms and standards for audio compression

7.3.3.1   MPEG-1 audio coding

7.3.3.1.1   MPEG-1 audio layers

The MPEG-1 audio compression algorithm [35], like all other MPEG family standards, is distinguished by the use of the psychoacoustic model of hearing. MPEG-1 Audio Standard, officially referred to as "ISO/IEC 11172-3—Information Technology—Coding of Moving Pictures and Associated Audio for Digital Storage media at up to about 1.5 Mbit/s. Part 3 Audio," defines three compression subsets: Layer I, Layer II, and Layer III. They have a lot in common and are downward-compatible, i.e., they understand the bottom layers. They assume the use of sampling frequencies: 32, 44.1, and 48 kHz.

Layer I provides high quality, but at a relatively high compressed data rate. The bit rate ranges from 32 (mono) to 448 kbits/s (stereo). Quality close to CD stereo is achieved at bit rates of 256–384 kbits/s.

Layer II is characterized by greater complexity of the algorithm used. It has a lower compressed data bit rate. The coded data stream bit rate ranges from 32–192 (mono) to 64–384 kbits/s (stereo). Quality close to CD stereo is achieved at bit rates of 192–256 kbits/s.

Layer III, commonly known as MP3, is the most complex and provides the lowest bit rate of encoded data. The range of compressed data stream bit rates defined by the standard is 32–320 kbits/s. Layer III provides significantly better quality for the same bit rate than Layers I and II. CD-quality audio is achieved with 128–192 kbits/s (i.e., compression ratios in the range of 12:1 to 8:I).

7.3.3.1.2 MPEG-1 audio layer I

The general scheme of the algorithm implemented by the MPEG-1 Layer I standard corresponds to Figure 7.28. Time-to-frequency mapping is implemented using a bank of 32 narrowband filters. Frequency filtering is subjected to an array of 32 samples of the audio signal. The frequency band of the audio signal is divided into 32 sub-bands with equal widths of frequency bands. An analysis sub-band filterbank is used to split the broadband signal with sampling frequency f_s into 32 equally spaced sub-bands with sampling frequencies $f_s/32$; 12 groups of 32 frequency coefficients (spectral coefficients) or sub-band samples from the filter outputs form a frame of Layer I.

Twelve samples from the output of each filter are used to determine a scalefactor. The calculation of the scalefactor for each sub-band is performed every 12 sub-band samples. There are a total of 63 scalefactors specified in the MPEG-1 Audio standard. The maximum of the absolute value of these 12 samples is calculated. The next largest value of the table for codes of the scalefactors is used as the scalefactor. The sub-band output is divided by the scalefactor before quantizing. The scalefactor is used to make sure that the spectral coefficients make use of the entire range of the quantizer.

The task of the bit-allocation block is to distribute the total number of bits allocated for encoding the sub-band samples, so as to make quantization noise inaudible or minimize the audibility of the quantization noise. This is achieved with the help of a psychoacoustic model of hearing. For each sub-band, a minimum masking threshold is calculated. The minimum masking threshold is found using a fast Fourier transform (FFT) of the encoded audio signal in several steps described below.

With the use of the psychoacoustic model, the threshold of audibility (absolute threshold) is found and significant masking signals—maskers are defined. It is determined by the type of masker, which can be tonal (or close to tonal, i.e., similar to a sine wave) or noise (or similar to noise). Based on these data, the minimum masking threshold and the SMR in each sub-band (in the band of each filter) are calculated. In Layer I, the psychoacoustic model uses only frequency masking.

A lookup table in the MPEG Audio standard also provides an estimate of the signal-to-noise ratio (SNR), assuming quantization to a given number of quantizer levels. Then the mask-to-noise ratio (MNR) is defined as the difference:

$$MNR_{dB} = SNR_{dB} - SMR_{dB}.$$

The allocation procedure is an iterative procedure, where in each iteration step, the number of levels of the sub-band samples of the greatest benefit is increased. The lowest MNR is determined, over all the sub-bands, and the number of code-bits allocated to this sub-band is incremented. Then a new estimate of the SNR is made, and the process iterates until no more bits are left to allocate.

Using the FFT in parallel with sub-band filtering allows us to compensate for the lack of frequency selectivity of sub-band filters in the low-frequency region. As noted, the bands of all 32 sub-band filters are the same, therefore in the low-frequency region, several critical bands fit into the band of one filter. The method used in the standard provides both sufficient temporal resolution (polyphase filter with optimized window for minimal pre-echoes) and sufficient frequency resolution (1 024-point FFT for sufficient spectral resolution for the calculation of the masking thresholds).

Then quantization and encoding of sub-band samples are performed. A linear midtread quantizer with a symmetric zero representation is used to quantize the sub-band samples. This representation prevents small value changes around zero from quantizing to different levels. Each of the sub-band samples is normalized by dividing its value by the scalefactor.

16 uniform quantizers are pre-calculated, and for each sub-band, the quantizer giving the lowest distortion is chosen. The index of the quantizer is sent as 4 bits of side information for each sub-band. The maximum resolution of each quantizer is 15 bits.

Then the bitstream formatting is performed. Quantized samples are sent to a stream formatting unit of compressed audio data. Elementary stream of compressed audio data is a sequence of frames. A frame is a part of a stream that is decoded without attracting additional data, i.e., using only the information contained in the frame. In Layer I, the frame contains information about 384 samples. The frame begins with a sync word and ends before the next sync word. A frame always consists of an integer number of slots (4 bytes).

The frame consists of four parts: header, error check, audio data, and ancillary data (Figure 7.32). The header is the first part of the frame. This is a string of 32 bits. The first eleven bits of a frame header are always equal to 1. They are called "frame sync". The header contains a number of identifiers.

ID—this bit is to indicate the algorithm (1—MPEG audio and 0—reserved).

Layer—2-bit code to indicate the layer used (Layers I, II, and III).

Protection—this bit is set to 1 if there is CRC protection.

Bit rate index—4-bit code of the bit rate in kbits/s (there are 14 specified bit rates from 32 to 448 kbits/s for Layer I, 32 to 384 for Layer II, and 32 to 320 for layer III).

Sampling frequency—2 bits that indicate the sampling frequency (32, 44.1, 48 kHz),

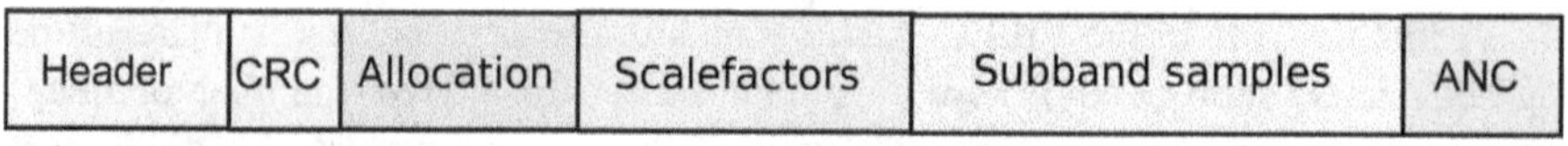

Header	CRC	Allocation	Scalefactors	Subband samples	ANC

Figure 7.32   MPEG-1 Layer I Frame

mode—2 bits to indicate the mode: single channel, dual channel, stereo, and joint stereo. The dual-channel mode consists of two channels that are not intended to be played together. The stereo mode consists of two channels that are encoded separately but played together. The joint stereo mode consists of two channels that are encoded together.

Mode extension—2 bits that are used in the joint stereo mode to indicate which sub-bands are in intensity stereo.

Copyright—this bit that is set to 1 if the material is copy righted.

Original—this bit that is set to for original bistreams.

Emphasis—2 bits to indicate the type of de-emphasis (no emphasis, 50/15 ms emphasis, CCITT J.17).

If the CRC bit is set, the header is followed by a 16-bit CRC parity-check word which is used for optional error detection within the encoded bitstream ("CRC"). The CRC parity-check word is followed by the bit allocation words named "Allocation." It is reasonable to remind that "allocation" is a 4-bit code word to indicate the number of bits used to code the samples in sub-bands (from sub-band 0 to sub-band 31). The allocation data are followed by the set of 6-bit "scalefactors." "Scalefactor" indicates the factor of a sub-band by which the requantized samples of this sub-band shall be multiplied. The scalefactor data are followed by the quantized 384 samples ("Sub-band samples"). The sub-band sample data are followed by ancillary data ("ANC").

Although most of this information may be the same for all frames, MPEG decided to give each audio frame such a header in order to simplify synchronization and bitstream editing.

7.3.3.1.3   MPEG-1 audio layer II
Layer II has much in common with Layer I. The main difference relates to the fact that three groups of 12 sub-band samples are included in the frame, which gives 1 152 samples in each frame. This means that the number of sub-band samples corresponding to 3 * 384 (1 152) input PCM samples is calculated for three blocks. When calculating the masking threshold, both frequency and time masking are taken into account. If the scalefactor has a close value for three groups of frame samples, then one scalefactor can be used for the entire frame. Resolution of quantizers is increased from 15 to 16.

7.3.3.1.4   MPEG-1 audio layer III (MP3)
A subset of compression standard MPEG-1 Audio named Layer III has become widely known and popular under the name MP3 [31]. Layer III is much more complex than Layers I and II but provides higher quality at the same coded data stream bit rate. It eliminates the problem of Layers I and II, associated with the decomposition of the audio signal using a block of 32 filters with equal frequency bands. At low frequencies, the bandwidth of the sub-bands is much wider than the critical bands, making it difficult to accurately calculate the mask-to-signal ratio. A simple way to improve the frequency resolution would be to increase the number of sub-bands of the filter

bank. However, one of the requirements for Layer III was backward compatibility with Layers I and II. In this case, the optimum was to perform the spectral decomposition in two stages. First, the spectrum of the sound signal is divided into 32 sub-bands as it is done in Layers I and II. Then the output of each sub-band is transformed using a modified DCT (MDCT) with a 50% overlap. Layer III defines two sizes for calculating MDCT: 6 and 18. It can be used to reduce the size of the filter.

Different bit rate control methods are used for Layers I–II and Layer III. In Layers I and II, a bit allocation process is used. This means that a number of bits is assigned to each sample (or group of samples) in each sub-band. A noise allocation is used for Layer III. It means that the actually injected noise is the variable to be controlled.

In Layers I and II, a fixed PCM code is used for each sub-band sample, with the exception that in Layer II, quantized samples may be grouped. In Layer III, Huffman codes are used to represent the quantized frequency samples. These Huffman codes are variable-length codes that allow for a more efficient bitstream representation of the quantized samples at the cost of additional complexity.

The process to find the optimum quantization step size values and scalefactors for a given block, bit rate, and output from the psychoacoustic model is usually done by two iteration loops. These are the rate loop (inner iteration loop) and noise control loop (outer iteration loop).

In either case, the result is a set of quantization parameters and quantized output samples that are given to the bitstream formatter. Quantization is done via a nonlinear power-law quantizer. In this case, smaller values are automatically coded with greater accuracy. It means that some noise shaping is built into the quantization process.

The bitstream formatter takes the quantized filterbank outputs, the noise allocation, and other required side information, and encodes and formats that information in an efficient fashion. The Huffman codes are also inserted at this point.

7.3.3.2   MPEG-2 audio coding

7.3.3.2.1   ISO/IEC 13818-3 audio coding

A standard on low bit rate coding for carrying of high-quality digital mono or stereo audio signals was established in MPEG-1 Audio (ISO/IEC 11172-3). MPEG-2 Audio standard, officially referred to as "ISO/IEC 13818-3—Information Technology—Generic coding of moving pictures and associated audio information—Part 3: Audio," is an extension of MPEG-1 Audio [36]. This standard is backward compatible to ISO/IEC 11172-3 coded mono, stereo, or dual channel audio programs.

Three additional sampling frequencies are provided for Layers I, II, and III: 16, 22.05, and 24 kHz. This extension is done to achieve better audio quality at very low bit rates (less than 64 kbits/s per audio channel). The syntax of the bit stream and coding techniques of MPEG-1 Audio are maintained except for a new definition of the sampling frequency field, the bit rate index field, and the bit allocation tables.

MPEG-2 Audio also provides the extension of MPEG-1 Audio to 3/2 multi-channel audio and an optional low frequency enhancement (LFE) channel. This extension allows using an additional center loudspeaker channel C and two surround loudspeaker channels LS and RS. These three speakers augment the front left and right loudspeaker channels L and R. Of course, this requires the transmission of five audio signals. An LFE channel can be added to this configuration. The LFE channel is capable of handling signals in the range from 15 to 120 Hz. It is similar to the LFE channel proposed in the film industry (5.1 system). MPEG-2 Audio multichannel audio systems provide enhanced stereophonic stereo performance.

7.3.3.2.2 ISO/IEC 13818-3 advanced audio coding (AAC)

MPEG-2 AAC, officially referred to as "ISO/IEC 13818-7—Information Technology—Generic coding of moving pictures and associated audio information—Part 7: Advanced Audio Coding (AAC)" uses the same basic coding concept such as MPEG-1 Audio Layer III: high-frequency resolution filterbank, non-uniform quantization, Huffman coding, and iteration loop structure [36, 37]. But new tools are used in AAC to enhance coding efficiency and improve quality at low bit rates.

The tools to enhance coding efficiency are higher frequency resolution, prediction, improved joint stereo coding, and improved Huffman coding.

The frequency decomposition is done by using an MDCT only instead of combination of 32-band filterbank and MDCT. In the AAC algorithm, a window length of 2 048 and 256 samples is used. The MDCT algorithm includes 50% overlap. That is why 1 024 or 128 spectral coefficients are generated (compared with up to 576 for Layer III). The long block length allows us to get enhancement in compression for stationary portions of the input audio signal. The short block length allows us to process sharp sound attacks.

An optional backward prediction can be used to increase higher coding efficiency for the signals that are very tone like ones.

Some tools allow us to get high quality for difficult-to-compress signals. AAC uses a switched MDCT filterbank. The impulse response is about 5 ms for short blocks at the sampling frequency of 48 kHz (Layer III has the impulse response of 18 ms) [38]. As a result, the pre-echo artifacts are reduced.

Thanks to many small improvements and enhancements, MPEG-2 AAC allows us to get on average the same quality as MPEG-1 Layer III but at 70% of the bit rate of the coded data.

The MPEG-2 AAC standard is widely used. On the one hand, it allows to deliver the audio component for theaters at 320 kbits/s for five channels: left, right, center, left-surround, and right-surround. The 5.1 system also has an LFE channel. On the other hand, MPEG-2 AAC can achieve high-quality stereo sound at the bit rates below 128 kbits.

7.3.3.3 MPEG-4 audio coding

Several different audio compression components are included in the MPEG-4 Audio standard officially referred to as "ISO/IEC 14496-3:2009 Information technology—Coding of audio-visual objects—Part 3: Audio" [39]. These are speech compression, perceptually based coders, and text to speech. The primary general audio coder, MPEG-4 AAC is similar to the MPEG-2 AAC standard with some new tools. These tools are a perceptual noise substitution tool, the interband prediction tool in the spectral coding block, transform-domain weighted interleave vector quantization, and bit sliced arithmetic coding [40].

7.3.3.4 Dolby AC-3 audio coding

Dolby AC-3 audio compression algorithm was developed as a part of standardization activities of the Grand Alliance, which was developing a standard for HDTV in the United States [41, 42]. But it was first released in the cinema industry with the movie "Batman Returns" in 1992. AC-3 became a part of the United States Advanced Television Systems Committee (ATSC) standard in 1995 under the name: ATSC Standard A/53, "Digital Television Standard for HDTV Transmission" [43].

Many details of the Dolby-AC3 concept are similar to details of the MPEG Audio algorithms. The actual audio data transmitted by the AC-3 bit stream are the quantized frequency coefficients. As in the MPEG Audio algorithms, the Dolby-AC3 scheme uses the MDCT with 50% overlap for time-to-frequency mapping. AC-3 uses two different sizes of windows. Like in the MPEG Audio schemes, there is a psychoacoustic model in AC-3. This model incorporates the hearing thresholds in the quiet and in the presence of tone-like and noise-like maskers. But the approaches to bit allocation are quite different.

In the MPEG Audio algorithms, the audio signal being encoded is the input to the bit allocation procedure. Then the bit allocation data are sent to the decoder as side information. In the AC-3 scheme, a representation of the spectral envelope is provided to the bit allocation procedure sent to the decoder.

The frequency coefficients are calculated in floating point form and in binary exponential notation. Each coefficient consists of an exponent and a mantissa. Exponents indicate the number of leading zeros in the binary representation of a frequency coefficient. They are estimates of the relative magnitude of the frequency coefficients. The exponent acts as a scale factor for each mantissa.

The Dolby-AC3 algorithm uses the exponents of the MDCT coefficients in the binary exponential form as the representation of the spectral envelope. The sequence of the exponents is sent to the bit allocation block, which works in conjunction with a psychoacoustic model. The spectral envelope is used to generate the number of bits to be used to quantize the mantissa of the binary exponential representation of the MDCT coefficients.

AC-3 is a flexible audio data compression technology. It is capable to encode a range of audio channel formats into a low rate bit stream. Channel formats range from monophonic to 5.1 channels.

7.4   Data packing and data transmission

7.4.1   Elementary video stream

7.4.1.1   Elementary video stream components

The most important feature of MPEG compression standards is the representation of television image and sound signals in a form that allows you to treat video and audio streams as computer data streams. These data can be recorded on a wide variety of information carriers, transmitted and received using communication channels and telecommunications networks that exist today and will appear in the future. The description of the principles of data stream formation, packaging, and transmission will be given for definiteness on the example of the MPEG-2 standard specifications, which became the basis for the creation of the widely used AVC/H.264 and HEVC/H.265 standards today. Among the constituent parts of the MPEG-2 standard, there are three main ones: 13818-1—Systems, 13818-2—Video, and 13818-3—Audio. Specification 13818-2 regulates the code representation and the decoding process for reproducing compressed television images. 13818-2 involves the compression of the video stream by eliminating the spatial and temporal redundancy inherent in the television image. The elimination of spatial redundancy is based on the use of DCT, temporal—on differential coding with motion compensation, as described above. Specification 13818-3 specifies the code representation of the audio signal. System Specification 13818-1 sets the rules for combining video and audio data into a single stream.

The video data stream defined by specification 13818-2 is a hierarchical structure, the elements of which are built and combined with each other in accordance with certain syntactic and semantic rules. There are six types of elements of this hierarchical structure:

- video sequence
- group of pictures
- picture
- slice
- macroblock
- block

Video sequence is a top-level video stream element. It is a series of consecutive frames of a television image. All MPEG standards allow for both progressive and interlaced sequences. Interlaced sequence is a series of television fields. In the process of compression, fields can be encoded separately. This gives a field type image. The two fields, encoded as a television frame, form a frame-like image. In a single interlaced sequence, both field images and frame image may be used. When progressively scanned image sequences are used, each image is a frame.

According to the differential encoding methods used, three types of images are distinguished: I, P, and B. A series of images containing one I-image is called a group of images. An example of a video sequence with different types of images is shown in Figure 7.26 (arrows indicate the direction of prediction within the same group of images). The larger the group of images, the greater the degree of compression that can be achieved.

From the informational point of view, each image consists of three rectangular image arrays of images: the luminance Y and two matrices of the color difference signals C_B and C_R. The MPEG-2 standard allows for various matrix structures. The ratio between the number of luma and chroma samples is determined by the sampling format. In 4:4:4 format, all matrices are the same. The 4:2:2 format is different in that all three matrices have the same vertical dimensions, but in the horizontal direction of the matrix of the color, difference signals have half the number of elements. In the case of the 4:2:0 format, the dimensions of the matrices C_B and C_R are two times smaller than Y, both in the horizontal and vertical directions (Figure 7.33).

As you can see, the samples of the luminance signal and the color difference signals in the 4:2:0 format have different positions. This is due to the desire to support both progressive and interlaced images. If the samples of the luminance signal and the color-difference signals were the same, then if the number of samples of the color-difference signals was reduced by two times in the vertical, all the samples of C_B and C_R would be in one and the same field of the interlaced image. The absence of color difference signals in one field means that the sampling rate in time for large colored parts is halved and becomes equal to the frame rate. This could lead to color distortions in the transmission of dynamic images. In order for the color to be present in each field, the samples of the color difference signals are between the lines of the luminance signal matrix. This, of course, does not mean that additional "inter-line" signals are needed that carry color information. In each field, samples of color difference signals between the lines are calculated by interpolation before compression. After decoding and before representing a color image, the values of the color difference signals should be calculated by interpolating at the same points as the luminance signal samples.

Each image is divided into slices, which consist of macroblocks (Figure 7.34). A macroblock is an area of an image with a size of 16 × 16 pixels. A macroblock is made up of blocks of 8 × 8 image elements (these blocks are applied to the DCT when encoding). Each macroblock contains a group of four blocks with luminance samples and a group of blocks with chrominance samples taken from the same

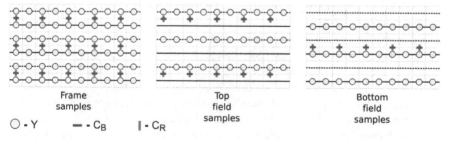

Figure 7.33   4:2:0 sampling structure (MPEG-2)

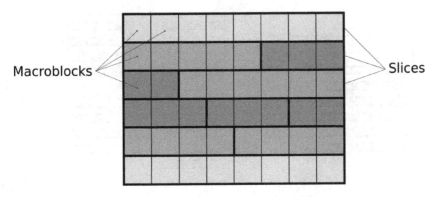

Figure 7.34   Slices and macroblocks

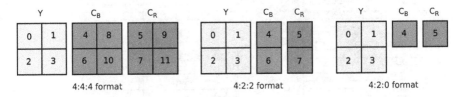

Figure 7.35   Macroblock structures (4:4:4, 4:2:2, and 4:2:0 formats)

image area as the brightness block (Figure 7.35). The number of blocks with chroma samples depends on the sampling format: four blocks of C_B and C_R in 4:4:4 format, two in 4:2:2 format, and one in 4:2:0 format.

In images of the "frame" type, in which both frame and field coding can be used, two variants of the internal organization of the macroblock are possible (Figure 7.36). In the case of frame coding, each block of brightness Y is formed from alternating lines of two fields (Figure 7.36a). During field coding, each block Y is formed from the lines of only one of the two fields (Figure 7.36b). Chroma blocks are formed according to the same rules in the case of 4:2:2 and 4:4:4 sampling formats. However, when using the 4:2:0 format, chroma blocks are organized to perform a DCT within the frame structure (Figure 7.36a).

All structural elements of the video data stream obtained as a result of intraframe and interframe coding (except for the macroblock and block) are supplemented with special and unique starting codes. Each element contains a header followed by data from lower level elements. The header of the video sequence (as the top-level element) provides various additional information, e.g., the size and aspect ratio of the image, the frame rate, the data rate, the quantization matrix, the image color sampling format, the coordinates of the primary colors and white color, the parameters of the brightness matrix and color difference signals, transfer characteristic parameters (gamma).

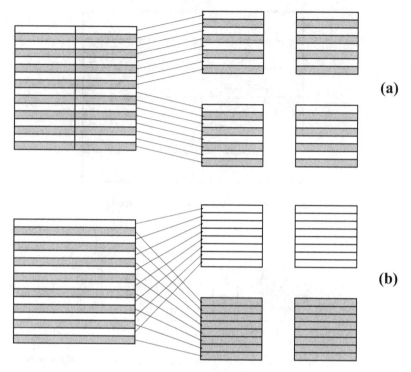

Figure 7.36   Structure of macroblock Y: frame (a) and field (b) coding

7.4.1.2   The order of transmission of images in the elementary stream of video data

As a result of compression, the amount of data representing the original images is compressed (Figure 7.37). But the MPEG-2 standard does not regulate the coding process itself, therefore the images (presentation blocks) in it are considered as the result of decoding compressed images (access blocks). The use of bidirectional prediction leads to the fact that the decoder can proceed to decoding a type B image only after both the preceding and the subsequent reference images with which the prediction has been calculated are already received and decoded. In order not to install huge buffer arrays in the decoder, in the data stream at the output of the encoder (this stream is called the elementary video data stream), the encoded images follow in decoding order. For example, instead of the sequence I-B-B-P, the series I-P-B-B is formed (Figure 7.38).

7.4.2   Packetized elementary video stream

The MPEG-2 system specification (ISO/IEC 13818-1) describes the integration of elementary streams of one or several television programs into a single data stream, convenient for recording or transmitting via digital communication channels. It should be noted that the MPEG-2 standard does not define protection against errors possible during recording or transmission, although, of course, it provides such an opportunity,

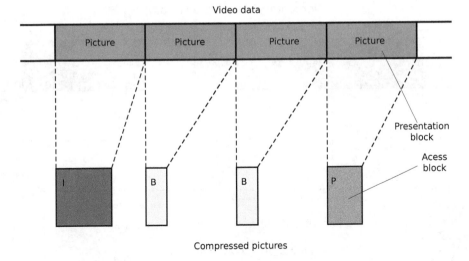

Figure 7.37 Conversion of presentation blocks to access blocks during compression

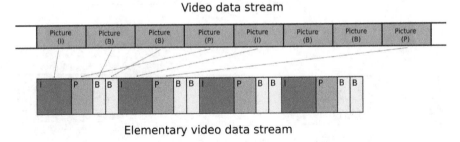

Figure 7.38 Formation of elementary stream of video data

making it easier to protect due to the optimal selection of stream parameters. MPEG-2 defines two possible forms of a single data stream—a program stream and a transport stream. The first step on the way to getting a single stream is the formation of a packetized elementary stream (PES), which is a sequence of PES packets (Figure 7.39). Each packet consists of a header and user data, or payload, which is a fragment of the original elementary stream. There is no requirement for matching the beginning of the packet payload and the start of the access blocks, so the start of the access block can be at any point of the PES packet, and several small access blocks can fall into one PES packet. PES packets can be of variable length. This freedom can be used in different ways. For example, you can simply set a fixed length for all packets, or you can coordinate the beginning of a packet with the beginning of an access block.

At the beginning of the PES-packet header (Figure 7.40), there is a 32-bit start code, consisting of a start prefix and a stream identifier. The stream identifier allows

Elementary video data stream

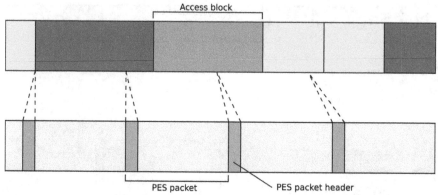

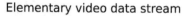

Figure 7.39   Formation of PES

PES packet

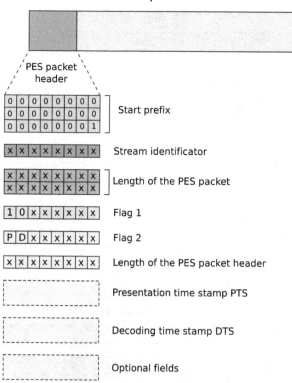

Figure 7.40   PES packet header

you to allocate PES packets belonging to one elementary stream of a television program. The specification defines the allowed values of numbers in the identifier field for 32 elementary audio streams and 16 elementary video data streams. Flags 1 and 2 are bits indicating the presence or absence of additional fields in the header, which are not mandatory. These fields are used to transfer additional information, such as copyright, scrambling, and priority. Of particular importance are the bits P and D of flag 2, indicating the presence of fields with time stamps PST (presentation time stamps) and DTS (decoding time stamps). Time stamps are a mechanism for synchronizing data streams in a decoder.

7.4.3   Elementary audio stream

As a result of compression, the bit rate of data representing the original audio is decreased. The MPEG standards do not regulate the coding process itself, therefore the presentation units in it are considered as the result of decoding access units. Audio and video elementary streams consist of access units. In the case of video, an access unit is a coded picture. In the case of audio, an access unit is the coded representation of an audio frame. A sequence of coded frames described in section 7.3 is the audio elementary stream.

A broadcast program can have several audio channels being coded into different elementary streams. Each elementary audio bit stream is segmented into PES. A PES stream consists of PES packets, all of whose payloads consist of data from a single elementary audio stream, and all of which have the same stream identifier. Then respective packets are multiplexed into program stream or transport stream. Packetized elementary audio streams are formed in the same way as packetized elementary video streams. The main difference is that a packetized elementary audio stream contains presentation time stamps (PTS) only. In the case of audio, if a PTS is present in PES packet header, it shall refer to the first audio frame commencing in the PES packet. The audio PTS and video PTS are both samples from a common time clock. The multimedia synchronization is achieved by PTS. Encoders save time stamps at capture time. Decoders use those time stamps to schedule presentations.

7.4.4   Program stream

The program stream combines elementary streams that form a television program (Figure 7.41). When forming a program stream, blocks of PES packets are formed. The block contains a block header, a system header (optional), followed by a number of PES packets. The length of a program stream block can be arbitrary, the only restriction is that block headers should appear no less than 0.7 s later. This is due to the fact that the header contains important information—the reference system time. The system header contains information about the characteristics of the program stream, such as, e.g., the maximum data transfer rate and the number of video and audio elementary streams. The decoder uses this information, e.g., to decide whether it can decode this program stream.

A program stream combines elementary streams of a single program that have a common time base. It is intended for use in an environment that does not introduce errors into digital data. The reason for this is relatively large blocks of variable

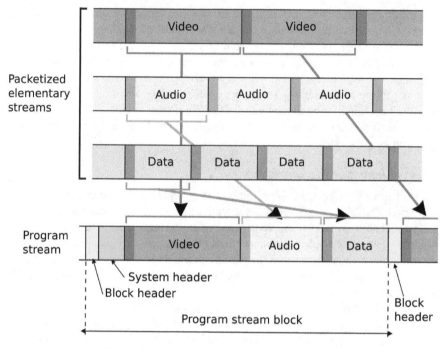

Figure 7.41   Formation of program stream

length. Distortion due to errors of one block can mean the loss of, e.g., a whole frame of the television image. Since the block length is variable, the decoder cannot predict the end time of one block and the beginning of another and is forced to rely only on the length information contained in the header. If the corresponding header field is hit by errors, the decoder will go out of sync and lose at least one block. The advantages obtained when using a program stream include the fact that the procedure for demultiplexing a program stream is relatively simple.

7.4.5   Transport stream

7.4.5.1   Structure

A transport stream may combine packet elementary streams carrying data from several programs with independent time bases. It consists of short packets of fixed length (188 bytes). Elementary video, audio, and additional data streams (e.g., teletext) are broken up into fragments equal in length to the payload of the transport packet (184 bytes) and multiplexed into a single stream (Figure 7.42). This process is subject to a number of restrictions.

- The first byte of each PES packet of the elementary stream shall be the first byte of the transport packet payload.
- Each transport packet may contain data of only one PES packet.

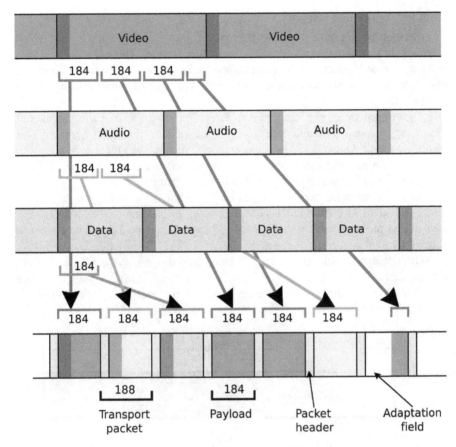

Figure 7.42   Formation of transport stream

If a PES packet does not have a length multiple of 184 bytes, then one of the transport packets is not completely filled with the PES packet data. In this case, the excess space is filled with the adaptation field (Figure 7.37). Transport packets carrying different elementary streams may appear in a random order, but packets belonging to the same elementary stream must follow the transport stream in chronological order, i.e., in the order of their "cutting" of the PES packages.

The structure of the transport stream is optimized for data transmission conditions in communication channels with noise. This is manifested primarily in the small length of the packages. Typical examples of protection against transport data errors are given by digital television broadcasting systems. In the DVB and ISDB systems, 16 check bytes of the Reed–Solomon code are added to 188 bytes of each transport packet, which allows us to correct up to 8 affected bytes in each packet. In ATSC, 20 check bytes are added to each packet, which allows for up to 10-byte errors to be corrected in one packet.

290 Digital television fundamentals

7.4.5.2 Transport packet

The transport packet starts with a 4-byte header (Figure 7.43), the first byte of which is the synchronization (the number 47 in hexadecimal code). This value is not unique and may appear in other fields of the transport packet. However, the fact that headers always follow with an interval of 188 bytes makes it easy to determine the beginning of a packet.

A transport stream may carry several television programs consisting of a set of elementary streams. To identify packets belonging to one elementary stream, a 13-bit identifier is used. Of the 8 192 possible values, 17 are reserved for special purposes, and the remaining 8 175 can be used for use as numbers of elementary streams. Thus, a single transport stream can carry up to 8 175 elementary streams.

An important component of the header is a continuity counter, which is incremented in successive transport packets belonging to the same elementary stream. This allows the decoder to detect the loss of the transport packet and take steps to conceal errors that may occur due to the loss.

The adaptation field is optional. It can be used not only to fill the "voids" (Figure 7.37). This field also carries important additional information about the use of packet data, e.g., the program reference time (PCR—Program Clock Reference).

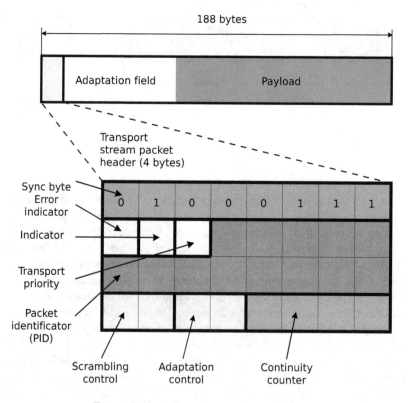

Figure 7.43   Transport packet structure

7.4.5.3 Hierarchical program identification

The identifier of a transport packet belonging to a certain elementary stream is the PID value (Figure 7.43). And for the recognition of elementary streams and their integration into television programs, the program information PSI (Program Specific Information) is used, which must necessarily be transmitted in the transport stream. The system specification MPEG-2 defines 4 types of tables with program information:

- program association table (PAT)
- program map Table (PMT)
- network information table (NIT)
- conditional access table

Each of these tables is transmitted as a payload of one or more transport packets. The PAT connection table is always carried by transport packets with a PID = 0. This table (Figure 7.44) reports the list of numbers of all programs that are contained in the transport stream and indicates the identifiers of the packets that contain the PMTs

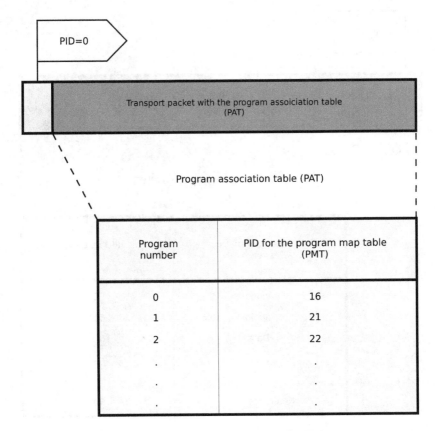

Figure 7.44   Presentation of PAT

with information about the programs and the elementary streams from which they are added. Program number 0 is reserved, it is used to indicate the PID of a packet with NIT network information about transport streaming networks, channel frequencies, modulation characteristics, etc. (in the example of Figure 7.39, the PID of a packet with NIT is 16, and the PID of a packet with PMT information about program 1 is 21).

The PMT (Figure 7.40 shows an example of a PMT for program 1 with a PID of 21) provides information about the program and the elementary streams from which it is composed. From the example of Figure 7.45, it follows that the elementary video data stream of this program is carried by packets with PID = 50, audio stream—by packets with PID = 51, additional data—by packets with PID = 52. The table also indicates the PID of the transport packets carrying the timestamps of this program (usually these packets have the same PID as the elementary video stream).

Tables with PSI form a hierarchical index mechanism. Figure 7.46 shows the principle of multiplexing elementary and transport streams in the process of obtaining a multi-program transport stream, and Figure 7.47 illustrates the demultiplexing

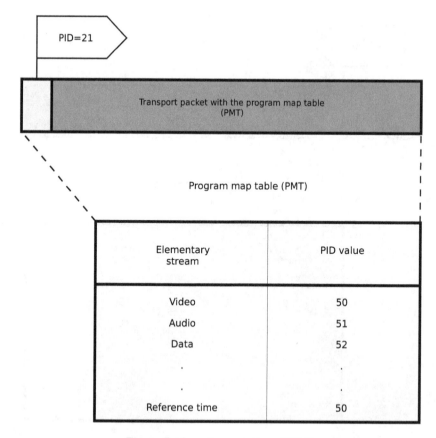

Figure 7.45   Presentation of PMT

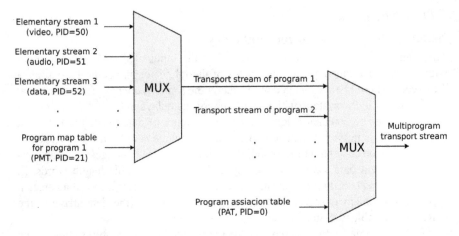

Figure 7.46   *Transport stream multiplexing model*

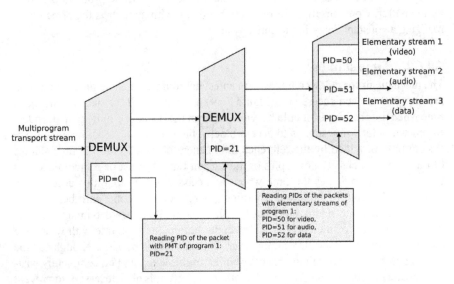

Figure 7.47   *Transport stream demultiplexing model (extracting elementary streams of program 1)*

of a transport stream to extract the elementary streams from which television program 1 is composed (the PID values in these figures correspond to the examples of the tables shown in Figures 7.39 and 7.40).

Due to the small packet length, the transport stream can easily carry several television programs with different time bases, but this has to be paid by a more a complicated multiplexing and demultiplexing scheme than in the case of a program flow.

7.4.6   Synchronization

7.4.6.1   The principle of constant delay

The frames of the television image are fed to the input of the MPEG-2 encoder with a constant frequency, and the frames of the television image at the output of the decoder should be reproduced with exactly the same frequency. This means that the total delay in the system, which is the sum of the delays in the individual circuit elements, must be constant (Figure 7.48). The amount of data needed to represent coded images is not a constant. It depends on the image detail, on the presence of fast-moving objects, and on the encoding method (I, P, and B images are character-ized by different data volumes). Entropy coding forms variable-length words. In order to load the communication channel evenly, the data must follow at a constant speed. The problem is solved by using the encoder buffer (the data arrive in the buffer at a variable rate, and go out at a constant one).

The coded images (access blocks) due to the noted coding features come to the decoder with variable frequency, but the decoded images should be reproduced with a constant frequency equal to the frame rate. In the decoder, the problem is solved by the buffer. Compensating for one variable delay with another is the principle of realizing a constant delay in the entire system.

7.4.6.2   Time stamps

The mechanism for delay compensation and synchronization is the time stamps that are assigned to each access unit (Figure 7.49) and which tell the decoder the exact time when the access unit should be removed from the decoder's buffer and decoded. In order to attach timestamps to access blocks, the encoder must know the current system time provided by the reference time generator. But the access time stamps of the access blocks are not copies of the current time. It must be remembered that the time stamp indicates the time when the decoder will decode this access unit, which should occur in the future. Therefore, there should be some shift between the current time and the stamp. How big this shift should be depends on many factors, including the size of the encoder and decoder buffers, and the rate with which the elementary stream enters the multiplexer. The shift must be large enough for the access unit to pass through the encoder buffer, multiplexer, and be completely writ-ten into the decoder buffer. When calculating the shift, it is also necessary to prevent possible overflow or complete emptying of the decoder buffer, because in this case,

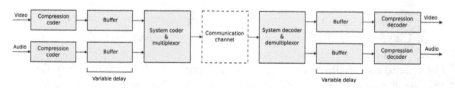

Figure 7.48   Principle of constant delay

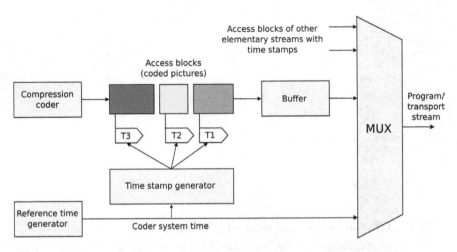

Figure 7.49   Use of time stamps in formation of program or transport streams

there will be a failure in the continuous reproduction of the decoded images. To do this, the encoder uses a hypothetical decoder, which is connected to the output of the encoder. Of course, this is not a real decoder, but a computational model, followed by a determinant of the degree of filling of the decoder buffer. The purpose of the models is to impose restrictions on the encoding process in order to ensure that there is no overflow or full release of the decoder buffer capacity. Data on the degree of filling of the buffer are reported to the real decoder so that it can compare the calculated values with the current values of similar parameters in the process of real decoding.

7.4.6.3   Adjusting the system clock

In order to correctly interpret the time stamps, the decoder must have its own system time, and the decoder "clock" must be adjusted to the encoder "clock" time. For this, the current time of the encoder is regularly transmitted to the decoder. The system time of each program is counted in units of the oscillation period with a frequency of 27 MHz. The samples of this time are transmitted in the program stream in one of the block header fields (they are called SCR—system clock reference) at least at every 0.7 s. In the transport stream, data of several television programs can be transferred, each of which can have its own independent time, called program time. The samples of program time (PCR) are carried in the adaptation field of the transport packet with the corresponding PID (usually it coincides with the identifier of the elementary video data stream, as illustrated in Figure 7.45). PCR stamps should appear at least once every 0.1 s. Despite the difference in names, the basic functions of PCR and SCR are the same. The principle of synchronization of the decoder with the encoder by using the samples of program time is illustrated in Figure 7.50.

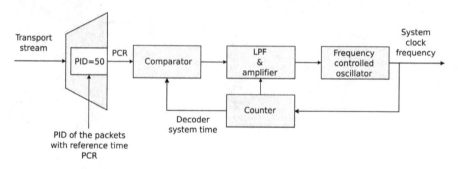

Figure 7.50   *Synchronization of decoder with encoder*

7.4.6.4   Time to decode and time to present

The time stamps associated with the access blocks are expressed in units of time of the oscillation period with a frequency of 90 kHz, obtained by dividing the frequency of 27 MHz. These stamps come in two types: PTS presentation timestamps and DTS decoding timestamps. PTS stamps determine the point in time at which the decoded access unit (encoded image or audio fragment) must be presented to the viewer. For all elementary streams, except video, PTS are the only tags that are needed. The video stream requires DTS decoding timestamps, which determine the times at which access blocks are extracted from the buffer and decoded, but not presented to the viewer. Decoded images are temporarily stored and presented at a later time designated by PTS stamps. DTS stamps are needed for I and P type images, which must be decoded earlier than B-images, for which I and P images were used as references for encoding. DTS stamps do not appear alone, but must be accompanied by PTS stamps.

The time stamps must not accompany every block of access. The limitation defined by the MPEG-2 standard is that stamps must appear in elementary video and audio streams at least once every 0.7 s. Stamps are transported in the headers of PES packets (Figure 7.40). If the stamp accompanies the access block, then it appears in the header of the PES packet in which this access block begins.

7.4.7   Switching compressed data streams

7.4.7.1   Is it possible to switch compressed data streams?

It is often considered that switching programs encoded in accordance with the MPEG compression standards is not possible. Such a judgment is explained by the fact that as a result of coding with prediction in the process of eliminating temporal redundancy, all frames are connected into a single chain, which supposedly cannot be broken without a failure in the reproduced image. It is argued that the only possible switching method requires decoding, i.e., converting compressed streams into the original form, after which you can perform a switching operation and re-encode the new program. Of course, this type of installation is possible, but it is associated with time-consuming and even with potential distortions and artifacts resulting from repeated compression–decompression cycles. However, switching of video streams compressed according to the MPEG-2 standard is also possible, although, of course, the compression system leaves a significant imprint on the switching methods.

First of all, it should be noted that all images are connected and form a chain with interdependent elements only in the case of using open image groups. Image frames within a closed group (it ends with an image of type P) do not depend on frames of other groups (the prediction is performed strictly within one group). Therefore, video streams of closed groups can easily be switched and even edited on the boundaries of groups. However, streams based on open groups of images can be switched in compressed form as well. For this, it is necessary to break the continuity chain at the selected switching point. However, the switched program must have all the properties of the MPEG-2 data stream.

7.4.7.2   Remarking frames in the transition area

One of the options for switching elementary flows is illustrated in Figure 7.46. This option is based on remarking frames—type B images (which are associated as a result of the prediction with both previous and subsequent I and P type frames) into P type images without changing the video data of the corresponding access unit. As shown in Figure 7.51, frames B_{15} and B_{16} of elementary stream 1 are renamed frames P'_{15} and P'_{16}. The continuous chain of predictions is breaking. When decoding P'_{15} and P'_{16}, only the P_{14} image will be used as a reference, therefore switching is possible.

The fact that the decoding strategy is transformed without changing the coded prediction error/residual is not associated with high visibility of potential

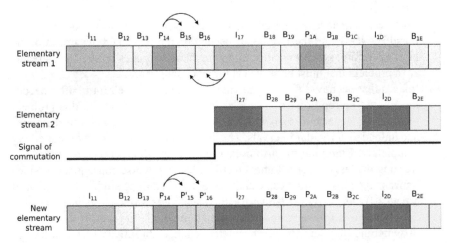

Figure 7.51   *Switching of elementary streams with alignment of groups of pictures*

distortions due to the short duration of the transition. More serious is the fact that such switching can lead to a potential overflow of the decoder buffer, which is associated with large distortions and malfunctions of the decoder. A possible solution to this problem is to insert a short pre-coded black data stream, which can be inserted between the first and second streams to normalize the state of the buffer. The fact that the method requires temporal alignment of groups of images of switched streams is not a serious complication. The method can be recommended for use in TV distribution systems, e.g., for inserting local advertising clips.

7.4.7.3   Recoding frames in the transition area

Another, and probably the best in terms of image quality method, is associated with the recoding of switched elementary streams in the transition area. This option can be recommended for systems in which television programs are stored compressed.

7.4.7.4   Seamless splicing of compressed data streams

The wide distribution of video compression makes it increasingly necessary to combine coded programs not only without decoding but also without changing the content of the access units. What should be the switching of data streams? In its external manifestation, it should be similar to the change of plot in a single program. In its internal essence, this, of course, is not just switching, but splicing of data streams, in which the resulting stream will correspond to the syntax and semantics of MPEG-2. In the standard, this is called splicing.

There are three reasons that impose restrictions on the switching of MPEG-2 data streams.

- P and B frames cannot be recovered without the reference pictures that were used to predict in the encoding process. Switching can leave P and B images without reference.
- Compressed images require different time intervals for transmission (I—more, P and B—less), and these intervals depend on the detail and dynamism of the plot. Therefore, synchronization and alignment of frames of switched streams are a problem that must be solved at the time of splicing.
- Images that occupy different time intervals in compressed form, after decoding, should be reproduced at regular intervals. The solution to this problem requires a decoder buffer, in which access blocks are loaded at different times and unloaded at regular intervals. The buffer should neither overfill nor empty completely. Emptying means no data to decode, which can be overcome by freezing the last decoded frame. Overflow leads to worse consequences, since it means data loss, which can cause the reproduced image to be distorted until a new I-frame arrives. Standard MPEG encoders work in such a way that both buffer overflow and underflow are eliminated. However, at the moment of switching, the data stream parameters change abruptly, which can lead to disruption of the normal operation of the buffer, at which its capacity is filled on average by 50%.

These problems lead to the fact that only separate data flow points are suitable for splicing without changing the coded data of access objects (Figure 7.52). In the MPEG specification, these points are called splicing points. The switching of two streams and the transition from the old stream to the new one are possible only if the splicing points of the two streams coincide in time.

The MPEG-2 syntax provides the means to ensure splicing even at the transport stream level. Among these means, the first place belongs to the packet counter to the splicing point. The counter is an 8-bit counter, which is decremented with each packet, and the state of which becomes zero at the nearest potential splicing point. The counter is located in the field of adaptation of the transport package. Its purpose is to inform the switching equipment about the possible splicing and indicate its exact position (Figure 7.53).

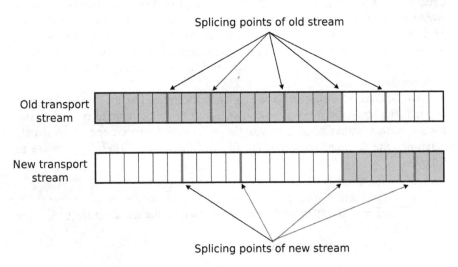

Figure 7.52   Splicing points of transport streams

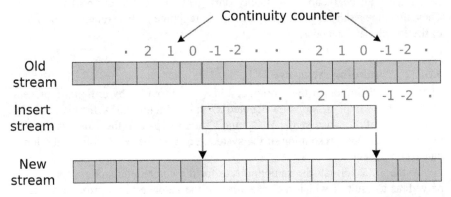

Figure 7.53   Principles of fragment insertion into the transport stream

7.5   Comparison of compression systems

7.5.1   *Criteria and parameters of comparison of compression systems*

7.5.1.1   Introduction

The use of video compression is almost always associated with some distortions in image quality. The compression algorithm is a compromise between the rate of the compressed data stream and the quality of the image reproduced at the output of the decoder. The development of a compression algorithm is always associated with finding a balance between the amount of discarded data (the degree of compression) and the distortions of image quality. Indeed, discarding a part of the video stream cannot pass without consequences. In the world of ideal technology, there would be no place for compression. But high and ultra-high-definition television requires the transmission of a huge stream of uncompressed video data, as noted above. Working with such streams, e.g., when recording and reproducing an image, is an extremely complex task. Simplification is achieved through the use of video compression systems.

An indicator of compression is efficiency determined as the rate of the compressed data stream at a given level of the quality of the reproduced image. But the parameters that are associated with the technological process and which largely determine the scope of the codec are of great importance. These options are as follows:

- sampling structure of image components and pixel format
- the number of quantization levels, determined by the length of the code word (bit depth)
- coding type (interframe, intraframe)

These parameters affect the image quality (and in a nonlinear manner!) and determine the possible technological process. The values of these parameters are associated with degradation of quality during re-coding and during transcoding. They affect the ability to perform operations in real time and the relative complexity of the encoder and decoder.

7.5.1.2   Sampling structure

The highest definition of the reproduced image is provided by a digital representation with a sampling structure denoted as 4:4:4 (Figure 7.35). In this case, the components of the video signal Y, C_R, and C_B are sampled in the same way, which provides a full color resolution of the system, equal to the resolution for the luminance component.

In the 4:2:2 format, the sampling rate for the color difference components is halved, as a result of which the resolution in the color in the horizontal direction is also halved. The admissibility of reducing the resolution in color is due to the properties of the human visual system, which has a lower sensitivity to small color details. This technique is used in analog television, in which the frequency band of color-difference signals is significantly less than the frequency band of the luminance signal. In digital television, using a 4:2:2 sampling structure can be considered as a form of compression with a ratio of 3:2 = 1.5:1.

A further reduction in the data rate of the digital video signal is possible due to the transition to the 4:2:0 sampling structure, at which the resolution in color is halved in both the horizontal and vertical directions (compared with the 4:4:4 structure). The 4:2:0 sampling structure used in the MPEG-2 video compression system is shown in Figure 7.33. The win is to reduce the flow rate of video data by half. The volume of transmitted data decreases twice with the 4:1:1 sampling structure, which means a fourfold decrease in the sampling frequency of the color-difference components of the signal (this leads to a fourfold decrease in color clarity in the horizontal direction in comparison with the 4:4 structure: four). Thus, the use of 4:2:0 and 4:1:1 sampling structures is a compression ratio of 2:1.

The following examples illustrate the significance and practical application of these considerations.

As shown above, a format of 1080/50/p, often simply referred to as 1080p, provides very high image quality. However, when the digital representation of the video signal is in accordance with the 4:4:4 sampling structure and with the number of bits per sample equal to 10, the digital video stream has a rate of 4.455 Gbits/s. If the quantization uses 8 bits per sample, then the bit rate becomes equal to 3.564 Gbits/s. The transition to the 4:2:2 sampling structure, in which the resolution of color-difference components in the horizontal direction is two times lower, reduces the flow rate to 2.376 Gbits/s.

Replacing the non-interlaced image decomposition with an interlaced image makes it possible to further reduce the flow rate of digital video data. To transmit a TV image in 1080/50/I format (interlaced image with a total number of active lines in a frame equal to 1 080, a field frequency of 50 Hz, and a frame rate of 25 Hz), it is necessary to set the stream rate at 2.227 Gbits/s with 4:4:4 sampling structure and 10 bits per sample. The transition to the 4:2:2 sampling structure makes it possible to reduce the flow rate to 1.485 Gbits/s.

The digital representation of standard definition TV images (576 active lines, 720 active elements per line, frame rate 25 Hz) at a 4:3 aspect ratio, 4:4:4 sampling structure, and quantization with a bit depth of 10 per sample corresponds to a stream rate of 405 Mbits/s. The transition to the 4:2:2 sampling structure makes it possible to reduce the speed to 270 Mbits/s when quantizing at a bit depth of 10 bits per sample and to 216 Mbits/s when quantizing at a bit depth of 8 bits per sample. Further reduction of the data rate of the digital video signal is possible due to the transition to the 4:2:0 sampling structure, at which the video data stream rate is 162 Mbits/s. The same value has a video stream with a sampling structure of 4:1:1. Eliminating structural redundancy can further reduce the bit rate of video data carrying the TV image. If, when sampling at 4:2:0 (only the active part of the raster) and quantizing at a bit depth of 8 bits, then the bit rate will be 124.416 Mbits/s.

7.5.1.3   Pixel format

Standard definition digital television uses a rectangular pixel (even with a 4:4:4 sampling structure). This was due to the desire to have one value of the sampling rate and one number of pixels in the active part of the line for two systems: 625/50 and 525/60. The format of a standard television image, or the ratio of the width of the image to its height, is 4:3. If we proceed from the requirement of the same definition horizontally and vertically, then when scanning a television image into 625 lines (576 active lines), each line would contain 576 * 4/3 = 768 pixels, and when scanning into 525 lines (480 active lines), 480 * 4/3 = 640 pixels. But the active part of the line in accordance with Recommendation 601 contains 720 pixels for both the 625/50 system and the 525/60 system. This means that the pixel is not square either in the 625/50 system or in the 525/60 system and that the definition in the horizontal and vertical directions is not the same. In the 625/50 system, the actual pixel is stretched out in the horizontal direction (its format can be estimated as 768:720 = 1.07) and the horizontal definition is worse than the vertical, even for the brightness component of the image. In the 525/60 system, the pixel is compressed in the horizontal direction (its format is 640:720 = 0.89) and the resolution in the horizontal direction is better than vertical. The same pixel aspect ratio is also preserved in the 16:9 image format, which assumes the same number of active elements in a row, equal to 960, and in the 625/50 system and in the 525/60 system.

High-definition digital television formats define the use of a square pixel, in which the horizontal and vertical spatial resolution are equal. However, to reduce the bit rate, a horizontal definition reduction can be used, achieved by introducing a rectangular pixel for the luminance component of the image. For example, in a format with 1 080 active lines, a pixel can be lengthened in the horizontal direction and take the form of a rectangle with an aspect ratio of 4:3. The number of pixels in the active part of the line, initially 1 920, becomes equal to 1 440 and the video data stream rate decreases. Using a rectangular pixel in this case is a form of compression with a ratio of 4:3 = 1.33:1.

7.5.1.4   Quantization

In the digital representation, the value of a video signal is expressed by a binary code number, the number of bits of which determines the number of quantization levels. For example, using a code word of 8 bits means representing a video signal with one of the 256 allowed values. Quantization is the rounding off of the actual value of the video signal to the nearest quantization level. Rounding is always associated with an error, the maximum value of which is equal to half the interval between adjacent levels—the quantization step. The discrepancy between the actual values of the video signal and its quantized values is the quantization noise. With an 8-bit representation, the ratio of the peak-to-peak video signal value to quantization noise is 48 dB, with a 10-bit representation—60 dB. The greater the number of quantization levels, the lower the quantization noise and, accordingly, the higher the accuracy of the digital representation of the image. On the other hand, increasing the length of the code word increases the rate of the stream of digital video data. Therefore, reducing the number of bits of a codeword can be considered as a form of compression. For example, the transition from 10-bit quantization to 8-bit, i.e., discarding two bits in each codeword reduces the amount of data transferred by 20% and is a video compression with a ratio of $10:8 = 1.25:1$.

7.5.1.5 Interframe and intraframe coding

Interframe coding involves the elimination of not only spatial but also temporal redundancy. Imagine a group of 12 images with the following structure: I-B-B-P-B-B-P-B-B-P-B-B. As noted above, the volume of the compressed image of type P for typical plots of broadcast television is about a third of the image of type I, and the B-image is about a quarter. The amount of data that, after compression, represents the entire group of 12 images, would be equal to four type I images. But if there was no reduction in temporal redundancy, then the required data volume would be $12/4 = 3$ times more. A factor of 3 gives an approximate decrease in the rate of data stream, achieved through the use of interframe coding with groups of 12 images, with approximately the same visibility of compression distortions. The larger the group of images, the greater the gain achieved by eliminating temporal redundancy. However, it should be noted that in the case of the use of compression systems with intraframe coding, there would be no artifacts associated with moving objects. However, on the other hand, the eye notices such distortion the less, the faster objects move in the image field. As you can see, these two factors to some extent compensate each other. The gain calculated above is, of course, indicative, it depends on many factors, but the fact that this gain is significant is not in doubt, therefore the use of interframe coding with the elimination of temporal redundancy is, of course, expedient in all links of the path where it is a desirable significant reduction in data rate.

Codec features with interframe coding can be formulated as follows.

- High efficiency (video data stream speed is on average three times less than in the case of intraframe coding with the same quality level of the decoded image).
- Difficulties with implementation of editing (due to the use of prediction, points of potential transitions are tied to groups of frames, i.e., separated from each other by about half a second when using groups of 12–15 frames).
- The compressed data stream rate may be constant, but the volumes of the individual encoded frames are not the same.
- When moving objects in the image field and panning, additional artifacts and distortions arise.
- Long recovery time after loss of data (due to errors, a whole group of frames may be distorted).
- Codec asymmetry (a decoder requires significantly less processor processing power than an encoder, which is essential for television broadcasting).

Intraframe coding allows us to reduce only the spatial redundancy of the digital representation of the TV image. Therefore, codecs with intraframe coding provide less efficiency than codecs with interframe coding. On the other hand, each frame is encoded and, accordingly, decoded without using data from other frames, which leads to significant advantages of such codecs.

The features of codecs with intraframe coding can be formulated as follows.

- Easy editing; individual compressed frames can be eliminated, replaced, or added without affecting other frames.
- All compressed frames can be of equal volume.
- The movement of an object in the television image field is not associated with the appearance of additional artifacts, which makes such codecs suitable for use, e.g., in sports programs.
- Fast recovery after the loss of part of the data (only those frames in which there were transmission errors) is distorted.
- Codec symmetry (encoders and decoders require data processors of roughly the same computing power).
- Low efficiency (in comparison with interframe coding at the same level of quality of the decoded image).

7.5.2   Requirements of the workflow

To date, a large number of video compression systems with various features and parameters have been developed. The choice of a compression system and a codec is a complex task that is the subject of research and discussion [44, 45]. Its decision depends, first of all, on the requirements determined by the stage of the workflow/technological process at which this codec will be applied. The main requirements of several typical workflows can be formulated as follows:

1. Digital intermediate process (Digital Intermediate)

    The transformation of a film frame image into a stream of video data should ensure the highest quality. Ideally, compression should not be used at all. However, a slight lossless compression or compression accompanied by visually imperceptible loss of image quality may be acceptable. Naturally, this compression is a compression of the first generation.

    Requirements:

    Intraframe coding.

    The sampling structure is RGB 4:4:4.

    Quantization—12–14 bits or more per sample of one image component (10 bits with a logarithmic representation of the signal).

    Spatial resolution—1 920 × 1 080 or better.

    The target bit rate is at least 300 Mbits/s when writing to the memory subsystem.

2. Shooting and delivery of program materials

    Video captured data can be compressed to simplify recording and storage devices and transmission systems. This compression is a first-generation compression.

Requirements:

Intraframe coding.

The sampling structure is YC_RC_B 4:2:2.

Quantization—8 or 10 bits per sample of one image component.

The spatial resolution is 1 280 × 720p, 1 920 × 1 080i, or 1 920 × 1 080p.

The target bit rate is 100–300 Mbits/s.

3. Electronic news gathering

Image quality must comply with the requirements of news programs broadcast. Such programs are usually formed within the framework of a simplified technological process of linking programs, so there is no need for multiple transcoding. A higher degree of compression can be achieved through the use of interframe coding and reduced color space, which reduces the cost of equipment. This compression is a first-generation compression.

Requirements:

Interframe coding.

The sampling structure is YC_RC_B 4:2:0.

Quantization—8 bits per sample image component.

The spatial resolution is 1 280 × 720p, 1 440 × 1 080i, 1 920 × 1 080i.

The target bit rate is 25–50 Mbits/s for standard definition television and 35–100 Mbits/s for high and ultra-high-definition television.

4. Program assembling and editing

When editing and assembling programs, the image quality that is present in the original program materials should be preserved; therefore, only slight compression is allowed, allowing for effective access to materials in real time, as well as efficient transfer of materials between different platforms. It would be possible to use one type of compression with a small degree in the whole workflow; however, this leads to additional losses during recoding. A more appropriate option is to support the editing systems of the timeline in any source format and retain references to the source materials for multilayer operations. Within this approach, it is possible to limit the number of transcoding to two. Input data are presented in the compressed format of the first generation, weekend—the third. The final coding must comply with the compression format set in the program distribution system.

Requirements:

Intraframe or interframe coding.

The sampling structure is YC_RC_B 4:2:2 or 4:2:0.

Quantization—8 or 10 bits per sample of one image component.

The spatial resolution is 1 280 × 720p, 1 440 × 1 080i, 1 920 × 1 080i.

The target bit rate is 25–150 Mbits/s.

5. Distribution and archiving

Communication channels and networks used in the distribution of programs, as a rule, have limited bandwidth. Therefore, compression systems with interframe coding are used in the average bit rate range of compressed video data. It must be possible to perform with a compressed data some limited set of operations, e.g., logo insertion, switching. Operations should not lead to distortion, visible to the viewer. The compression used is a second- or third generation compression.

Requirements:
Interframe coding.
The sampling structure is YC_RC_B 4:2:2 or 4:2:0.
Quantization—8 bits per sample image component.
The spatial resolution is 1 280 × 720p, 1 440 × 1 080i, 1 920 × 1 080i.
The target bit rate is 25–50 Mbits/s.

6. Broadcasting

    Transmission of a full-resolution television image to television receivers over narrow-band channels requires the use of high-performance compression, therefore interframe coding must be used. The compression used is a fourth-generation compression.

    Requirements:
    Interframe coding.
    The sampling structure is YC_RC_B 4:2:0.
    Quantization—8 bits per sample image component.
    The spatial resolution is 1 280 × 720p, 1 920 × 1 080i.
    The target bit rate is not higher than 20 Mbits/s.

7. Recording movies and programs to disks.

    Writing movies and other audiovisual materials to disk in full resolution requires the use of efficient video compression systems with interframe coding due to limited disk capacity. The compression used is a fourth generation compression.

    Requirements:
    Interframe coding.
    The sampling structure is YC_RC_B 4:2:0.
    Quantization—8 bits per sample image component.
    The spatial resolution is 1 280 × 720p, 1 920 × 1 080i.
    The target bit rate is not higher than 8–25 Mbits/s.

8. IPTV and content download via the Internet

    Maintaining full resolution image quality at a satisfactory level when transmitting over data networks using the TCP/IP protocol stack is an achievable goal, although this will require solving a number of difficult tasks. Downloadable videos can be watched on the TV screen connected to the computer. There are already TVs with interfaces for direct playback of media data received from network sources (e.g., YouTube). Applied compression systems must allow software implementation and have the highest efficiency, this is the fifth generation compression.

    Requirements:
    Interframe coding.
    The sampling structure is YC_RC_B 4:2:0.
    Quantization—8 bits per sample image component.
    The spatial resolution is 1 280 × 720p, 1 440 × 1 080i, 1 920 × 1 080i.
    The target bit rate is 1–14 Mbits/s.

9. Mobile TV

    Mobile TV playback devices (mobile phones and handheld computers) have small dimensions and not always high resolution. The compression systems used must have very high efficiency, operate at low video data flow rates, and support scaling. Decoders should be simple and work with low resolution. The compression used is a fifth generation compression.

Requirements:

Interframe coding.

The sampling structure is YC_RC_B 4:2:0.

Quantization—8 bits per sample image component.

Reduced spatial resolution.

The target bit rate is 1 Mbits/s.

7.5.3 Comparison of parameters and prospects for the development of compression systems

Data on the support of different image presentation formats and the use of different types of coding in the codecs of various video compression systems are given in Table 7.4. The following video compression systems were considered: DV, MPEG-2, MPEG-4-ASP (advanced simple profile), AVC/H.264, HEVC/H.265, AVS, and JPEG2000. Comparing the parameters of the compression systems and the requirements of the above processes, estimate the scope of codecs of different systems Table 7.5. As can be seen from the table, all systems can be used in several technological processes. Obviously, there is no compression system that could be used absolutely in all areas of modern digital television and cinema. Note that the naib H.264/AVC and HEVC/H.265 codecs have a longer range of possible application areas.

The above consideration can only be considered an example of the initial technical stage. Many other considerations should be taken into account, including financial and economic considerations, the issue of intellectual property rights, and others.

Over the past three decades, a number of video compression systems have been developed and standardized, the main ones of which were considered in this article. None of these systems meets the requirements of all applications, so it is not surprising that there have been occasional reports of new proprietary codecs that promise a huge increase in efficiency. However, only some of them have achieved international recognition and standardization. Many failures are explained by the fact that devices that work well in the laboratory were unable to work in real studio and field conditions with real images. The reason for some failures was the underestimation of the fact that the complexity of the hardware or software implementation of codecs operating in real time is one of the most important factors that should be taken into account when developing compression systems in the real world.

New codecs allow you to achieve higher video compression efficiency. However, it can be stated that higher efficiency is achieved due to higher demands on the performance of data processors. New codecs almost always require large computing power, which increases the cost of hardware and software tools for encoding and decoding. For example, the efficiency of AVC compression is two times higher compared with MPEG-2. However, the AVC hardware encoder requires about 16 times more computational resources than the MPEG-2 encoder, and the AVC decoder is about four times harder when compared with the number of gates. The compression

Table 7.4   Codec support of different types of representing and coding

Name	DV	MPEG-2	MPEG-4 ASP	AVC/ H.264	AVS	HEVC/H.265	JPEG-2000
Quantization (bit depth)	8	8	8	8/10/12/14	8	8/10/12/14/16	8/10/12
Intraframe coding	+	+	+	+		+	+
Interframe coding		+	+	+	+	+	
Lossless coding							+
Near lossless coding					+	+	+
Sampling structure 4:2:0, 4:1:1	+	+	+	+	+	+	+
Sampling structure 4:2:2	+	+	+	+	+	+	+
Sampling structure 4:4:4				+		+	+

Table 7.5   *Codec support of requirements of different workflows*

Name	DV	MPEG-2	MPEG-4 ASP	AVC/ H.264	AVS	HEVC/H.265	JPEG-2000
Digital Intermediate							+
Shooting and delivery		+	+	+		+	+
Electronic news gathering	+	+		+		+	+
Assembling and editing	+	+		+		+	
Distribution and archiving		+		+	+	+	
Broadcasting		+	+	+	+	+	
Film recording		+			+	+	
IPTV			+	+	+	+	
Mobile TV			+	+		+	

efficiency of JPEG2000 is higher in comparison with JPEG by 20–40%, but the hardware implementation of JPEG2000 is more difficult (6–10 times) than JPEG [44, 45].

It can be stated that the times of compression codecs with only the hardware implementation are over. Any compression system that is intended to be used in the future should be implemented in software in real time. This will allow the use of accumulated audio-visual materials in the future. The program codec can be saved as a file along with archived audiovisual materials that have been encoded using it. Software implementation of codecs is provided using universal computers.

The main hardware element of the computer is a microprocessor. Microprocessor developers improve their products, and give them the opportunity to work effectively with multimedia materials. These enhancements have a strong impact on improving the performance of codecs in the professional and broadcast market. The emergence of real-time codecs that can be implemented without the use of specialized hardware contributes to the rapid development of consumer electronics. Algorithms previously implemented using specialized and expensive hardware are now increasingly performed on universal computers. Undoubtedly, this trend will continue in the future, and the rate of implementation of software codecs will be further accelerated.

In recent years, two concepts of developing new compression systems have come together. Over the 30 years of its existence, the MPEG group has developed about 180 international standards that have given impetus to the birth of the digital media industry and are now contributing to its development. As Leonardo Chiariglione, the founder of the MPEG group, notes, this was made possible by the MPEG business model, which has always considered the main goal to achieve the best performance, irrespective of the IPR involved. But patent holders who allowed us to use patented patents in the standards received hefty money as royalties, which was used to develop new technologies for the next generation of MPEG standards [46]. Recently, however, this business model began to falter. The slow implementation of HEVC is largely due to the pricey and complicated licensing situation of HEVC. The challenge was the announcement of the Alliance for Open Media on the development of the new AV1 video compression codec, which was claimed to perform better than HEVC and was said to be offered royalty free [47]. MPEG group may lose out in this rivalry if it does not change its business model when developing new compression standards. The future of development and standardization of compression systems promises to be exciting!

References

[1]   Shannon C.E. 'A mathematical theory of communication'. *Bell System Technical Journal*. 1948, vol. 27(4), pp. 623–56. Available from http://doi.wiley.com/10.1002/bltj.1948.27.issue-4
[2]   Huffman D.A. 'A method for the construction of minimum-redundancy codes'. *Proceedings of the IRE*. 1987, vol. 40(9), pp. 1098–101.

[3] Sayood K. *Introduction to data compression*. Third Edition. The Morgan Kaufmann Series in Multimedia Information and Systems Series. Fox, Virginia Polytechnic University; 2005.

[4] Li Z.-N., Drew M.S. *Fundamentals of multimedia*. Upper Saddle River, NJ: Pearson Education International; 2004.

[5] Witten I.H., Neal R.M., Cleary J.G. 'Arithmetic coding for data compression'. *Communications of the ACM*. 1987, vol. 30(6), pp. 520–40.

[6] Recommendation ITU-R BT.656-4. *Interfaces for digital component video signals in 525-line and 625-line television systems operating at the 4:2:2 level of recommendation ITU-R BT.601 (Part A)*.

[7] Recommendation ITU-R BT.709-5. *Parameter values for the HDTV standards for production and international programme exchange*.

[8] Recommendation ITU-R BT.799-3. *Interfaces for digital component video signals in 525-line and 625-line television systems operating at the 4:4:4 level of recommendation ITU-R BT.601 (Part A)*.

[9] Recommendation ITU-R BT.1120-4. *Digital interfaces for HDTV Studio signals*.

[10] *ISO/IEC 11172-2 – Information Technology – coding of moving pictures and associated audio for digital storage media at up to about 1,5 Mbit/s – Part 2: video*.

[11] Recommendation ITU-T H.262, (ISO/IEC 13818-2). *Generic coding of moving pictures and associated audio: video*.

[12] ISO/IEC 13818-3. *Generic coding of moving pictures and associated audio: audio*.

[13] Recommendation ITU-T H.222.0, (ISO/IEC 13818-1). *Generic coding of moving pictures and associated audio: systems*.

[14] *IEC 61834 – Recording Helical-scan digital video cassette recording system using 6,5 MM magnetic tape for consumer use (525-60, 625-50, 1125-60 and 1250-50 systems)*.

[15] *ISO/IEC 15444-1 – Information technology – JPEG 2000 image coding system: core coding system*.

[16] *ISO/IEC 14496-2 – Information technology – coding of audio-visual objects – Part 2: Visual*.

[17] Recommendation ITU-T H.264, (ISO/IEC 14496 Part 10). *Advanced Video Coding (AVC)*.

[18] Recommendation ITU-T H.265, (23008-2 (MPEG-H Part 2)). *High efficiency video coding (HEVC)*.

[19] *Information technology, advanced audio video coding, part 2: video (AVS1, standard No.: GB/T 20090.2-2006)*.

[20] *Information technology, advanced audio video coding part 16: radio television video (AVS+, standard No.: GB/T 20090.16-2016)*.

[21] *AVS2 – the second generation AVS standards. What is AVS2?*. Available from www.avs.org.cn/AVS2/en/index.asp

[22] Ackerman E. *Biophysical science*. Englewood Cliffs, N.J: Prentice Hall, Inc; 1962.

[23] Вахитов Я.Ш. *Теоретические основы электроакустики и электроакуст-ическая аппаратура.* М.: Искусство; 1982.

[24] Zwicker E., Feldtkeller R. *Das Ohr als Nachrichtenempfaenger.* S. Hirzel Verlag Stuttgart; 1967.

[25] Chedd G. *Sound: from communication to noise pollution.* Doubleday Science Series. Garden City, N.Y: Doubleday and Co., Inc; 1970.

[26] Fletcher H. 'Uditory patterns'. *Reviews of Modern Physics.* 1940, vol. 12(1), pp. 47–65.

[27] Greenwood D.D. 'Critical bandwidth and the frequency coordinates of the basilar membrane'. *The Journal of the Acoustical Society of America.* 1961, vol. 33(10), pp. 1344–56.

[28] Robinson D.J.M. *Perceptual model for assessment of coded audio.* Department of Electronic Systems Engineering, University of Essex; 2002 Mar.

[29] Thiede T. *Perceptual audio quality assessment using a non-linear filter bank.* Berlin: Technische Universität; 1999.

[30] Zwicker E., Zwicker U. 'Audio engineering and psychoacoustics: matching signals to the final receiver, the human auditory system'. *Journal of the Audio Engineering Society. Audio Engineering Society.* 1991, pp. 115–26.

[31] Brandenburg K., Popp H. 'An introduction to MPEG layer-3'. *EBU Technical Review.* 2000.

[32] Cave C.R. *Perceptual modelling for low-rate audio coding.* Montreal, Canada: Department of Electrical & Computer Engineering McGill University; 2002 Jun.

[33] Ferreira A.J.S. 'Optimizing high quality audio coding: advantages of full system observability'. *In IEEE International Conference on Acoustics Speech and Signal Processing*; 1995.

[34] Ehliar A., Eilert J. 'A hardware MP3 decoder with low precision floating point intermediate storage. LiTH-ISY-EX-3446-2003'. *Dept. of Electrical Engineering at Linkopings Universitet. Linkoping.* 2003.

[35] *ISO/IEC 11172-3 – Information technology – coding of moving pictures and associated audio for digital storage media at up to about 1,5 Mbit/s. Part 3 audio.*

[36] *ISO/IEC 13818-3 – Information technology – generic coding of moving pictures and associated audio information – Part 3: Audio.*

[37] *ISO/IEC 13818-7 – Information technology – generic coding of moving pictures and associated audio information – Part 7: advanced audio coding (AAC).*

[38] Brandenburg K. 'MP3 and AAC explained'. *In Proceedings of the AES 17th International Conference on High Quality Audio Coding*; Florence, Italy, 1999. pp. 99–111.

[39] *ISO/IEC 14496-3:2009 information technology – coding of audio-visual objects – Part 3: Audio.*

[40] Sayood K. *Introduction to data compression.* Third Edition. The Morgan Kaufmann Series in Multimedia Information and Systems Series Editor, Edward A. Fox, Virginia Polytechnic University; 2005.

[41] Todd C., Davidson G.A., Davis M.F, *et al.* 'AC-3: flexible perceptual coding for audio transmission and storage'. *AES 96th Convention*; 1994.

[42] *ATSC standard: digital audio compression (AC-3, E-AC-3). Doc. A/52:2012.*

[43] *ATSC digital television standard, part 5-AC-3 audio system characteristics. Doc. A/53:2014.*

[44] Roth T. 'Practical applications of compression standards'. *Proceedings for the 62nd annual NAB Broadcast Engineering Conference*; 2008. pp. 573–91.

[45] Naylor J.R., Bancroft D. 'Understanding and implementing JPEG2000 compression for long-form EFP acquisition'. *Proceedings for the 62nd Annual NAB Broadcast Engineering Conference*; 2008. pp. 604–11.

[46] Chiariglione L. *A crisis, the causes and A solution.* Available from http://blog.chiariglione.org/2018/01/28/

[47] *Alliance for open media (AOMedia). AV1 show demos.* Available from https://aomedia.org/av1-show-demos/

Chapter 8

Fundamentals of broadcast systems

Michael Isnardi[1]

This chapter will provide historical background on the Moving Picture Experts Group (MPEG), and describe the MPEG standards that apply to digital television.

8.1 Life before MPEG

Up until the early 1980s, consumers worldwide had only experienced media in analog form: radio and television broadcasts, audio, and video cassette recordings [1]. In 1982, things began to change with the commercial introduction of compact discs (CDs) and CD players by Sony and Philips. For the first time, consumers were exposed to digital media: high-quality digital audio, albeit in uncompressed format. The advantages of digital media were immediately apparent: fast search, perfect copying, and no generational loss.

Digital video, on the other hand, would take a little longer to arrive in consumers' homes. The audio bit rate on audio CDs is about 1.4 Mbps. Uncompressed standard-definition video (e.g. 525-line National Television Sytems Committee (NTSC) or 625-line Phase Alternating Line (PAL)/Séquentiel de couleur à mémoire, French for color sequential with memory (SECAM)) requires at least 100× higher bit rate than uncompressed audio, and at the time only television studios could afford the expensive hardware and infrastructure required to store and transmit uncompressed digital video. Television studios began making the transition to digital in the early 1980s, and were aided by the Comité Consultatif International pour la Radio, a forerunner of the ITU-R (CCIR) 601, the first digital video standard, ratified in 1982. Other standards development organizations, such as the Society of Motion Picture and Television Engineers (SMPTE), were creating digital video and audio standards that allowed equipment from different vendors to interoperate in digital television studios.

The use of digital media was growing but its voracious appetite for bits presented a great technical challenge. In order for digital media to become affordable and ubiquitous, both in the commercial and consumer worlds, digital audio and video content must be greatly reduced in terms of their representation. In other words, digital media must be *compressed* to about 1–10 per cent of their uncompressed rate in order to enable inexpensive storage and efficient transmission.

Fortunately, audio and video compression had been recognized as a need decades earlier, and sophisticated compression algorithms were starting to emerge

[1]SRI (Stanford Research Institute), USA

from academic and commercial research labs. By 1988, a large number of audio compression systems had been developed and demonstrated, nearly all of them using perceptual techniques and frequency-domain analysis [2].

Video compression techniques relied on a large amount of redundancy in neighboring spatial and temporal pixel values. Good prediction can transform typical video signals into a form that benefits from statistical coding techniques. The earliest digital video compression systems used differential pulse-code modulation to achieve compression ratios on the order of 2–4 [3]. Nippon Electric Company, Ltd. (NEX) introduced the first commercial broadcast video codec in 1982 [4]. The first commercial use of a digital fiber optic line for broadcast television was at the 1984 Summer Olympic Games in Los Angeles, CA. The data rate was 90 Mbps [4]. Also, there was interest around the world in developing video compression systems that could interface with the digital transmission infrastructure being developed by telecommunication companies. Leaders in the field of video compression systems in the mid- to late-1980s were PictureTel and CLI.

Still image coding was being perfected to the point where international standardization was required in order to foster economic viability and technical interoperability. The Joint Photographic Experts Group (JPEG) was formed in 1986 by the International Telegraph and Telephone Consultative Committee (CCITT) and International Organization for Standardization (ISO) to set worldwide standards for image compression [5].

Thus, by 1988 the following factors were present:

- Digital video and audio compression were being used successfully in commercial broadcast and video teleconferencing applications
- Consumers were adopting and becoming accustomed to some forms of digital media: audio CDs, CD- Read-Only Memory (ROMs), Digital Video Interactive (DVI), and Personal Computer (PCs)
- Interoperability of digital devices was generally limited to devices made by the same company or organization
- There was a growing need for multimedia compression standards

It is against this backdrop that the MPEG Committee was formed [6].

8.2   MPEG's birth: the need for standardization

MPEG (pronounced M-peg), is the nickname given to a family of international standards used for coding audio-visual information in a digital compressed format. The MPEG family of standards includes MPEG-1, MPEG-2, and MPEG-4, formally known as ISO/International Electrotechnical Commission (IEC)-11172, ISO/IEC-13818, and ISO/IEC-14496. MPEG is originally the name given to the group of experts that developed these standards. The MPEG working group (formally known as ISO/IEC Joint Technical Committee (JTC1)/Subcommittee (SC29)/Working Group (WG11) is part of JTC1, the Joint ISO/IEC Technical Committee on information technology. The convenor of the MPEG group was Leonardo Chiariglione, who founded the group in January 1988 with the first meeting consisting of about 15 experts on compression technology.

MPEG was established with the mandate to develop standards for coded representation of moving pictures, audio, and their combination. Starting from its first meeting, MPEG has grown to become a very large committee. Usually, some 350 experts from some 200 companies and organizations from about 20 countries take part in MPEG meetings. Traditionally, MPEG meets four times a year but may meet more frequently when the workload so demands.

A large part of the MPEG membership is made of individuals operating in research and academia. Even though the MPEG environment looks rather informal, the standards it creates must be of high strategic and economic relevance. It should be no surprise that the operation of ISO standards committees is carefully regulated by "Directives" issued by ISO/IEC and "Procedures for the Technical Work" issued by JTC1.

8.3   MPEG's standardization process

MPEG exists to produce standards. Those currently produced by ISO are indicated by five digits (e.g., the ISO number for MPEG-1 is 11172 and for MPEG-2 is 13818). Published standards are the last stage of a long process that starts with the proposal of new work within a committee. These proposals of work (New Proposal) are approved at the Subcommittee and then at the Technical Committee level (SC29 and JTC1, respectively, in the case of MPEG).

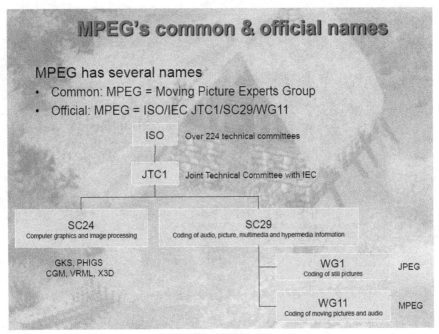

Source: mpeg-43dgraphics-120706063942-phpapp01.pdf

When the scope of new work has been sufficiently clarified, MPEG usually makes open requests for proposals. Examples of proposals include:

- MPEG-1 Audio and Video (July 1989)
- MPEG-2 Audio and Video (July 1991)
- MPEG-4 Audio and Video (July 1995)
- Synthetic/Natural Hybrid Coding (March 1996)

Depending on the nature of the standard, different documents are produced. For Audio and Video coding standards, the first document that is produced is called a Verification Model (VM) or Test Model (TM). In MPEG-1 and MPEG-2 this was called Simulation and Test Model, respectively. The VM describes, in some sort of programming language, the operation of the encoder and the decoder. The VM is used to carry out simulations to optimize the performance of the coding scheme. When MPEG has reached sufficient confidence in the stability of the standard under development, a Working Draft (WD) is produced. This is already in the form of a standard but is kept internal to MPEG for revision. At the planned time, the WD has become sufficiently solid and becomes Committee Draft (CD). It is then sent to National Bodies (NB) for the ballot. If the number of positive votes is above the quorum, the CD becomes Final Committee Draft (FCD) and is again submitted to NBs for the second ballot after a thorough review that may take into account the comments issued by the NBs. If the number of positive votes is above the quorum the FCD becomes FD International Standard (FDIS). ISO will then hold a yes/no ballot with NBs where no technical changes are allowed. The document then becomes IS.

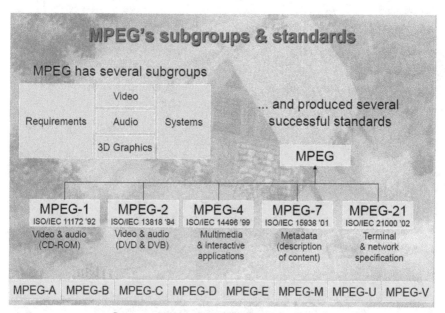

Source: mpeg-43dgraphics-120706063942-phpapp01.pdf

A WD usually undergoes several revisions before moving to the Committee Draft stage. A key role is played by "Core experiments" where different technical options are studied by at least two different committee members. Each revision involves a large number of experts who draw the committee's attention to possible errors contained in the document. Moreover, depending on the nature of comments that usually accompany NB votes, important changes may have to be made on documents when they progress from Committee Draft to Draft International Standard (DIS) to International Standard (IS).

The net result is that standards produced by MPEG are of very high quality. No single error was discovered in MPEG-1 and only small items of dubious interpretation were found in MPEG-2 Video and MPEG-2 Audio. In MPEG-2 Systems, Video and Audio it was found useful to introduce some new features that built upon the standards as originally released. This was done using the "amendment" procedure defined by ISO. For MPEG-2 Audio it was found useful to produce a new revision of the standard.

A large part of the technical work is done at MPEG meetings, usually lasting one full week. Several hundred contributions are submitted by members by electronic means to the MPEG File Transfer Protocol (FTP) site. Delegates are then able to come to meetings without having to spend precious time to study other delegates' contributions at the meeting. The meeting is structured in plenaries and subgroup meetings.

Usually, over 100 documents are produced at every meeting that captures the agreements reached. Of particular importance are "Resolutions" which document the outline of each agreement and make reference to the documents produced, and "Ad-hoc groups," groups of delegates working on some specified area of work, usually until the following meeting. Ad-hoc groups work by e-mail and in some exceptional cases, they are authorized to hold physical meetings. Output documents, too, are stored on the MPEG FTP site. Access to input and output documents, however, is restricted to MPEG members.

As a testament to the importance of interoperability, MPEG standards have been adopted by leading multimedia technology firms like Thomson, Philips, Samsung, Intel, and Sony for their products.

The National Academy of Television Arts and Sciences awarded its 1995–1996 Engineering Emmy for Outstanding Achievement in Technological Development to ISO/IEC for the development of the MPEG and JPEG standards. It was a recognition of the MPEG compression technology as a cost-effective means of delivering high-quality audio and video programming to consumers by reducing the bandwidth necessary to carry the signal to homes and businesses across the globe. Other success stories for MPEG standards include:

- Today's computer systems support numerous MPEG audio and video standards for the creation, editing and playback of multimedia files.
- MPEG-2 is the video standard for DVD players and MPEG-4 AVC is a supported video standard for Blu-ray players.
- For DVD multichannel audio, MPEG-2 is the standard audio format for PAL/SECAM countries and an option for NTSC countries.

- A large number of broadcasting applications are based on MPEG technology, e.g., Digital Satellite System, Digital Audio Broadcast (DAB), Digital Video Broadcast (DVB), Astra Digital Radio, satellite feeds to cable networks, etc.
- MPEG-1 Audio Layer 3 (MP3) and MPEG Advanced Audio Coding (AAC) compression standards are supported by most of the world's smart phones and music players.

In summary, MPEG technology has enabled the introduction and evolution of digital storage and transmission products and services that continues to this day.

8.4   The MPEG family of standards

8.4.1   MPEG-1

The MPEG-1 standard, established in 1992, is designed to produce reasonable quality images and sound at low bit rates. MPEG-1 consists of four parts:

- IS 11172-1: System describes synchronization and multiplexing of video and audio.
- IS 11172-2: Video describes compression of non-interlaced video signals.
- IS 11172-3: Audio describes compression of audio signals using high-performance perceptual coding schemes.
- CD 11172-4: Compliance Testing describes procedures for determining the characteristics of coded bit-streams and the decoding process, and for testing compliance with the requirements stated in the other parts.

MPEG-1, IS 11172-3, which describes the compression of audio signals, specifies a family of three audio coding schemes, simply called Layer-1, -2, and -3, with increasing encoder complexity and performance (sound quality per bit rate). The three codecs are compatible in a hierarchical way, i.e., a Layer-N decoder is able to decode bit-stream data encoded in Layer-N and all Layers below N (e.g. a Layer-3 decoder may accept Layer-1, -2, and -3, whereas a Layer-2 decoder may accept only Layer-1 and -2). MPEG-1 Layer-3 is more popularly known as MP3 and has revolutionized the digital music domain.

MPEG-1 is intended to fit the bandwidth of CD-ROM, Video-CD, and CD-i. MPEG-1 usually comes in Standard Interchange Format, which is established at 352×240 pixels NTSC at 1.5 Mbits/s, a quality level about on par with Video Home System (VHS). MPEG-1 can be encoded at bit rates as high as 4–5 Mbits/s but the strength of MPEG-1 is its high compression ratio with relatively high quality. MPEG-1 is also used to transmit video over digital telephone networks such as asymmetrical digital subscriber lines, VOD, video kiosks, and corporate presentations and training networks. MPEG-1 is also used as an archival medium or in an audio-only form to transmit audio over the Internet.

8.4.2   MPEG-2

The MPEG-2 standard, established in 1994, is designed to produce higher quality images at higher bit rates. MPEG-2 is not necessarily better than MPEG-1, since MPEG-2 streams at lower MPEG-1 bit rates would not look as good as MPEG-1. But at its specified bit rates between 3 and 10 Mbits/s, MPEG-2 at the full Comité Consultatif International pour la Radio, a forerunner of the ITU-R (CCIR)-601 resolution of 720 × 486 pixels NTSC delivers true broadcast quality video. MPEG-2 was engineered so that any MPEG-2 decoder will play back an MPEG-1 stream, ensuring a backwards-compatible path for users who enter into MPEG with the lower-priced MPEG-1 encoding hardware. MPEG-2 video is the original standard for standard- and high-definition television (HDTV) and is supported by DVD and Blu-ray players. The primary users of MPEG-2 are broadcast and cable companies who demand broadcast quality digital video and utilize satellite transponders and cable networks for delivery of cable television and direct broadcast satellite.

8.4.3   MPEG-3

MPEG-3 was initially intended to cover HDTV, providing larger sampling dimensions and bit rates between 20 and 40 Mbits/s. It was later discovered that MPEG-2 met the requirements of HDTV, so the MPEG-3 standard was dropped.

8.4.4   MPEG-4

The MPEG-4 Visual standard was initiated in 1995, reached a committee draft status in March 1998 (ISO 14496-2) and was finalized by the end of 1998. This standard was initially specified for very low bit rates but now it supports up to 4 Mbits/s. MPEG-4 specifies sampling dimensions up to 176 × 144 pixels at comparatively low bit rates between 4,800 and 64,000 bits/s (not megabits but bits). It has six parts:

- Systems
- Visual
- Audio
- Conformance testing
- Software
- Delivery multimedia integration framework

MPEG-4 is designed for use in broadcast, interactive, and conversational environments. The way MPEG-4 is built allows MPEG-4 to be used in television and Web environments, not just one after the other but also facilitates the integration of content coming from both channels in the same multimedia *scene*. Its strong points are inherited from the successful MPEG-1 and -2 standards (broadcast-grade synchronization and the choice of online/offline usage) and Virtual Reality Modeling Language (VRML) (the ability to create content using a *scene description*).

MPEG-4 adds to MPEG-1 and -2:

- Integration of natural and synthetic content, in the form of *objects*. Such objects could represent 'recorded' entities (a person or a chair) or synthesized material (a voice, a face, or an animated 3D model)
- Support for 2D and 3D content
- Support for several types of interactivity
- Coding at very low rates (2 kbit/s for speech and 5 kbit/s for video) to very high ones (5 Mbit/s for transparent quality video and 64 kbit/s per channel for CD quality audio)
- Support for management and protection of intellectual property

MPEG-4 adds to VRML:

- Native support for natural content and real-time streamed content, using Uniform Resource Locator (URLs)
- Efficient representation of the scene description

Several forms of scalability support usage over networks with a bandwidth that is unknown at the time of encoding.

MPEG-4 preserves compatibility with major existing standards: MPEG-1, MPEG-2, International Telecommunications Union – Telecom Sector (ITU-T) H.263, and VRML.

While the full MPEG-4 toolbox is very rich and powerful, it is generally too expensive to implement in full for many applications. That is why MPEG has defined 'profiles,' which group the capabilities into useful subsets. This means that the standard can be used for both simple applications now and for future applications containing rich Web content.

8.4.5   Other MPEG video standards

MPEG video compression has seen three major developments and standardization efforts, namely MPEG-4 AVC (ISO/IEC 14496-10), MPEG-4 HEVC (ISO/IEC 23008-2, MPEG-H Part 2) and VVC (ISO/IEC 23090-3, MPEG-I Part 3), which are more efficient than MPEG-2 video and are, of this writing, being used or evaluated in broadcast and streaming applications. Figure 8.1 is a timeline of MPEG video standards in relation to standards developed by other organizations.

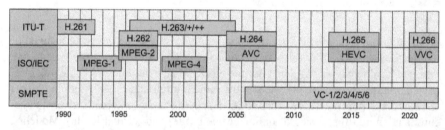

Figure 8.1   Evolution of video coding standards

8.4.6   MPEG-7

MPEG-7 is a standardized description of various types of multimedia information. This description is associated with the content itself, to allow fast and efficient searching for material that is of interest to the user. MPEG-7 is formally called Multimedia Content Description Interface.

The increasing availability of potentially interesting audio/video material makes its search more difficult. This challenging situation led to the need for a solution to the problem of quickly and efficiently searching for various types of multimedia material interesting to the user. MPEG-7 provides a solution.

The people taking part in defining MPEG-7 represent broadcasters, equipment manufacturers, digital content creators and managers, transmission providers, publishers, and intellectual property rights managers, as well as university researchers.

MPEG-7 will not replace MPEG-1, MPEG-2, or MPEG-4. It is intended to provide complementary functionality to these other MPEG standards, representing information about the content, not the content itself ("the bits about the bits"). This functionality is the standardization of multimedia content descriptions. MPEG-7 can be used independently of the other MPEG standards - the description might even be attached to an analog movie. The representation that is defined within MPEG-4, i.e., the representation of audio-visual data in terms of objects, is however very well suited to what will be built on the MPEG-7 standard. This representation is basic to the process of categorization. In addition, MPEG-7 descriptions could be used to improve the functionality of previous MPEG standards.

There are many applications and application domains that will benefit from the MPEG-7 standard. A few application examples are:

- Digital libraries (image catalog and musical dictionary)
- Multimedia directory services (e.g. yellow pages)
- Broadcast media selection (radio channel, TV channel)
- Multimedia editing (personalized electronic news service and media authoring)

8.5   MPEG's place in the ecosystem of digital television (DTV) standards

MPEG is not an independent organization but an ISO/IEC/International Telecommunication Union (ITU) working group. All three of these international organizations are involved in this area. The ISO collaborates with its international standardization partner, the IEC, whose scope of activities complements ISO's. In turn, ISO and the IEC cooperate on a joint basis with the ITU. Like ISO, the IEC is a non-governmental body, while the ITU is part of the United Nations Organization and its members are governments. The three organizations have a strong collaboration on standardization in the fields of information technology and telecommunications.

ISO promotes the development of standardization and related activities in the world with a view to facilitating the international exchange of goods and services, and developing cooperation in the spheres of intellectual, scientific, technological, and economic activity.

IEC is the world organization that prepares and publishes international standards for all electrical, electronic, and related technologies. It promotes international cooperation on all questions of electrotechnical standardization and related matters, such as the assessment of conformity to standards, in the fields of electricity, electronics, and related technologies.

The ITU is an international organization within which governments and the private sector coordinate global telecom networks and services.

MPEG also has a liaison subgroup, which maintains communications between MPEG and other standards bodies on topics of common interest. A few such bodies are:

- ITU-T
- ITU-Radio (ITU-R) Sector
- European Broadcast Union (EBU)
- Advanced Television Systems Committee (ATSC)
- SMPTE
- ISO
- JPEG, aka ISO/IEC JTC1 SC29/WG1
- VRML
- World Wide Web Consortium
- DVB
- International Federation of Film Producers Associations
- International Telecommunications Satellite Organization
- Audio Engineering Society

8.6   ATSC

The Advanced Television Systems Committee (ATSC) has developed a set of standards for digital television transmission over terrestrial, cable and satellite networks. ATSC standards are used in the U.S., Canada, Mexico and South Korea.

The ATSC standards were developed in the early 1990s by the Grand Alliance, a consortium of electronics and telecommunications companies that assembled to develop a specification for what is now known as HDTV. The ATSC formats also include standard-definition formats, although initially only HDTV services were launched in the digital format.

8.6.1   *Audio*

Dolby Digital AC-3 is used as the audio codec, though it was standardized as A/52 by the ATSC. It allows the transport of up to five channels of sound with a sixth channel for low-frequency effects (the so-called "5.1" configuration). In contrast,

Japanese Integrated Services Digital Broadcasting (ISDB) HDTV broadcasts use MPEG's Advanced Audio Coding (AAC) as the audio codec, which also allows 5.1 audio output. DVB (see below) allows both.

8.6.2 Video

ATSC system supports a number of different display resolutions, aspect ratios, and frame rates. The video formats are described in more detail below.

For transport, ATSC uses the MPEG systems specification, known as an MPEG transport stream, to encapsulate data, subject to certain constraints. ATSC uses 188-byte MPEG transport stream packets to carry data. Before decoding of audio and video takes place, the receiver must demodulate and apply error correction to the signal. Then, the transport stream may be demultiplexed into its constituent streams.

8.6.3 MPEG-2

There are three basic display sizes for ATSC. Basic and enhanced NTSC and PAL image sizes are at the bottom level at 480 or 576 lines. Medium-sized HDTV images have 720 scanlines and are 1280 pixels wide. The top tier has 1080 lines and 1920 pixels wide. A 1080-line video is actually encoded with 1920 × 1088 pixel frames but the last eight lines are discarded prior to display. This is due to a restriction of the MPEG-2 video format, which requires the number of coded luma samples (i.e. pixels) to be divisible by 16.

Most resolutions can operate in the progressive scan or interlaced mode, although the highest 1080-line system cannot display progressive images at the rate of 50, 59.94, or 60 frames per second, because such technology was seen as too advanced at the time and the image quality was deemed to be too poor considering the amount of data that needs to be transmitted. The standard also demands 720-line video be progressive-scan.

A terrestrial (over-the-air) transmission carries 19.39 Mbits/s (a fluctuating bandwidth of about 18.3 Mbit/s left after overhead such as error correction, program guide, closed captioning, etc.), compared to a maximum possible MPEG-2 bit rate of 10.08 Mbit/s (7 Mbit/s typical) allowed in the DVD standard and 48 Mbit/s (36 Mbit/s typical) allowed in the Blu-ray disc standard.

Although the ATSC A/53 standard limits MPEG-2 compression formats to those listed in Table 6.2 of [14], the US Federal Communications Commission declined to mandate that television stations obey this part of the ATSC's standard. In theory, television stations in the US are free to choose any resolution, aspect ratio, and frame/field rate, within the limits of Main Profile at High Level. Many stations do go outside the bounds of the ATSC specification by using other resolutions – e.g. 352 × 480 or 720 × 480.

The "Enhanced-Definition Television (EDTV)" displays can reproduce progressive scan content and frequently have a 16:9 widescreen format. Such resolutions are 704 × 480 or 720 × 480 [7] in NTSC and 720 ×576 in PAL, allowing 60 progressive frames per second in NTSC or 50 in PAL.

The ATSC A/53 specification imposes certain constraints on MPEG-2 video stream:

- The maximum bit rate value in the sequence header of the MPEG-2 video stream is 19.4 Mbit/s for broadcast television and 38.8 Mbit/s for the "high data rate" mode (e.g. cable television). The actual MPEG-2 video bit rate will be lower since the MPEG-2 video stream must fit inside a transport stream.
- The amount of MPEG-2 stream buffer required at the decoder (the vbv_buffer _size_value) must be less than or equal to 999,424 bytes.
- In most cases, the transmitter cannot start sending a coded image until within a half-second of when it is to be decoded (vbv_delay less than or equal to 45,000 90-kHz clock increments).
- The stream must include colorimetry information (gamma curve, the precise RGB colors used, and the relationship between RGB and the coded YCbCr).
- The video must be 4:2:0 (chrominance resolution must be 1/2 of luma horizontal resolution and 1/2 of luma vertical resolution).

The ATSC specification and MPEG-2 allow the use of progressive frames coded within an interlaced video sequence. For example, some DTV stations transmit a 1080i60 video sequence, meaning the formal output of the MPEG-2 decoding process is 60 540-line fields per second. However, for prime-time television shows, those 60 fields can be coded using 24 progressive frames as a base – actually, an 1080p24 video stream (a sequence of 24 progressive frames per second) is transmitted, and MPEG-2 metadata instructs the decoder to interlace these fields and perform 3:2 pulldown before display, as in soft telecine.

The ATSC specification also allows 1080p30 and 1080p24 MPEG-2 sequences; however, they are not used in practice because broadcasters want to be able to switch between 60 Hz interlaced (news), 30 Hz progressive or Progressive Segmented Frame(PsF) (soap operas), and 24 Hz progressive (prime-time) content without ending the 1080i60 MPEG-2 sequence.

The 1080-line formats are encoded as 1920 × 1088 luma frames; however, the last eight lines are discarded by the MPEG-2 display process.

8.6.4   *MPEG-4 AVC/H.264*

In July 2008, ATSC was updated to support the ITU-T H.264 video codec. The new standard is split in two parts:

- A/72 part 1: video system characteristics of AVC in the ATSC DTV system [8]
- A/72 part 2: AVC video transport subsystem characteristics [9]

The new standards support 1080p at 50, 59.94, and 60 frames per second; such frame rates require H.264/AVC *High Profile Level 4.2*, while standard HDTV frame rates only require Levels 3.2 and 4, and Standard-Definition Television (SDTV) frame rates require Levels 3 and 3.1.

8.6.5 MPEG-2 transport stream

An MPEG-2 transport stream, which typically has a ".TS" file extension, is a media container format. It may contain a number of streams of audio or video content multiplexed within the transport stream. Transport streams are designed with synchronization and recovery in mind for potentially lossy distribution (such as over-the-air ATSC broadcast) in order to continue a media stream with minimal interruption in the face of data loss in transmission. When an over-the-air ATSC signal is captured to a file via hardware/software, the resulting file is often in a .TS file format.

8.7 ISDB

ISDB, which stands for Integrated Services Digital Broadcasting, is a Japanese standard for DTV and digital radio used by the country's radio and television network. ISDB replaced NTSC-J analog television system and the previously used *MUSE Hi-vision* analog HDTV system in Japan. Digital Terrestrial Television Broadcasting services using ISDB-T started in Japan in December 2003 and in Brazil in December 2007 as a trial. Since then, many countries have adopted ISDB-T as their terrestrial DTV standard.

ISDB is maintained by the Japanese organization Association of Radio Industries and Businesses (ARIB). The standards can be obtained for free at the Japanese organization DiBEG website and at ARIB.

The core standards of ISDB are ISDB-S (satellite television), ISDB-T (terrestrial), ISDB-C (cable), and 2.6 GHz band mobile broadcasting which are all based on MPEG-2 or MPEG-4 standards for multiplexing with transport stream structure and video and audio coding (MPEG-2 or H.264), and are capable of HDTV and standard definition television. ISDB-T and ISDB-Tsb is the terrestrial digital sound broadcasting specification for Integrated Services Digital Broadcasting (ISDB-Tsb) is for mobile reception in TV bands. 1seg is the name of an ISDB-T service for reception on cell phones, laptop computers, and vehicles.

The concept was named for its similarity to ISDN because both allow multiple channels of data to be transmitted together (a process called multiplexing). This is also much like another digital radio system, Eureka 147, which calls each group of stations on a transmitter an ensemble; this is very much like the multi-channel DTV standard Digital Video Broadcasting - Terrestrial (DVB-T). ISDB-T operates on unused TV channels, an approach that was taken by other countries for TV but never before for radio.

Table 8.1 shows the main characteristics of the three main terrestrial HDTV transmission sysytems.

Table 8.1   Main characteristics of the three main terrestrial HDTV systems

System	ATSC 8-VSB	DVB COFDM	ISDB BST-COFDM
Source coding			
Video	Main profile syntax of ISO/IEC 13818-2 (MPEG-2 – video)		
Audio	ATSC standard A/52 (Dolby AC-3)	ISO/IEC 13818-3 (MPEG-2 layer II audio) and Dolby AC-3	ISO/IEC 13818-7 (MPEG-2 – AAC audio)
Transmission system			
Channel coding			
Outer coding	R-S (207, 187, t = 10)	R-S (204, 188, t = 8)	
Outer interleaver	52 R-S block interleaver	12 R-S block interleaver	
Inner coding	Rate 2/3 trellis code	Punctured convolution code Rate 1/2, 2/3, 3/4, 5/6, and 7/8 constraint length = 7, polynomials(octal) = 171, 133	
Inner Interleaver	12 to 1 trellis code interleaver	Bit-wise interleaving and frequency interleaving	Bit-wise interleaving frequency interleaving and selectable time interleaving
Data randomization	16-bit PRBS		
Modulation	8-VSB and 16-VSB	COFDM QPSK, 16QAM and 64 QAM Hierachical modulation: multi-resolution constellation (16QAM and 64QAM) Guard interval: 1/32, 1/16, 1/8, and 1/4 of OFDM symbol 2 modes :2k and 8k FFT	BST-COFDM with 13 frequency segments DQPSK, QPSK, 16QAM and 64QAM Hierachical modulation: choice of three different modulation on each segment Guard interval 1/32, 1/16, 1/8, and 1/4 of OFDM symbol 3 modes: 2k, 4k, and 8k FFT

MPEG, the ISO, IEC, and ITU govern the standards but not the patents for the technology used to apply these standards. There can be competition between various standards, for instance, this is the case in audio compression. MPEG-2, e.g., must compete with audio standards such as AC-3, developed by Dolby Digital. MPEG also incorporates outside technologies such as AudioMP3 for its own MPEG-1 Layer-3 standard. AudioMP3 was mainly developed by Fraunhofer, an organization specialized in applied research.

Many companies try to differentiate themselves after adopting MPEG standards by adding special operating features or customizing the look and feel of the hardware and software. The strength of the MPEG standards is that any country or company with an interest in the development of this technology can participate.

8.8   DVB

DVB, which stands for Digital Video Broadcasting, is a suite of internationally accepted open standards for digital television. DVB standards are maintained by the DVB Project, an international industry consortium with more than 270 members, and are published by JTC of the European Telecommunications Standards Institute (ETSI), European Committee for Electrotechnical Standardization, and EBU. The interaction of the DVB sub-standards is described in the *DVB Cookbook* [10]. Many aspects of DVB are patented, including elements of the MPEG video coding and audio coding (Table 8.2).

8.9   Digital Terrestrial Multimedia Broadcast (DTMB)

DTMB is the TV standard for mobile and fixed terminals used in the People's Republic of China, Cuba, Hong Kong, and several other countries. It is a merger of several digital multimedia broadcast standards. It supports both fixed reception (indoor and outdoor) as well as mobile DTV reception.

8.10   Digital Multimedia Broadcasting (DMB)

DMB is a digital radio transmission technology developed in South Korea [11] as part of the national Information Technology (IT) project for sending multimedia such as TV, radio, and datacasting to mobile devices such as mobile phones, laptops, and Global Positioning System (GPS) navigation systems. This technology, sometimes known as mobile TV, should not be confused with DAB which was developed as a research project for the European Union. DMB was developed in South Korea as the next generation digital technology to replace FM radio. The world's first official mobile TV service started in South Korea in May 2005, although trials were available much earlier. It can operate via satellite (S-DMB) or terrestrial (T-DMB) transmission. DMB has also some similarities with the

Table 8.2 Main technical specifications for DVB family of DTV standards

	DVB-S2	DVB-T2	DVB-C2
Input interface	Multiple transport stream and generic stream encapsulation (GSE)	Multiple transport stream and generic stream encapsulation (GSE)	Multiple transport stream and generic stream encapsulation (GSE)
Modes	Variable coding and modulation and adaptive coding and modulation	Variable coding and modulation [4]	Variable coding and modulation and adaptive coding and modulation
Forward Error Correction (FEC)	Low-Density Parity-Check Code (LDPC) + BCH 1/4, 1/3, 2/5, 1/2, 3/5, 2/3, 3/4, 4/5, 5/6, 8/9 and 9/10	Low-Density Parity-Check Code (LDPC) + BCH 1/2, 3/5, 2/3, 3/4, 4/5, and 5/6	Low-Density Parity-Check Code (LDPC)+ Bose–Chaudhuri–Hocquenghem code) a code in coding theory (BCH) 1/2, 2/3, 3/4, 4/5, 5/6, 8/9, and 9/10
Modulation	Single carrier QPSK with multiple sStreams	OFDM	Absolute OFDM
Modulation schemes	QPSK, 8-Phase-Shift Keying (PSK), 16-Amplitude and Phase-Shift Keying (APSK), 32-Amplitude and Phase-Shift Keying (APSK)	QPSK 16-QAM, 64-QAM, and 256-QAM	16- to 4096-QAM
Guard interval	Not applicable	1/4, 19/256, 1/8, 19/128, 1/16, 1/32, and 1/128	1/64 or 1/128
Fourier transform size	Not applicable	1k, 2k, 4k, 8k, 16k, and 32k, Discrete Fourier Transform (DFT)	4k inverse FFT
Interleaving	Bit-interleaving	Bit-, time-, and frequency-interleaving	Bit-, time-, and frequency-interleaving
Pilots	Pilot symbols	Scattered and continual pilots	Scattered and continual pilots

main competing mobile TV standard, Digital Video Broadcasting – Handheld (DVB-H) [12].

T-DMB is made for terrestrial transmissions on band III Very High Frequency (VHF) and L Ultra High Frequency (UHF) frequencies. DMB is unavailable in the US because those frequencies are allocated for television broadcasting (VHF channels 7 to 13) and military applications. The US is adopting ATSC-M/H for free broadcasts to mobiles, and Qualcomm's proprietary MediaFLO system was used there. In Japan, 1seg is the standard, using ISDB.

T-DMB uses MPEG-4 Part 10 (H.264) for the video and MPEG-4 Part 3 Bit Sliced Arithmetic Coding (BSAC) or HE-AAC v2 for the audio. The audio and video is encapsulated in an MPEG transport stream (MPEG-TS). The stream is forward error corrected by Reed Solomon encoding and the parity word is 16 bytes long. There is convolutional interleaving made on this stream, then the stream is broadcast in data stream mode on DAB. In order to diminish the channel effects such as fading and shadowing, the DMB modem uses OFDM-DQPSK modulation. A single-chip T-DMB receiver is also provided by an MPEG transport stream demultiplexer. DMB has several applicable devices such as mobile phones, portable TV, Personal Digital Assistant (PDAs), and telematics devices for automobiles.

T-DMB is a (ETSI) standard (TS 102 427 and TS 102 428). As of late 2007, ITU formally approved T-DMB as a global standard, along with three other standards like DVB-H, 1Seg, and MediaFLO.

8.11   Emerging MPEG standards and the road ahead

Since 2010, MPEG has begun to develop standards beyond traditional television broadcasting. These include standards for streaming (MPEG-DASH), immersive media and virtual/augmented reality (MPEG-I) and internet-connected devices (MPEG-IoMT). The use of Artificial Intelligence (AI) is also being studied within MPEG and by other organizations, such as MPAI [13].

MPEG is currently powerful as a standards body because of the technical superiority of its standards and the broad applicability of the standards. The standards' extensive reach generates widespread participation among industry members. MPEG draws upon the broad range of technical expertise from its members for ideas and inputs on proposed standards. Widespread participation leads to a high probability that new ideas for new standards will be captured by the MPEG standards body, generating positive feedback, which results in increased power for MPEG over time.

As MPEG standards continue to proliferate, the issue of backward compatibility will also grow in importance. The first MPEG standards were minimally encumbered by compatibility issues. The compatibility of new emerging standards with old will require both more technical and political considerations.

Convergence of all forms of digital data in the future can easily bring new players to compete with MPEG's position as a standards body. Perhaps the greatest threat is for a strong market leader or group of powerful companies attempting to establish their standard independent of MPEG or in direct conflict with MPEG.

MPEG also must watch out for market forces such as initially high prices for hardware for new technologies. These hinder the adoption of standards. MPEG must make sure new standards are still economically feasible over the short term, even if there are significant benefits over the long term. The ability for MPEG-4 to be implemented in partial form, rather than always in full, shows how this issue might be addressed.

Finally, MPEG patent pool terms and royalties can complicate the path to acceptance of MPEG standards. This has led to the development of standards such as MPEG-5 Efficient Video Coding (EVC) which consists of a royalty-free subset and individually switchable enhancements. The Baseline profile only contains video coding technologies that are older than 20 years and freely available for use in the standard. As patents begin to expire or as organizations freely donate intellectual property, more royalty-free codecs are expected to enter and become an important part of the digital media ecosystem.

References

[1]   Jurgen R.K. 'Digital consumer electronics handbook' in Jurgen R.K. (ed.). *Digital consumer electronics handbook*. New York: McGraw-Hill; 1997. pp. 1–45.
[2]   Available from https://en.wikipedia.org/wiki/Data_compression#Audio
[3]   Available from https://en.wikipedia.org/wiki/Differential_pulse-code_modulation
[4]   Available from http://s3.amazonaws.com/itp_archive/documentation/349/gianetti.christopher.1992.thesis.doc.PDF
[5]   Available from https://en.wikipedia.org/wiki/JPEG
[6]   Available from http://courses.ischool.berkeley.edu/i224/s99/GroupG/report1.html
[7]   Available from https://en.wikipedia.org/wiki/ATSC_standards
[8]   Part 1. Available from https://prdatsc.wpenginepowered.com/wp-content/uploads/2023/04/A72-Part-1-2023-04.pdf
[9]   Part 2. Available from https://prdatsc.wpenginepowered.com/wp-content/uploads/2015/03/A72-Part-2-2014-1.pdf
[10]   Available from https://www.etsi.org/deliver/etsi_tr/101200_101299/101200/01.01.01_60/tr_101200v010101p.pdf
[11]   Available from https://en.wikipedia.org/wiki/Digital_multimedia_broadcasting
[12]   Available from https://en.wikipedia.org/wiki/DVB-H
[13]   Available from https://mpai.community/
[14]   Available from https://www.atsc.org/wp-content/uploads/2021/04/A_53-Part-4-2009.pdf

Chapter 9

Digital video broadcasting primer

Edward Feng Pan[1]

9.1 Introduction

Digital television (DTV) is the new type of broadcasting technology that has globally transformed the television industry and the viewing experience since its inception in the late 1990s. DTV refers to the complete digitization of the TV signal (video, audio, and data) from transmission to reception. By transmitting TV signals in digital format and compressing them with the state-of-the-art video compression scheme, a digital broadcaster can carry much more information than is possible with analog broadcasting technology. This allows for the transmission of TV signal with HDTV resolution for dramatically better picture and sound quality or several SDTV programs concurrently. The DTV technology can also provide high-speed data transmission, including fast Internet access.

There are several DTV broadcasting systems that have been developed and are in operation in different parts of the world, notably the Advanced Television Systems Committee (ATSC), Digital Video Broadcasting (DVB), Integrated Service Data Broadcasting (ISDB), and Digital Multimedia Broadcasting. This article focuses on DVB standards and introduces the principles behind DVB.

DVB is a suite of standards for DTV. DVB standards are maintained by the DVB project, an industry consortium with more than 270 members, and are published by a Joint Technical Committee of the European Telecommunications Standards Institute, European Committee for Electrotechnical Standardization, and European Broadcasting Union.

DVB standards specify a variety of approaches to distribute data using satellite (DVB-S), cable (DVB-C), terrestrial (DVB-T), and handheld systems (DVB-H). These standards define the physical layer of the distribution system. All data are transmitted in MPEG-2 transport streams (TSs) with some additional constraints. These distribution systems differ mainly in the modulation schemes and error correction codes to cater to noise/reflections/distortions in the physical transmission media. The DVB-x2 (DVB-T2, DVB-S2, and DVB-C2) is a new suite of standards that were published by ETSI in September 2009. The DVB-x2 standard gives more robust TV reception and increases the possible bit rate by up to 30% for single transmitters and should increase the maximum bit rate by over 50% in large single-frequency networks (SFNs).

[1]Advanced Micro Devices, Inc., Canada

The rest of the article is organized as follows. Section 9.2 gives an overview of the DTV system. Section 9.3 describes the source data encoding techniques used in DTV systems, which include audio/video coding and their multiplexing, data scrambling, and conditional access (CA). Section 9.4 describes forward error correction (FEC) techniques such as convolutional coding and Reed–Solomon coding. Channel modulation techniques are introduced in section 9.5. Finally, this article discusses the various DVB delivering formats such as DVB-S, DVB-C, DVB-T, and the latest DVB-T2.

9.2   DTV system overview

The block diagram of a DTV broadcasting system is shown in Figure 9.1. The video, audio, and other service data are compressed to form elementary streams and then are multiplexed to form the MPEG-2 TS. These streams may be multiplexed again with the source data from other programs to form a TS with multiple programs. A TS consists of transport packets that are 188 bytes in length.

The FEC encoder takes preventive measures to protect the TSs from errors caused by noise and interference in the transmission channel. It includes Reed–Solomon coding, outer interleaving, and convolutional coding. This process mainly introduces some redundancy in data, so that the original information could be recovered even when some of the data bits are compromised by the channel distortions, and the capability of error correction largely depends on the amount of redundancy it introduces.

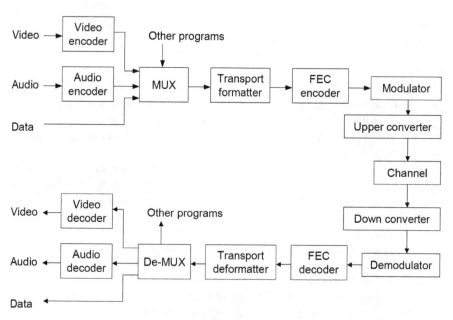

Figure 9.1   Diagram of the DTV system

The modulator then converts the FEC-protected transport packets into digital symbols that are suitable for transmission in the terrestrial channels. This involves quadrature amplitude modulation (QAM) and orthogonal frequency-division multiplexing (OFDM) in DVB-T and ISDB-T systems, or PAM and vestigial sideband (VSB) in ATSC-T. The final stage is the upper converter, which converts the modulated digital signal into the appropriate RF channel. The sequence of operations in the receiver side is a reverse order of the operations as in the transmitter side.

9.3 Video compression

The compression of digital video signal [1,2] is indispensable as the raw video data are extremely bulky. It is impossible to transmit uncompressed video signals in most applications. For example, an uncompressed high-definition television (HDTV) program has a bit rate of over 1.4 Gbps, and even a standard-definition television (SDTV) program needs a transmission bit rate of up to 240 Mbps. These are far higher than the typical transmission bit rate that a digital multiplex can accommodate, which is typically around 20 Mbps. In the early 1990s, the MPEG-2 video compression standard has been developed, which was designed for the application of DVB. With the use of MPEG-2, one HDTV program or up to six SDTV programs nicely fit into one multiplex. More recently, new compression standards have been considered for video broadcasting as well. Those include AVC/H.264 and High-Efficiency Video Coding (HEVC)/H.265, which are more efficient in compression efficiency compared to their predecessors, so more TV programs or even ultra-high definition television broadcasting is possible.

All video-coding standards mentioned above can be classified into a category called block-based video compression standards, which exploit certain characteristics of video signals to remove redundant information. In a typical video signal, redundancy exhibits many ways, such as spatial redundancy inside a frame and temporal redundancy in-between frames. The video compression mechanism also removes the psycho-visual redundancy based on the characteristics of the human vision system (HVS) such that HVS is less sensitive to error in detailed texture areas and fast-moving pictures. Finally, the video compression techniques also use entropy coding to remove statistical redundancy in the data to increase the data-packing efficiency.

9.3.1 Transform-based intra-frame coding

The intra-frame coding algorithm (Figure 9.2) begins by calculating the transform (DCT or integer transform) coefficients over small non-overlapping image blocks (variable in size, from 4×4 to 64×64 pixels). This block-by-block processing takes advantage of the image's local spatial correlation properties. The transform process produces many 2D blocks of transform coefficients that are quantized to discard some of the trivial coefficients that are likely to be perceptually masked. The quantized coefficients are then zigzag scanned to output the data in an efficient way. The final step in this process uses variable length coding to further reduce signal entropy.

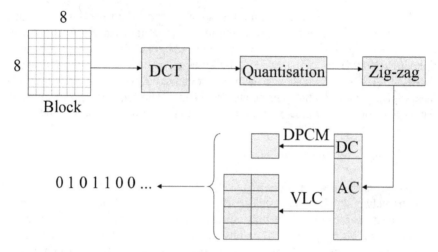

Figure 9.2   DCT-based intra-frame coding

9.3.2   Prediction-based inter-frame coding

Inter-frame coding (Figure 9.3), on the other hand, exploits temporal redundancy by predicting the frame to be coded from a previous reference frame. The motion estimator searches previously coded frames for areas similar to those in the prediction blocks of the current frame. This search results in motion vectors (represented by x and y components in pixel lengths), which the decoder uses to form a motion-compensated prediction of the video. The motion-estimator circuitry is typically the most computationally intensive element in a video encoder (Figure 9.4). With motion-compensated inter-frame coding, the compression system only needs to

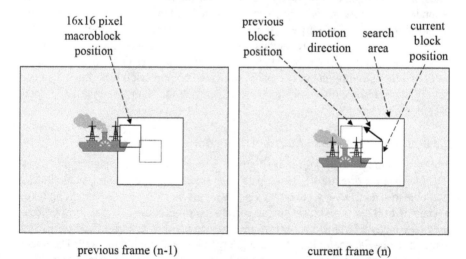

Figure 9.3   Motion-compensated inter-frame coding

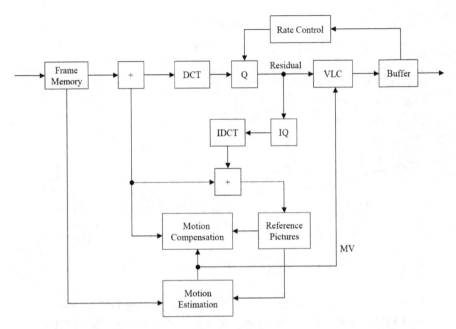

Figure 9.4   Block diagram of an MPEG-2 video compression system

convey the motion vectors required to predict each block to the decoder, instead of conveying the original block data, which results in a significant reduction in bit rate.

9.3.3   Overview of video coding standards

There are two major international standard bodies working on creating video-coding standards. One of them is Video-Coding Experts Group (VCEG) under International Telecommunication Union-Telecommunication, and the other is MPEG, a group formed under the auspices of the International Standards Organization (ISO) and the International Electro-Technical Commission (IEC). These two standards committee has been jointly developing video compression standards for various applications since the 1990s. The three most successful standards are MPEG-2, AVC, and HEVC, which are widely used in modern video broadcasting systems.

9.3.3.1   MPEG-2

The MPEG-2 standard (ISO/IEC 13818), established in 1994, is designed to support a wide range of applications and services of varying bit rate, resolution, and quality. MPEG-2 standard defines four profiles and four levels for ensuring the interoperability of these applications. The profile defines the color space, resolution, and scalability of the bit stream. The levels define the range of frame resolution, luminance sample rate, the number of video and audio layers supported for scalable profiles, and the maximum bit rate per profile.

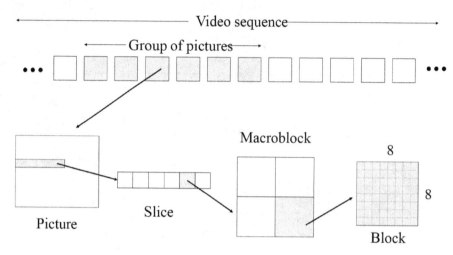

Figure 9.5   MPEG-2 video bit stream hierarchy

MPEG-2 video-coding standard is primarily a specification of bit stream semantics. The video bit stream is usually arranged in several layers; each corresponds
to a hierarchy, as shown in Figure 9.5. The functions of each of the hierarchies are
explained in the following.

- Video sequence. This is the highest layer that defines the contexts that are applicable to the whole sequence. A video sequence consists of a sequence header,
  the number of groups of pictures (GOP), and the end-of-sequence code. The
  sequence header initializes the state of the decoder. This allows a decoder to
  decode any sequence without being affected by past decoding history.
- GOP. A GOP consists of a series of one or more video frames intended to allow
  random access to the sequence. A GOP always starts with an I-frame.
- Picture. The picture is the primary coding unit of the video sequence, which
  consists of three rectangular matrices representing one luminance (Y) and two
  chrominances (Cb and Cr) components.
- Slice. A slice is a layer for intra-frame addressing and (re)synchronization. It
  consists of one or more contiguous macroblocks (MBs).
- MB. A typical MB consists of a 16 × 16 pixels area, with four Y blocks and one
  or two Cb and Cr blocks depending on whether it is 4:2:0 or 4:2:2 sub-sampled.
  MB is the basic unit for motion prediction and compensation.
- Blocks. A block is an 8 × 8 pixel section of luminance or chrominance components. It is the basic unit for DCT-based intra-frame coding.

MPEG-2 supports both progressive scanning and interlaced scanning. Interlaced
scanning is the scanning method used in analog television systems where odd lines
of a frame are scanned first as one field (odd field), and even lines (field) are then
scanned after the odd field. An interlaced video coding can use two different picture

structures: frame structure and field structure. In the frame structure, lines of two fields are coded together as a frame, the same way as in sequential scanning. One picture header is used for two fields. In the field structure, the two fields of a frame are coded independently, and the odd field is followed by the even field. Cross-field predictive coding is possible as the two fields can be treated as two independent frames. The interlaced video sequence can switch between frame structure and field structure on a picture-by-picture basis.

9.3.3.2   H.264/AVC

H.264 advanced video coding (ISO/IEC14496) [3] is also the result of a joint effort between MPEG and VCEG. It was developed in response to the growing need for higher compression of video signal for various applications including television broadcasting, videoconferencing, digital storage media, Internet streaming, and wireless communication. H.264/AVC started as an H.26L project in 1997 by VCEG, with the aim to define a new video-coding standard that would have improved coding efficiency by 50% bit rate reduction compared to any previous standards, simple syntax specification, simple and clean solutions without an excessive quantity of optional features or profile configurations (a lesson learned from MPEG-4), as well as network friendliness. In the meantime, MPEG issued a call for proposals in July 2000 to look for the possibility of a new video-coding standard after MPEG. VCEG proposed H.26L to this call, and among all the proposals submitted to MPEG, H.26L achieved the best performance. Therefore, in December 2001, MPEG and VCEG formed a joint video team (JVT) to continue H.26L project, which was renamed the JVT project. After that, the new video-coding standard was renamed H.264 or MPEG-4 part 10 or AVC.

H.264/AVC contains a rich set of video-coding tools that are organized into different profiles. The original three profiles were defined in the H.264 specification that was completed in May of 2003: the baseline profile, the extended profile, and the main profile. For applications such as high-resolution entertainment-quality video, content distribution, studio editing, and post-processing, it is necessary to develop some extensions of the tool set for professional video applications. This effort has resulted in a new set of extensions that are named the fidelity range extensions. These extensions include four new profiles: the high profile, the high 10 profile, the high 4:2:2 profile, and the high 4:4:4 profile.

H.264/AVC represents a major step forward in the development of video-coding standards. It typically outperforms prior standards by a factor of two or more, particularly in comparison to H.262/MPEG-2 and H.26XX. The improvements enable new applications and business opportunities to be developed. However, the improvement in coding efficiency is at the cost of coding complexity, which is about three to five times more complex than any of its predecessors. Furthermore, H.264 is a collaboratively designed open standard with the aim of very low licensing cost. This will help create a competitive market, keeping prices down and ensuring that products made by a wide variety of different manufacturers will be fully compatible with each other.

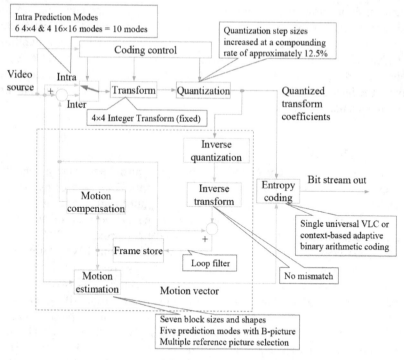

Figure 9.6   H.264/AVC encoder block diagram

Some new techniques that H.264/AVC has implemented:

- spatial prediction in intra-coding with two block sizes of 16 × 16 and 4 × 4;
- adaptive block size motion compensation with variable block sizes of 4 × 4, 4 × 8, 8 × 4, 8 × 8, 8 × 16, 16 × 8, and 16 × 16;
- 4 × 4 integer transformation;
- multiple reference frames;
- content adaptive binary arithmetic coding.

The increased coding efficiency of the H.264/AVC has led to new application areas and business opportunities and has found many successful applications in various industries such as DTV broadcasting, Blu-ray movie, and Internet video, as well as in mobile communication.

Figure 9.6 shows the block diagram of the H.264/AVC encoder. In the diagram, we have specified the new features of the various function blocks in comparison to the MPEG-2 standard.

9.3.3.3   HEVC

The HEVC or the H.265 standard (ISO/IEC 23008-2) is the most recent joint video-coding standard ratified in 2013 by MPEG and VCEG. Compared to the state-of-the-art

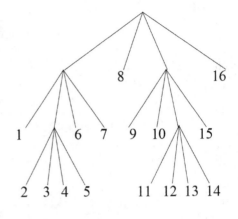

Figure 9.7 HEVC quadtree coding unit

H.264/AVC standard, HEVC can achieve the same PSNR with 40% of bit reduction and the same subjective quality at half the bit rate. Note that the underlying compression algorithms used in HEVC are very similar to previously adopted standards such as MPEG-2 and H.264. However, there are two main improvements of HEVC compared to its predecessors.

9.3.3.3.1 Variable block size

Instead of using a fixed MB size of 16 × 16, HEVC employs the concept of coding units with sizes ranging from 4 × 4 to 64 × 64. It even uses non-square block size, from 4 × 8/8 × 4 to 8 × 32/32 × 8. This is done with a more efficient but also complex set of coding units that can be efficiently partitioned using a quadtree. Depending on the complexity of the blocks, it could choose to use large partitions when they predict well and small partitions when more detailed predictions are needed. This leads to higher coding efficiency. Figure 9.7 shows one example of quadtree coding unit.

9.3.3.3.2 Much more prediction modes

To further improve the coding efficiency, HEVC describes the video contents in a much finer detail compared to that of previous standards. This is at the cost of much high encoder complexity. For example, HEVC uses 35 directional modes for intra-prediction, while AVC only uses up to nine different modes. Also, HEVC uses 24 different block types for inter-prediction, while AVC uses only seven block types.

9.4 DVB audio compression

Unlike video, the three current DTV standards use three different audio-coding schemes: Dolby AC-3 for ATSC, MPEG audio and Dolby AC-3 for DVB, and MPEG-AAC for ISDB. However, these audio standards use a similar technique called perceptual coding and support up to six channels—right, left, center, right

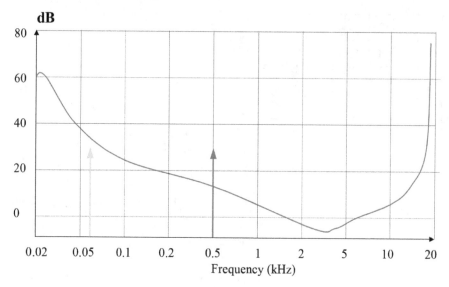

Figure 9.8   Audio perceptual masking

surround, left surround, and subwoofer—often designated as 5.1 channels. A perceptual audio coder [5] exploits a psycho-acoustic effect known as masking, (Figure 9.8). This psycho-acoustic phenomenon states that when sound is broken into its constituent frequencies, those sounds with relatively lower energy adjacent to others with significantly higher energy are masked by the latter and are not audible. Therefore, the masked audio frequency can simply be ignored in the encoding process to the same bitrate.

AC-3 is one of the most popular audio compression algorithms used in DTV, movie theater, and home theater systems. AC-3 makes use of the psycho-acoustic phenomenon to achieve great data compression. In the encoding process, the modified DCT algorithm transforms the audio signal into the frequency domain, which generates a series of frequency coefficients that represent the relative energy contributions to the signal of those frequencies.

By analyzing the incoming signal in the frequency domain, psycho-acoustically masked frequencies are given fewer (or zero) bits to represent their frequency coefficients; dominant frequencies are given more bits. Hence, besides the coefficients themselves, the decoder must receive the information that describes how the bits are allocated so that it may reconstruct the bit allocation. In AC-3, all the encoded channels draw from the same pool of bits, so channels that need better resolution can use the most bits.

The output coefficients generated by the time-domain to frequency-domain transformation are typically represented in a block floating-point format to maintain numeric fidelity. Using the block floating-point format is one way to extend the dynamic range in a fixed-point processor. It is done by examining a block of (frequency) samples and determining an appropriate exponent that can be associated

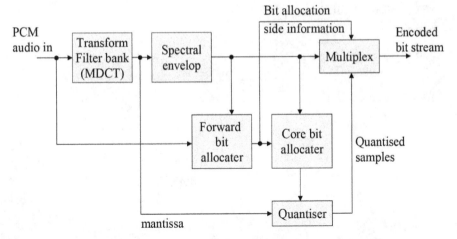

Figure 9.9   Dolby AC-3 audio-coding block diagram

with the entire block. Once the mantissas and exponents are determined, the mantissas are represented using the variable bit-allocation scheme described above; the exponents are DPCM coded and represented with a fixed number of bits (Figure 9.9).

MPEG audio is a type of forward adaptive bit allocation, while AC-3 uses hybrid adaptive bit allocation, which combines both the forward and backward adaptive bit allocation. The main advantage of MPEG audio is that the psycho-acoustic model resides only in the encoder. When the encoder is upgraded, legacy decoders continue to decode the newly coded data. However, the disadvantage is that it could have a heavy overhead for complicated music pieces.

9.5   MPEG-2 transport stream and multiplex

Audio and video encoders deliver elementary stream outputs. These media streams, as well as streams carrying other private data, are multiplexed in an organized manner and supplemented with additional information to allow demultiplexed in the decoder, synchronization of picture and sound (lip sync), and program selection by the end user. This is done through packetization specified in the MPEG-2 system layer. The elementary stream is divided into packets to form a packetized elementary stream (PES). A PES starts with a header, followed by the content of the packet (payload) and its descriptor. Packetization provides the protection and flexibility for transmitting multimedia steams across different networks. In general, a PES can only contain the data from the same elementary stream.

9.5.1   Elementary streams, packetized elementary streams, and transport streams

In broadcasting applications, a multiplex usually contains different data streams (audio, video, and other data) that might even come from different programs.

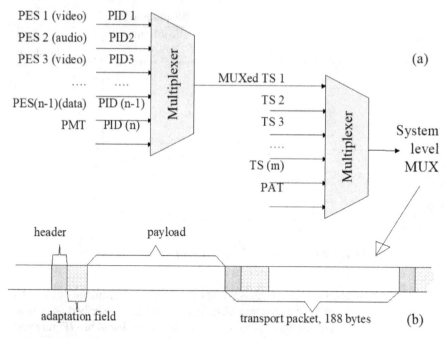

Figure 9.10   *(a) The process of multiplexing. (b) The structure of a transport packet.*

Therefore, it is necessary to multiplex them into a single stream—the TS [6]. Figure 9.10 shows the process of multiplexing. A TS consists of fixed-length transport packets. The header contains important information such as the synchronization byte and the packet identifier (PID). PID identifies a PES within the multiplex.

To create a TS, each of these PESs is packetized again and the data from the stream to form *transport packets*. Each transport packet has a length of 188 bytes, which is much smaller than a PES packet, and so a single PES packet will be split across several transport packets. This extra level of packetization allows the stream to support much more powerful error correcting techniques.

It is important to ensure that all the packets within the stream come in the right order—MPEG defines a strict buffering model for MPEG decoders, and so we must take care that each elementary stream in our TS is given a data rate that is constant enough to ensure that the receiver can decode that stream smoothly, with no buffer underflow or overflow. We must also note that video streams will use a much larger proportion of the final TS than audio streams, and so a ratio of 10 video packets to every audio packet is a good way to start with.

It is necessary to include additional program-specific information (PSI) within each TS to identify the relationship between the available programs and the PID of their constituent streams. This PSI consists of four tables: program associate table (PAT), program map table (PMT), network information table (NIT), and CA table (CAT).

Within a TS, the reserved PID of 0 indicates a transport packet that contains a PAT. The PAT associates a PID value with each program that is currently carried in the transport multiplex. This PID value identifies the PMT for that program. The PMT contains details of the constituent elementary streams for the program. Program 0 has a special meaning within the PAT and identifies the PID of the transport packets that contain the optional NIT. The contents of the NIT are private to the broadcaster and are intended to contain network-specific information. The CAT is identified by a PID of 1 and contains information specific to any CA or scrambling schemes that are in use.

9.5.2   Navigating an MPEG-2 multiplex

MPEG-2 PSI tables only give information concerning the multiplex. The DVB standard adds complementary tables (DVB-SI) to allow the user to navigate the available programs and services by means of an electronic program guide (EPG). DVB-SI has four basic tables and three optional tables to serve this purpose. As shown in Figure 9.11, the decoder must perform the following main steps in order to locate a program or a service in an MPEG-2 transport multiplex.

1. As soon as the new channel is acquired (synchronized), the decoder must filter the PID 0 packets to acquire the PAT sections and construct the PAT to provide the available choice (services currently available on the air) to the user.

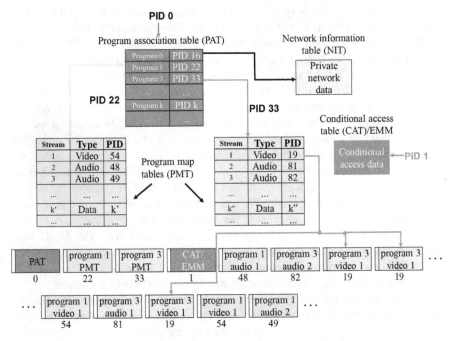

Figure 9.11   Navigating an MPEG-2 multiplex

2. Once the user choice is made, the decoder must filter the PID corresponding to the PMT of this program and construct the PMT from the relevant sections. If there is more than one audio or video stream, the user should be able to make another choice.

3. The decoder must filter the PID corresponding to this choice.

The audio/video decoding can now start. The part of this process that is visible to users is the interactive presentation of the EPG associated with the network, which can be built by means of the PSI and DVB-SI tables in order to allow them to easily navigate the available programs and services.

9.6   Conditional access in DVB

Except free to air broadcasting, most DTV services will either be pay-per-view or at least include some elements that are not freely available to the public. DVB defined a set of standards for CA, the common scrambling algorithm, and the common interface (DVB-CI). These standards [7] collectively define a method by which one can encrypt a digital-television stream so that it can only be accessed by those with valid decryption systems.

CA is achieved by a combination of scrambling and encryption. The data stream is scrambled with a 48-bit secret key, called the control word. The control word is generated automatically by the content providers, and it is being updated continuously at very short intervals, typically a few times per minute. To protect the control word, it is further encrypted to form an entitlement control message (ECM).

For the receiver to access the data stream, it must be continuously informed about the current value of the control word. It can only decrypt the control word when authorized to do so. The authorization is sent to the receiver in the form of an entitlement management message (EMM). The EMMs are specific to each subscriber, as identified by the smart card in his receiver, or to groups of subscribers. Note that EMMs are also being updated continuously, but at much less frequency than ECMs, usually at weekly or even monthly intervals. Figure 9.12 shows an example of a DVB content protection system using ECM and EMM and controls words.

DVB-CI defines a common interface for CA to enable the TV receiver to descramble programs broadcast in parallel, using different CA systems. By way of inserting a smart card module into the common interface, the receiver can sequentially address different CA systems at the same time. The simultaneous operation of several CA systems in a single receiver is called MultiCrypt. The MultiCrypt approach has the additional advantage that it does not require agreements between networks, but it is more expensive to implement. Other applications, such as Ethernet connection or electronic commerce, may also utilize the DVB-CI connector.

SimulCrypt is another way of providing the viewer with access to programs. In this case, commercial negotiations between different service providers have led to a contract that enables the viewer to use the one specific CA system built into the IRD to watch all the programs, irrespective of the fact that these programs were

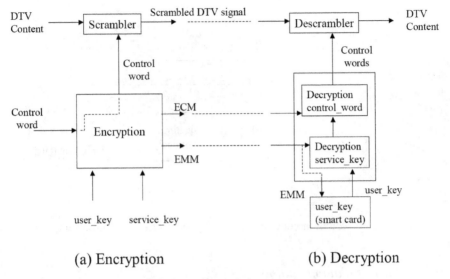

Figure 9.12   Example of DVB content protection

scrambled under the control of different CA systems. Note that DVB supports both MultiCrypt and SimulCrypt, while ATSC only supports the SimulCrypt.

9.7   Forward error correction

The transmission channels used for DTV broadcasting are, unfortunately, rather error-prone due to a lot of disturbances (such as noise, interference, and echoes). However, a digital TV signal, after almost all its redundancy is removed, requires a very low bit error rate (BER) for good performance. A BER of the order of 10^{-10} corresponds to an average interval of some 30 minutes between errors. Therefore, it is necessary to take preventive measures before modulation in order to allow detection and, as far as possible, correction in the receiver of most errors introduced by the physical transmission channel. These measures are called, collectively, FEC. FEC [8] requires that redundant data are added to the original data prior to transmission, allowing the receiver to use these redundant data to detect and recover the lost data caused by the channel disturbance.

Figure 9.13 illustrates the various steps of the FEC encoding process used in DTV broadcasting. Strictly speaking, energy dispersal is not part of the error correction process. The main purpose of this step is to avoid long strings of 0s or 1s in the TS, so that the energy is uniformly distributed across the frequency spectrum. Broadcasting standards often use the terms inner coding and outer coding. Inner coding operates just before the transmitter modulates the signal and just after the receiver demodulates the signal. Outer coding applies to the extreme input and output ends of the transmission chain. Inner coding is usually convolutional in nature, with optimal performance under conditions of steady noise interference. Outer

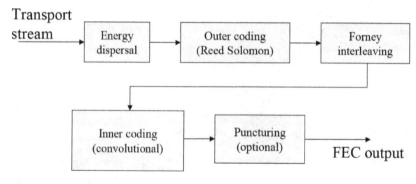

Figure 9.13   FEC coding process

coding is a Reed–Solomon code that is usually more effective for correcting burst errors.

9.7.1   Reed–Solomon coding

Outer coding is a Reed–Solomon code that is a subset of BCH cyclic block codes. As its name implies, in block coding, a block of bits is processed together to generate the new coded block. It does not have system memory, such that coding of a data word does not depend on what happens before or after that data occur. Reed–Solomon code, in combination with the Forney convolutional interleaving that follows it, allows the correction of burst errors introduced by the transmission channel. It is applied individually to all the transport packets in Figure 9.10b, excluding the synchronization bytes. R-S codes have been recently proved to operate at the theoretical limit of correcting efficiency. Therefore, it has been chosen for all DTV standards as outer coding. An R-S code is characterized by three parameters (*n, k, t*) where *n* is the size of the block after coding, *k* is the size of the block before coding, and *t* is the number of correctable symbols. Whether the received codeword is error-free could be checked through a division circuit corresponding to the generated polynomial *g(x)*. For a proper codeword, the remainder is zero. If the remainder is non-zero, a Euclidean algorithm is used to decide the two values needed for error correction: the location of the error and the nature of the error. However, if the size of the error exceeds half the amount of redundancy added, the error cannot be corrected.

In DVB and ISDB standards, we use the R-S(204,188,8) code. It adds 16 parity bytes and can correct up to 8 erroneous bytes per packet. On the other hand, the ATSC standard uses the R-S(207,187,10) code. It adds 20 parity bytes and can correct up to 10 erroneous bytes per packet.

9.7.2   Interleaving

The purpose of data interleaving is to increase the efficiency of the Reed–Solomon coding by spreading the burst errors (which is typically continuous) over a longer time and into several data packets. This can significantly increase the error correction capability of the Reed–Solomon coding. Interleaving is normally implemented

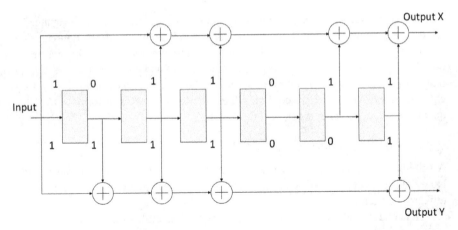

Figure 9.14   DVB convolutional coding

by using a two-dimensional array buffer, such that the data enter the buffer in rows and then read out in columns. The result of the interleaving process is that a burst of errors in the channel after interleaving becomes a few scarcely spaced single-symbol errors, which are more easily correctable.

The interleaver employed in the DVB standard is also called Forney interleaving and has an interleaving depth of 12, i.e., any two consecutive bits in the input will be separated to 12 bits apart. Note that only data bytes are interleaved. The interleaver is always synchronized to the first data byte of the data field.

9.7.3   Convolutional coding and puncturing

The inner coding used in DVB is convolutional coding. It is very powerful in correcting random errors, and thus an efficient complement to the Reed–Solomon coding.

DVB convolutional coding consists of two finite impulse response (FIR) filters that are implemented using polynomial generators with values of 0×79 and $0 \times 5B$, respectively. The input bit stream convolves with the two FIRs to produce two outputs simultaneously, which represent two different parity checks on the input data. Clearly, the code rate is 1/2. Figure 9.14 shows the diagram of convolutional coding.

Note that a ½ coding rate means that the error correction is very good, at the cost of halved bit rate. Convolutional code has the flexibility for the user to puncture the code to produce the variable coding rate, to achieve the required trade-off between error correction and bit rate or transmission efficiency. In DVB systems, 1/2, 2/3, 3/4, 4/5, 5/6, and 7/8 are possible code rates.

9.8   Digital modulations in DVB

Modulation is the technique that modifies the form of an electronic signal, so the signal can carry information on a communication media. There are typically three fundamental methods of modulating an analog carrier signal by digital binary signals, and they are amplitudes-shift keying, frequency-shift keying, and phase-shift keying. To increase the number of states per symbol or the efficiency of information-carrying capability, modern modulation schemes often combine multiple signal keying methods. The most popular combination is amplitude keying and phase keying. Depending on the required transmission bit rate and the desired C/N at a serviced receiver, different amplitude levels can be used during the modulation process to form 4-QAM [or quadrature phase shift keying (QPSK)], 16-QAM, 64-QAM, and 256-QAM [9].

9.8.1   QPSK modulation for DVB-S

A satellite transponder has a typical bandwidth of 27 MHz. This is a fairly wide bandwidth when compared with the 7 or 8 MHz allotted to a channel on a cable or terrestrial system.

Hence, a digital modulation technique used for satellite broadcasting (DVB-S) can use a fairly large bandwidth but should be capable of preserving the signal and maintaining a low BER even for very low signal strength.

The QPSK modulation system provides an ideal solution for this. QPSK is a very simple but robust form of digital modulation. The word quadrature simply means out of phase by 90° or orthogonal. Hence, as shown in Figure 9.2, QPSK provides four different states or possibilities for encoding a digital bit. This is because two components are used—one in phase (I) and the other out of phase or quadrature (Q). QPSK is a special case of QAM with only two amplitude levels and two orthogonal phases thus equivalent to 4-QAM. The QPSK system is now universally used, for all satellite DVB broadcasts.

9.8.2   QAM modulation for DVB-C

Similar to QPSK, QAM systems utilize changes in both phase and amplitude in the modulation process. Instead of using only two amplitude levels (or binary), QAM typically combines more bits to form multiple amplitude levels, and a large number of possible states can be created to provide dense digital modulation. It is obvious that each time the number of states per symbol is increased, the bandwidth efficiency also increases. As higher density modulation schemes are adopted, the modulation and de-modulation get progressively more complex. Therefore, this requires a rather high signal-to-noise ratio channel, which is exactly the case of cable broadcasting or DVB-S.

9.8.3   OFDM modulation for DVB-T

Both QPSK and QAM work well in applications where multi-path transmission does not exist or at least is not a big concern. This is the case of DVB-S and DVB-C. However, in terrestrial broadcasting, the distance between the signal source (transmitter) and the TV receiver is comparable with the many reflection paths caused by the high-rise buildings on the signal path. These multiple signals add up at the TV antenna, creating multiple images or "Ghosts" on the TV screen. This distortion is most pronounced in densely populated cities particularly those with high-rise buildings.

Modulation schemes for terrestrial television broadcasting are the main distinguisher among different DTV systems. DVB uses coded OFDM modulation and supports hierarchical transmission, ATSC uses eight-level VSB, and ISDB utilizes OFDM and two-dimensional interleaving. It supports hierarchical transmission of up to three layers and uses MPEG-2 video and advanced audio coding. Finally, digital terrestrial multimedia broadcasting adopts time-domain synchronous OFDM technology with a pseudo-random signal frame to serve as the guard interval (GI) of the OFDM block and the training symbol.

A European consortium, the Digital Video Broadcasting Project—developed the DVB-T OFDM system. The system uses a larger number of carriers per channel modulated in parallel via an FFT process, a technique referred to as OFDM. In case of multi-path interference, echoes could cause severe interference to the main signal. Therefore, a long symbol duration is necessary to suppress the echo interference. OFDM can achieve a long symbol duration (up to 1 ms) within the same bandwidth using parallel modulation. Thus, the receiver can wait until channel conditions have become stable before starting to decode each individual symbol. This means that all echoes have been collected. COFDM therefore turns echoes from destructive signals that cause inter-symbol interference into constructive signals adding to the energy of the direct transmission path.

In OFDM, symbols are de-multiplexed to modulate many different carriers (a few thousand), each of which occupies a much narrower bandwidth. These carriers are chosen to be orthogonal to each other so that they are separable in the decoder. The modulated symbols are frequency multiplexed to form the OFDM baseband signal, which is then up-converted to RF signal for transmission.

The OFDM transmission system can work QAM modulation, and different after levels of QAM modulation, such as 16-QAM and 64-QAM, can be selected. Moreover, a GI with selectable width (1/4, 1/8, or 1/16 of the symbol duration) separates the transmitting symbols, which gives the system an excellent capability for coping with multi-path distortion. OFDM modulation therefore can work well in a single-frequency network, such that in the single coverage area, multiple transmitters are used to transmit the same data using the same frequency at the same time. This gives the OFDM system another advantage of mobile reception—robust reception can be achieved even when the receiver is put on a vehicle running at over 100 km/h. The DVB-T system can operate in either a 2k mode or 8k mode. The 2k mode uses a maximum of 1 705 carriers, while in 8k mode the carrier number is 6817. The 2k mode system has a short symbol duration, so it is suitable for a small SFN network with limited distance between transmitters. The 8k mode is used in a large SFN network where the transmitters could be up to 90 km apart. Figure 9.15 shows an example of the OFDM frequency spectrum for 2k and 8k DVB-T systems.

In summary, with the use of the OFDM technique, the DVB-T system has performance advantages with respect to high-level (up to 0 dB), long-delay multi-path distortion. It works extremely well for services requiring large-scale, SFNs and for mobile reception. Hierarchical channel coding and modulation, which uses multi-resolution constellation on OFDM carriers, is also available to provide two-tier services within one DTV channel.

The standards of DVB-S and DVB-C were finalized in 1994, and that of DVB-T was ratified in early 1997. The first commercial DVB-T broadcasting was launched

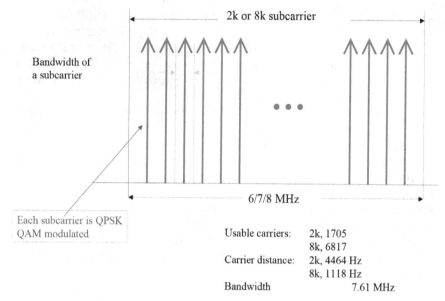

Figure 9.15   *OFDM frequency spectrum*

by the United Kingdom's Digital TV Group in late 1998. Actual DTV services are now offered in many countries thereafter. DTV also brings about new services and applications, such as home shopping and home banking, and Interactive TV. Therefore, the DTV revolution spells a technical progression similar to, but much profound than, the transition from black-and-white to color television.

9.9   DVB-T2 and the second-generation DVB systems

DVB-T2 [10] is the world's most advanced digital terrestrial television (DTT) system, offering more robustness, flexibility, and 50% more efficiency than any other DTT system. It supports up to 4k video broadcasting. The first version of the DVB-T2 was published in September 2009 by ETSI. The latest version is v1.3.1, which introduced the T2-Lite profile targeted for mobile services.

The aim of DVB in developing the second-generation DVB system was to significantly increase the system robustness and spectrum efficiency by taking advantage of the latest developments in IC technology and advanced signal processing algorithms, especially the cutting-edge video-coding techniques since the first digital standards conceived in 1993.

Like its predecessor, DVB-T2 uses OFDM modulation with a large number of sub-carriers delivering a robust signal, and offers a range of different modes, making it a very flexible standard. DVB-T2 uses the same error correction coding as used in DVB-S2 and DVB-C2, i.e., low-density parity check (LDPC) coding combined with Bose-Chaudhuri-Hocquengham (BCH) coding, offering a very robust signal. The

Table 9.1   Summary of differences between DVB-T and DVB-T2

	DVB-T	DVB-T2
FEC	Convolutional coding + Reed–Solomon 1/2, 2/3, 3/4, 5/6, 7/8	LPDC + BCH 1/2, 3/5, 2/3, 3/4, 4/5, 5/6
Modes	QPSK, 16-QAM, 64-QAM	QPSK, 16-QAM, 64-QAM, 256-QAM
GI	1/4, 1/8, 1/16, 1/32	1/4, 19/256, 1/8, 19/128, 1/16, 1/32, 1/128
FFT size	2k, 8k	1k, 2k, 4k, 8k, 16k, 32k
Scattered pilots	8%	1, 2, 4, 8%
Continual pilots	2.6%	0.35%
Bandwidth	6, 7, 8 MHz	1.7, 5, 6, 7, 8, 10 MHz
Typical data rate	24 Mbit/s	40 Mbit/s
Max. data rate (@20 dB C/N)	31.7 Mbit/s (using 8 MHz)	45.5 Mbit/s (using 8 MHz)
Required C/N ratio (@24 Mbit/s)	16.7 dB	10.8 dB

number of carriers, GI sizes, and pilot signals can be adjusted, so that the overheads can be optimized for different target transmission channel and applications.

Additional new technologies used in DVB-T2 are as follows.

- Multiple physical layer pipes allow separate adjustment of the robustness of each delivered service within a channel to meet the required reception conditions (e.g., in-door or roof-top antenna). It also allows receivers to save power by decoding only a single service rather than the whole multiplex of services.
- Alamouti coding is a transmitter diversity method that improves coverage in small-scale SFNs.
- Constellation rotation provides additional robustness for low-order constellations.
- Extended interleaving includes bit, cell, time, and frequency interleaving.
- Future extension frames allow the standard to be compatibly enhanced in the future.

As a result, DVB-T2 can offer a much higher data rate and a much more robust signal than DVB-T. Table 9.1 lists some major differences between DVB-T and DVB-T2 systems. For comparison, the two bottom rows show the maximum data rate at a fixed C/N ratio and the required C/N ratio at a fixed (useful) data rate.

9.10   Conclusions

The DVB-T may be one of the best terrestrial broadcasting systems because it was easily scalable with different noise and bandwidth environments, especially multi-path transmission channels. The key technology, COFDM, used in the DVB-T system makes it very suitable for mobile transmission.

A digital terrestrial system, DVB-T2, has lately been developed, which is more flexible and robust. DVB-T2 has advantages over DVB-T such as 30–60% more bandwidth; independent and flexible operation with multiple service providers; mobile and fixed services in the same bandwidth; and simple structure with constant coding and modulation for all programs. The implementation of the DVB-T2 standard will provide broadcasters the opportunity to deliver existing services using less frequency spectrum or more services using the existing frequency spectrum.

One of the main shortcomings of DVB-T2 terrestrial system is that the consumers will need to purchase new television receiving equipment. For example, in the case of HDTV services, consumers will need to change or upgrade their aerials and purchase a DVB-T2 receiver as well as an HD display. This is not very acceptable if the consumers have recently upgraded their television equipment as part of digital switchover. Manufacturers must ensure that their DVB-T2 devices comply with the receiver specifications determined by national administrations and industry groups. It is highly recommendable that countries to adopt common receiver specifications. This will help to reduce the need to produce different DVB-T2 receivers for each market, to alleviate the cost of transition from DVB-T to DVB-T2.

References

[1]   Jack K. *Video demystified – A handbook for the digital engineer*. 5th Ed. Elsevier; 2007.
[2]   Gao W., Ma S. *Advanced video coding systems*. Cham: Springer; 2014. Available from https://link.springer.com/10.1007/978-3-319-14243-2
[3]   Wiegand T., Sullivan G.J., Bjontegaard G., Luthra A. 'Overview of the H.264/ AVC video coding standard'. *IEEE Transactions on Circuits and Systems for Video Technology*. 2003, vol. 13(7), pp. 560–76.
[4]   Sullivan G.J., Tescher A.G., Ohm J.-R. 'Recent developments in standardization of high efficiency video coding (HEVC)'. Tescher AG. (ed.): *SPIE Optical Engineering + Applications*; San Diego, California, USA, 2010. pp. 30–7798.
[5]   Tozer E.P.J. *Broadcast engineer's reference book*. Focal Press; 2013.
[6]   *Digital Video Broadcasting (DVB); Specification for the Use of Video and Audio Coding in Broadcasting Applications Based on the MPEG-2 Transport Stream*. ETSI TS 101 154, v1.9.1, 2009.
[7]   *Conditional-access systems for digital broadcasting*. ITUR-R BT.1852-1, 2016.
[8]   Declercq D. *Channel coding: theory, algorithms and applications*. 1st Ed. Academic Press; 2014.
[9]   *Framing structure, channel coding and modulation for digital terrestrial television*. DVB document A012, DVB, 2015.

[10]   *Framing structure, channel coding and modulation for A second-generation digital terrestrial television broadcasting system (DVB-T2).* DVB document A012, DVB. 2015.

[11]   Eizmendi I., Velez M., Gomez-Barquero D., *et al.* 'DVB-T2: the second generation of terrestrial digital video broadcasting system'. *IEEE Transactions on Broadcasting.* 2014, vol. 60(2), pp. 258–71.

Chapter 10

Advanced Television System Commitee (ATSC)

Guy Bouchard[1]

10.1 Foreword

When I was asked to prepare this chapter on ATSC, I wonder for sometimes on what angle should I cover ATSC standards. I started reading the specifications available on the ATSC websites, the trade show presentations, etc., looking for what wasn't done yet.

The specifications show you the closest detail of what is achieved, giving me the feeling that I was trying to drink from a fire hydrant; the trade show presentation focused on what ATSC 3.0 can do, but I haven't seen an author trying to explain how ATSC and ATSC 3.0 works in terms that an unspecialized human can understand. This is the angle that I chose to take.

Choosing this angle implies that I can't cover the entire subject in detail and that I can't be 100% accurate, so some shortcuts have indeed been taken to ease the comprehension. I have the greatest respect for the brilliant people who came up with these standards, and apologize, in advance, for the technical inaccuracies they may find.

This tutorial approach demands that a logical flow is used; section 10.1 "The television broadcast environment, the RF channel" will cover what is the environment that a television signal has to evolve into, the radio frequency (RF) channel; section 10.2 "Tools required to use the RF channel as a communication media" will define the tools used to achieve communication in the so-defined RF channel; section 10.3 "Putting the blocks together" will put the blocks together that would build both ATSC 1.0 and 3.0 standards.

Today's television systems are defined in multiple layers; just as the other modern service delivery platforms, this chapter will focus its energy on physical layers.

Table 10.1 defines the frequency bands used in the television broadcast environment.

In North America, the National Television System Committee (NTSC) standard has basically ruled the airwaves for almost half a century. It was primarily based on a 6 MHz channel located in three different bands with their own characteristics as defined in Table 10.1.

---

[1] Télé-Québec, Canada

Table 10.1   Broadcast frequency bands characteristics

Band	Frequencies	Physical characteristics
Low VHF	52–88 MHz	Outstanding propagation characteristics, reaches beyond radio horizon thanks to diffraction, prone to impulse noise, and poor in-building penetration
High VHF	152–218 MHz	Good propagation characteristics, reaches beyond radio horizon thanks to diffraction, prone to impulse noise, and lousy in-building penetration
UHF	454–700 MHz	Fair propagation characteristic's limited to radio horizon and good in-building penetration

Radio horizon: The radio horizon represents the farthest point that an electromagnetic (EM) direct wave, issued from a communication tower, can touch the earth's surface (earth curvature accounted for). In geometrical form, it is the tangent of the line that links the top of the proposed tower to the earth's surface. The radio horizon (Figure 10.1) is deeply dependent on the local topography; however, a rough estimate ion can be derived via the following rule-of-thumb formula:

Radio horizon (km) = $3.57\sqrt{(\text{Tower height}(m))}$

So a 100 m tower would deliver an approximate radio horizon of some 36 km.

10.1.1   Gaussian noise

Gaussian noise is basically the thermal noise generated by the collision of electrons in any devices; as its name suggests, it follows Gaussian statistics, which mean that its amplitude variation is statistically limited to two standard deviations of its average value, 95% of the time.

It can be approximated by = KTB, where K is the Boltzmann constant, T is the temperature in K°, and B is the bandwidth in Hz.

where K = 1.380649×10^{-23} J/K

In practical terms, it equates to −114 dBm/MHz.

Applied to a 6 MHz television channel, the magic number becomes −106 dBm; in a system overview, the main factor degrading this number is the noise figure of the receivers input circuit.

Figure 10.1   Radio horizon

10.1.1.1 Impulse noise

Impulse noise is intrinsically impossible to predict.

Impulse noise is generally caused by electrical arcs in high-power devices such as utility companies' electrical distribution systems.

It is generally very high in amplitude and very short in duration (few microseconds to few tenths of microseconds); however, its repetition rate is randomly annoying for communication systems, from 3 hits per second to 3 per day. Trying to engineer the immunity against impulse noise is a real challenge as you must invest resources for a problem that cripples your service but will be used 0.000000001% of the time.

To make matter worse, in practice, it is next to impossible to see impulse noise with a spectrum analyzer, as a standard spectrum analyzer sweeps only one resolution bandwidth at any given moment; the probability to catch impulse noise on the fly ranges from slim to none, a real-time spectrum analyzer is required for such an undertaking.

On the high note, impulse noise is almost specific to the low-VHF band, and the systems performing in other bands don't need to build much immunity to impulse noise. For this reason, telecommunications authorities in North America granted very few licences for ATSC in low band VHF.

10.1.2   Static multipath

Static multipath (as shown in Figure 10.2) is part of human support infrastructure, buildings, hills, and vehicles that will transform direct propagation paths into several paths with diverse amplitudes and phase relationships.

As Figure 10.2 shows, the bunches of unwanted propagation components of unknown amplitude* and phase† will degrade the signals in an almost irreparable manner.

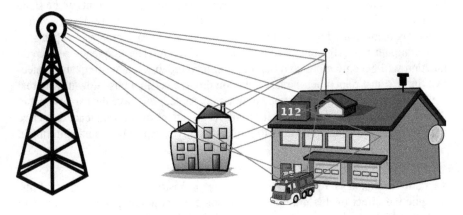

Figure 10.2   Multipath propagation

*Amplitude is a function of transmitted power, length of propagation path, reflection loss, and other losses.
†Phase is a direct function of the length of the propagation path

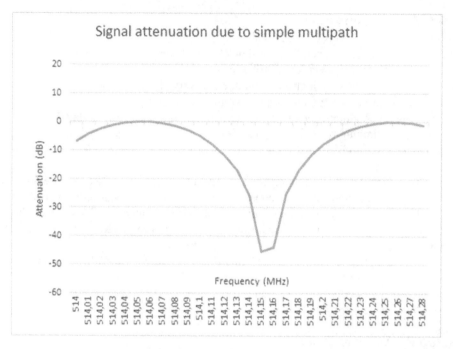

Figure 10.3   Selective multipath

In analog television, static multipath was known as a ghost.

Is multipath wideband or frequency selective?

Multipath is definitively frequency selective. Figure 10.3 represents the impact on the received signal of a single reflection arriving at the receive points at the same amplitude as the incoming signal between 514 MHz and 514.25 MHz. The frequency selectivity being caused by the difference in wavelength in each propagation paths.

Multipath is rarely composed of a single reflection. The frequency selectivity function will become more complex but will unavoidably remain frequency selective. In fact, just remember the last time you drove a car in a busy downtown area; the FM radio signal fades up to the point it's unusable; you move the car 1 m in any direction, and the signal is back. It is indeed the effect of the frequency selective multipath. Directive antennas tend to limit the receiver exposure to static multipath.

10.1.3   Dynamic multipath

As static multipath is a part of human support infrastructure, buildings, and hills. Imagine the effect of this multipath when some of these components are moving such as vehicles and airplanes, the results will look like:

- Constantly changing blend of received EM waves
- Doppler effect will start to kick in

A scenario that will make matter worse is that the receiving antenna is in a moving vehicle; this antenna is exposed to a different received environment every second.

Simple Doppler frequency shifts compounded by hundreds of propagation paths varying every second will create a Doppler spread that looks much like phase noise.

10.1.4 Man-made noise

The average household has dramatically changed in the last decades. Today, the average house has dozens of switching power supplies. Just start enumerating the number of devices: led light bulbs, cell phones chargers, radios, TV sets, computers, smart thermostats, smart speakers, coffee machine, smart toys, watches, and fitness monitors.

Any smart devices involve computing, computing involves DC (Direct Current), and today, conversion from AC to DC involves switching power supplies.

Switching power supplies involves the creation of EM waves in the high 100 KHz to a few MHz. Although they are not typically in the TV broadcast spectrum, the compounded effects of dozens of devices load today's virtually wide-open RF front end and degrade its performance in a hard to quantify manner.

10.1.5 Today's home environment

To make matter worse, today's building code suggests the use of metallic vapor barriers which reflect EM waves, transforming the average house into a punctured Faraday cage, thus keeping the interference indoor by reflecting the EM back in the house (Figure 10.4),

In the UHF band, the building penetration power (Figure 10.4) margin is defined as the average power margin required to go through the walls of the average house, It is about 13 dB; however, the standard deviation is 9 dB. So if a television signal aimed at an indoor reception and aim to cater to 95% of the population, it has to have 13+9+9 = 31 dB of margin.

Furthermore, as shown in Figure 10.5, once the signal has entered the house, through a window or else, it then bounces on all the walls, floors, and ceilings before

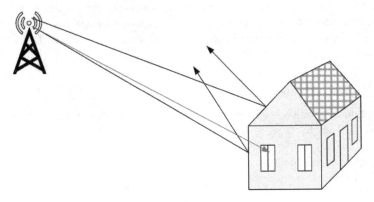

Figure 10.4   'In building penetration

Figure 10.5   Micro reflections

it reaches the antenna; these micro reflections (in the picoseconds range) are not something easy to handle for the receivers.

Negative delays: Television receivers are indeed exposed to negative delays. The receivers will lock up on the strongest signal, if this strong EM wave is not the wave that takes the shortest path, and it will lock on it regardless; therefore, the receiver will perceive waves that took a shorter path as negative delays.

10.2   Tools required to use the RF channel as a communication media

This section will define the tools most digital communication systems leverage to achieve communications in over-the-air broadcast and other systems. Although most of those techniques are not specific to ATSC, they are tools used to build both ATSC and ATSC 3.0 systems.

10.2.1   Modulation

Modulation science has not changed much since the inception of the digital technology. We are back to the same old techniques:

- Amplitude modulation
- Phase modulation
- Frequency modulation

One of the only different aspects that differ from analog to digital modulation is the concept of time gating. A modulated analog signal has no beginning and no defined ends. It is just broadcasted for as long as it is on-air. A digitally modulated time is defined, just as the bitstream it carries. The time period over which a parallel bit pattern is broadcasted is called the symbol duration. The RF modulation arrangement valid over that period is called the symbol. The speed of digital clock that paces the production of digital symbols is called the symbol rate.

As Figure 10.6 shows, modulations intrinsically produce two sidebands, an upper sideband, a lower sideband, and a carrier. The upper and lower sidebands carry exactly the same information, a the carrier carries none. For spectral efficiency sake, single sideband modulators have emerged and made their ways in digital modulation. In the analog days, older receivers had great difficulty of locking-up a SSB (Single Sideband) signal with a suppressed carrier; however, the carrier doesn't need to use half of the transmitter power as a traditional AM broadcast used to do; a partially suppressed carrier (or pilot) may be used.

When a digitally modulated EM wave is exposed to the multipath propagation interference, we define as intrasymbol interference the interference components received during the current symbol, the interference received before or after the current symbol is defined as intersymbol interference.

Modulation nomenclature (Analogue or Digital)

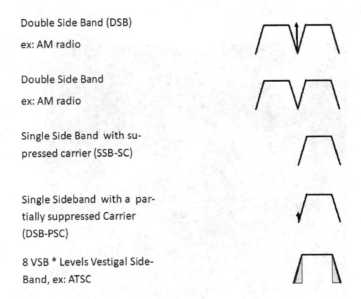

Double Side Band (DSB)

ex: AM radio

Double Side Band

ex: AM radio

Single Side Band with suppressed carrier (SSB-SC)

Single Sideband with a partially suppressed Carrier (DSB-PSC)

8 VSB * Levels Vestigal Side-Band, ex: ATSC

Figure 10.6   Modulations and their occupied bandwidth

10.2.2   *Digital modulation characteristics*

10.2.2.1   Receiver thresholds

Unlike analog modulation, digital modulations provide a quasi-error-free service at any carrier-to-noise ratio until about 1 dB from the infamous receiver's thresholds.

So, the receiver threshold is a measurement of the lowest amount of carrier-to-noise ratio, a receiver can decode virtually free of errors based on a Gaussian channel.

Figure 10.7 shows an ATSC carrier-to-noise measurements. Note that the value of the delta between the top of the modulated signal envelope and the noise floor is read directly as the C/N ratio. In this case, we are comparing dispersed energy to disperse energy. The pilot is ignored representing undispersed energy.

AWGN (Additive white Weighted Noise) threshold becomes the benchmark that qualifies most modulation techniques. The choice of modulation becomes a density/noise performance tradeoff; in the ATSC 1.0 specific case, the AWGN threshold is 15.6 dB.

10.2.2.2   Peak-to-average ratio

Digital modulation rarely end-up having a constant envelope. Depending on input signals, the amplitude of the modulated signal will vary. Statistically, they are compounded into an envelope that is the average amplitude + their peak-to-average ratio. Typically, the peak-to-average ratio of most digital modulations is between 3 and 7 dB. Figure 10.8 illustrates the instantaneous envelope power variations

So the signal peak-to-average ratio represents the delta between the average amplitude of a digitally modulated signal and the peaks observed in a given

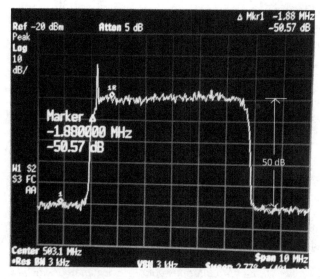

Figure 10.7   ATSC signal in the frequency domain

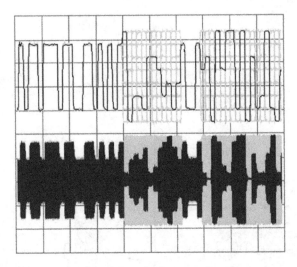

Figure 10.8   In the time domain

percentage of the time (typically 99.9%). ATSC 1.0 peak-to-average ratio is around 6 dB.

In their respective transmitter design, sufficient headroom has to be built in both power amplifier stages and passive components to make room for the peaks. To alleviate the issue, a noise-like signal is added to the signal; the latter is called a peak-to-average power ratio reduction.

Modulated signals are often represented as constellation.

10.2.2.2.1   Elementary constellations drawing

A quadrature phase modulated keying of QPSK signal is a signal that carries two bits as $2^2 = 4$ states signal.

The constellation represents the amplitude and phase relationships of each symbols (Figure 10.9).

An eight PSK signal carries three bits in a $2^3 = 8$ states constellation and so on.

QAM signals combine amplitude and phase modulation.

10.2.3   Interleaving

Interleaving is a technique that is aimed at helping the signal to survive in a real-life scenario. This specific scenario happens in virtually all aspect of human life, the bursty nature of errors. Figure 10.10 shows the error happening in brusts rather that occuring in a random manner.

All communication systems suffer from bursty errors, why? The RF channel conditions are changing all the times, and even the most sophisticated communication will take time to adapt to its new reality. Furthermore, all digital transmission

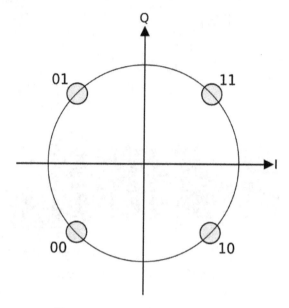

Figure 10.9   QPSK constellation

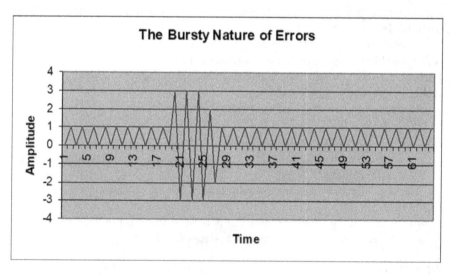

Figure 10.10   The bursty nature of errors

standards rely on a clock. Any error leading to the loss of the clock will indeed cause a long burst of errors due to the time required to rebuild that clock. Even an old analog television used to be affected by the burst of interference that was affecting vertical or horizontal synchronization. Impulse noise is notorious to cause bursts of interference.

10.2.4   Time interleaving

Interleaving is one of the oldest data integrity management tool; it is based on one principle: don't put all your eggs in the same basket...This technique consists of an alteration of the bits flow sequence.

Many of you have played the solitaire card game (specially easy since the inception of computers); this game consists of stacking the cards in order from ace, 2, 3...10, jacks, queen, and king in four stacks of each kind.

Scenario A: Imagine that I take a brand new stack of 52 cards right of the box. The entire stack represents your transport mechanism; it will be remarkably easy since the cards are already in the correct order in the stack. As it is already in order, I call this stack as the un-interleaved stack.

Scenario B: Imagine that I take a randomly shuffled stack of cards; it will be remarkably longer since the cards are not in the correct order in the stack. I call this stack as the interleaved stack.

Now imagine as in Figure 10.11, that I use a permanent ink marker and make a dot on the side of the stack of cards. In the transport process, now each stained card represents an error (that becomes all aces by example).

Stacks of cards of the scenario A (un-shuffled) will suffer consecutive errors, something like (ace,2,3,4,5,6,7,8,9, as consecutive cards were stained.

Stacks of cards of the scenario B (shuffled) will not suffer consecutive errors, as the errors (stained cards) will be spread in a random manner.

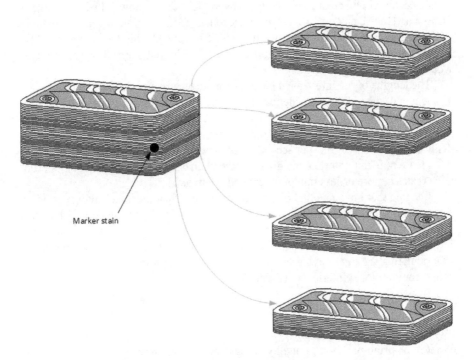

Marker stain

Figure 10.11   Classic example of interleaving

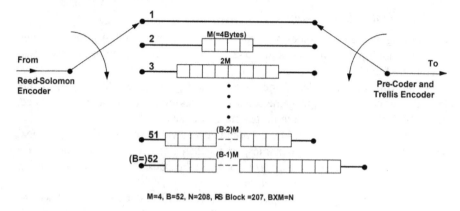

M=4, B=52, N=208, RS Block =207, BXM=N

Figure 10.12   Convolutional interleaver (ATSC A53)

As you may have guessed, each card of this game represents a bit, so the created randomly shuffled bitstream is less likely to feature consecutive errors.

Now producing a randomly shuffled stream may be challenging, so some digital-friendly shuffling techniques were required. One relies on the use of a memory chip. This RAM (Random Access Memory) chip is loaded in rows, read in columns, and serialized, and it will create a perfectly recoverable shuffled (now called interleaved) bitstream. How can you recover it at the other end using an inverse transform (de-serialize data, load in columns and read it in rows)? This technique is called linear interleaving, under this example a memory chip that stores 8 words of 8 bits would create an interleaving depth of 64 bits.

The lengths of the interleaved bitstream are driven by the size of the interleaver. The length of the interleaver is defined in bits; however, in practice, we tend to define interleaving length in time, for example, if I have an interleaver that is 10 kb wide, running on a system that carried 1 mb per second, my interleaving length will be defined as: 10^4 bits/10^6 bits per second $= 10^{-2}$ s or 10 ms interleaving period.

Please note that interleaving doesn't correct error; it breaks a long burst of error and creates a more or less randomly errored bitstream.

Imagine that I have a burst of error that is longer than the interleaving period or that we are dealing with a periodic noise that happens to be the same length as the interleaver (e.g. interference produced by snowmobile engines ignition system). Our interleaved bitstream will still suffer from consecutive error. A technique consisting of modulating the interleaving length in a pseudo-random manner is used, and it is called convolutional interleaver (Figure 10.12).

10.2.5   Frequency interleaving

10.2.5.1   Frequency

What if the principle "don't put all your eggs in the same basket" also applies in the frequency domain?

ATSC 1 carrier @ 19.3 MB/s (10.76 MS/s)   Symbol Duration 92 nS

OFDM ATSC 3.0 8k, 6919 carriers @2.91 kb/s
GI 111us   Symbol Duration 344 uS

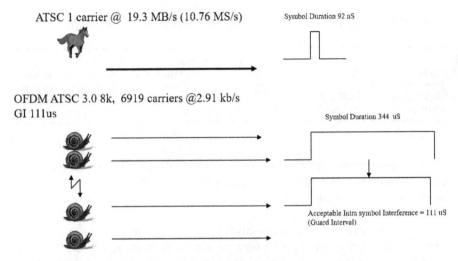

Acceptable Intra symbol Interference = 111 uS
(Guard Interval)

Figure 10.13   Comparison between single carrier and multicarriers modulations

Indeed it does: in order to better understand what frequency interleaving can do, may I attract your attention to two phenomena that were covered earlier:

1. Multipath is frequency selective
2. Interference causing errors are bursty in nature

We just created the environment that allows us to transform intersymbol interference to intrasymbol interference: intrinsically easier to manage.

Take for example a 10.76 MS/S carrier carrying 3 bits/symbol; the latter will carry a payload of 30 Mb/s (20 Mb/s after forward error correction (FEC)). Its symbol duration is 92 ns. Imagine, just like in Figure 10.13, that I could replace it by 6913 carriers carrying 2.91 kb/s.

This trade-off between inter-symbol-interference (ISI) and intra symbol-interference (ASI) is one of the complexity to multipath performance. In the absence of multipath interference, both the systems would work in a similar fashion.

When the multicarrier signal is looked at the macro level, multipath can easily kill a significant number of adjacent carriers. However, if the time/frequency interleaving is well chosen, FEC can come to the rescue by correcting a whole series of interleaved consecutive errors (that interleaving has transformed bursted errors into almost random errors).

In practice, what is the real benefit of frequency interleaving? By enlarging the symbol duration, the period where a multipath interference can be handled is then multiplied.

Table 10.2   *Comparison of monocarrier and multicarrier modulations*

Mode	Payload	Number of carriers	Symbol duration (s)	20% of SD or GI (s)	Impact in km
Single carrier	19.39 Mb/s	1	2 e-7 s	6 e-8	0.018
Multicarrier	20.1 Mb/s	6 913	344 µs	1.11 e-4	33

If the performance of the input equalizer permits the handling of multipath during a given portion like 30% of the symbol duration (SD), the period where the interference is admissible for both single and multicarrier proposals is defined in Table 10.2.

Just to put interference in perspective, a multipath coming from a reflection at 18 m of your antenna is probably strong and damaging to the signal. On the other hand, a reflection coming from an object 33 km from the antenna is probably faint or no longer existent, thus the superiority in multipath resistance of the multicarrier system.

A sophisticated technique called orthogonal frequency division multiple (OFDM) has been created, and the latter exploits both time and frequency diversity in an optimal manner.

OFDM systems are called by parameters:

* B: Bandwidth
* N: Number of carriers
* F: FFT (Fast Fourier Transform) window size
* GI: Guard interval (period over which the multipath improves the signal reception instead of destroying it)

10.2.6   Forward error correction

Earlier in this chapter, we mentioned that the error correction process creates miracles at the receiver level. FEC is a process by which additional codes are added to a bitstream in the transmitter to allow the receivers to detect and correct errors, bearing in mind that the communication channel between the transmitter and receiver is unidirectional.

The following example (Table 10.3) is a fictional FEC system used solely for tutorial purposes. Imagine that a block of (8×8) 64 bits has to be transmitted. The proposed FEC scheme just adds the sum of each rows and columns.

The 8×8 = 64 bits matrix is then loaded with 3×8 bits for the row sums and 3×8 bits for the columns sums so a total of 64+48 = 112 bits. The then created matrix has almost one-third of bits that are overhead bits, as FEC terminology would read two-third FEC (meaning that there is about 2 bits of useful data for each 3 bits transmitted).

Table 10.3   Simplistic FEC scheme

	Data									Sum	
	1	1	0	0	1	1	1	1	1	1	
	1	0	0	0	0	0	0	1	0	1	0
	1	0	1	0	1	1	0	0	1	0	0
	1	1	1	0	0	0	1	0	1	0	0
	0	0	1	0	1	1	1	1	1	0	1
	0	1	1	1	0	0	1	0	1	0	0
	0	0	0	0	0	0	0	1	0	0	1
	1	1	1	0	1	1	1	1	1	1	1
s	1	1	1	0	1	1	1	1			
u	0	0	0	0	0	0	0	0			
m	1	0	1	1	0	0	1	1			

If any one of these bits is altered during the transmission process, the sums of columns and rows will be false. So the error is identified and correctable. Even if the error takes place in one of the sums, one of the sums will be wrong in only one axis, telling the receiver that the error took place in the overhead bits and can be ignored. This system can correct all the errors if there are no more than one error per 112 bits. The system will work well as long as the interleaver ensures that there are no consecutive errors. The threshold of this system is defined as a probability of error of $1/112 = 8.9$ e-3.

Obviously, there are much smarter FEC systems in use, Viterbi, Reed-Solomon, Trellis Code Modulation (TCM), BCH (Bose–Chaudhuri–Hocquenghem), and LDPC (Low Density parity Codes) to name a few.

10.2.7 Compression

There are two types of compression one can use in digital communications, lossless and lossy. Lossless is defined as compression that has zero impact on picture and sound quality.

In a nutshell, lossless compression is entropy-based coding. Entropy coding may sound like absolute magic. Remember the first you used a PKzip, winZip, or winRAR session, you probably asked yourself where the bits went. It all began with the Huffman code and other probabilistic coding. The principle behind it is to exploit the variable density function of the incoming signal and remap the symbols to minimize data rate, just like if you draw a probability map of the average teenager's vocabulary, you will probably notice that the words "cool," "computer," and "nice" comes back very often, and the word "probabilistic" barely gets used. If I code the speech in a variable length symbol maps, and cleverly assign a long code to "probabilistic" and a short code to "nice," the end result will take significantly less space; however, this variable length code technique is highly dependent on incoming signal density function. Anyhow, its free compression (comes at the expense of computer processing power) and digital communication systems will use it even if the results are sometimes short of breathtaking.

Lossy compression is specially efficient in television as a ton of redundancy is built in the signal. There is basically two axis of compression, spacial and temporal. Spacial compression takes advantage of the fact that a picture of a given resolution doesn't use the full resolution across the entire screen. Repetitive pixels of pixel patterns are present in the vast majority of the picture that the latter can be compressed using various techniques.

In order to make picture data manageable, the picture is divided into pixel blocks of various sizes as shown in Figure 10.14.

The latter is going through a perfectly reversible time to frequency transform (typically a discrete cosine transform (DCT)). The latter doesn't really compress anything; however, it classified contents in the frequency domain, in most pictures that won't really demand the full resolution (e.g. cartoons). The last coefficient of the equation will be null or very low. The coefficients quantizing process will result in a string of coefficients scaled using an eight or ten bit resolution.

The system can then just shorten the equation and compress the picture. When more compression is required, the scale factor of the DCT coefficients can be altered to represent the pixels with obviously less bits.

We are referring to this compression technique as lossy. What exactly are we loosing?

- The non-carriage of all DCT coefficients will alter effective resolution
- The rescaling of DCT coefficients will show-up as increased picture noise

Temporal compression: When a television signal is looked at frame by frame, it shines by its repetitive nature, and it is the nature itself of television. Nobody would watch a newscast where the anchor constantly moves in the four quadrants of the screen. Even sporting events don't create a huge amount of motion. A mechanism has been created to handle television pictures efficiently, the latter is called, the temporal compressor.

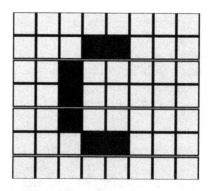

Figure 10.14   Pixel block

Defining motion: Motion is defined in most flavor of MPEG (Moving Picture Expert group) by a pixel block and a motion vector.

In this case, Figure 10.15, from frame 1 to frame 2, the pixel block highlighted has moved from (eight pixels the horizontal axis (H)-four pixels in the vertical axis (V)).

From frame 2 to frame 3, the pixel block has moved from (ten pixels H-five pixels V).

Predictive coding: MPEG loves short numbers, to achieve a higher rate of compression, MPEG uses a predictive approach. In a nutshell, MPEG has a prediction model that lives in both encoder and decoder, indeed both encoders are trying to predict the motion from each images to the next. MPEG carries the difference between the exact position of a pixel block and the prediction rather than carrying the entire number. Table 10.4 will illustrate the process.

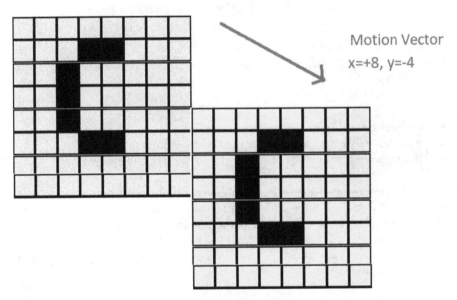

Motion Vector

x=+8, y=-4

Figure 10.15   Motion vector

Table 10.4   Predictive coding

Time ↓	X	Y	Vector Length = $\sqrt{a^2 + b^2}$
Original position	0	0	
At + 1 frame	+8	−4	
Prediction for frame 2	+16	−8	8.94
Actual position at frame 2	+18	−9	17.88
Number that MPEG would carry	+2	−1	2.23

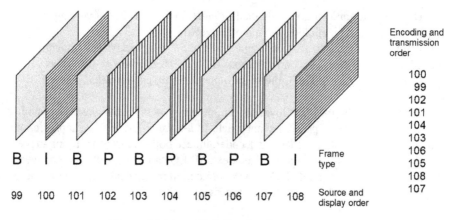

Figure 10.16   MPEG video frame structure

MPEG frame structure: MPEG frame structure is a clever structure that permits to borrow elements from each other in an orderly manner as shown in Figure 10.16.

It defines three types of frames:

- I frames are real complete pictures
- P frames are frames based on predicted elements from the past
- B frames are frames based on elements from the past and the future

Alert: Borrowing elements from the future, can the prediction model predict the future? Not really. However, by altering the transmission order of the frames, MPEG can send a future frame ahead of the B frame, so elements are available for the borrowing process, please note that in Figure 10.16 the frame transmission order has been altered so frame 100 (I) is transmitted before frame 99 (B) so it can borrow elements from frame 100 .

10.2.8   Adaptive tap equalizer

The adaptive tap equalizer is a near magic device that permits to eliminate/alleviate multipath interference in communications systems.

In most systems, the modulation scheme includes training signals of a known waveform and impulse response. The adaptive tap equalizer reads the training signal and defines one by one all the multipath components (ghost) it can handle by delay, amplitude, and phase.

It then generates the conjugate of each of the interfering signals and adds it to the incoming signal at the desired delay value (through a chain of time staggered taps). Figure 10.17 shows the result is the original signals free of the handled multipath interference signals. Figure 10.18 demonstrates the process.

Limitation of this system:

- The quality of the equalization achieved is limited by the available processing power present in the equalizer chip.

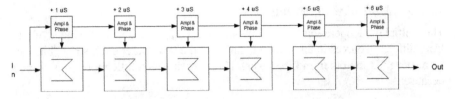

Figure 10.17   Adaptive tap equalizer

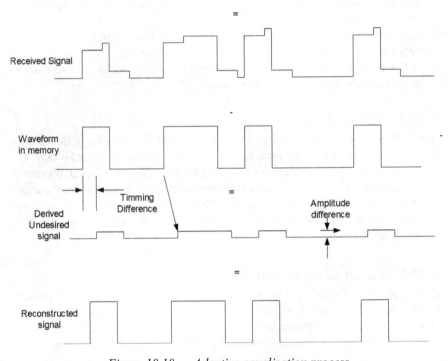

Figure 10.18   Adaptive equalization process

- The more taps are open in the equalizer, the noisiest the signals gets.
- The equalizer is limited in the time window, typically −20 μs to +60 μs.
- Great difficulty to handle micro reflections (inside the house).

The noise figure degradation caused by the number of taps open creates an altered phenomenon: Multipath no longer kills the signal, it causes a loss of margin so a receiver that features a 15.6 dB threshold under a Gaussian noise conditions will underperform in severe multipath conditions and deliver a receiver threshold of 20 or 21 dB.

Impact of Moore's law: The firsts ATSC decoders dated from the 1990s and their adaptive tap equalizer had access to the top processing power available in the days; 2019 receivers are built with an order of magnitude more processing power; therefore, today's tuner does a much better job of canceling multipath than their predecessors.

10.2.9   Multiplexing techniques

The multiplexing operation is all about building an aggregate bitstream made of bits from different sources in an orderly manner (at the transmitter end), so the bits can be retrieved at the other end (receiver). However, it gets a little more complicated because:

- Bits from various sources vary in debits (we are trying to multiplex a streams that occupies megabits with other services that occupy few kilobits).
- Bit rates of some services may vary in time (e.g. a video stream encoded at various bit rate mode).
- Bits from some sources may have more importance than bits from other sources.
- Bits from some sources have to be synchronized with bits from other sources (e.g. video and audio bitstreams have to be synchronized within milliseconds to avoid the annoying lip synch phenomenon).

10.2.9.1   Time division multiplex

Under this technique, different time slots are assigned for bits from different services. The schedule is re-made periodically to allow services to vary in bit rate.

Figure 10.19 shows a typical TDM multiplex where five services (represented in five different colors) are carried in different fixed duration time slots, the number of time slots used by each services varies with the payload it has to carry. Obviously, the whole process is made possible with a strong synch and schedule system (green & red).

10.2.9.2   Packet-based multiplex

In this case, the chunks of bits are serialized in an orderly manner in packets (Figure 10.20). Fixed length packets are then serialized at the transmitter end and routed to different components based on the information included in the header of the packets. This is exactly what takes place on the public Internet, where

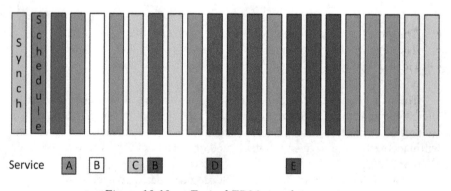

Service   A   B   C B   D   E

Figure 10.19   Typical TDM signal time map

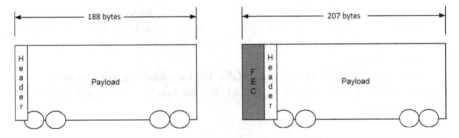

Figure 10.20   Packet-based multiplex

variable-length packets are used. Television services are heavily reliant on fixed-length called MPEG (Moving Pictures Expert Group) packets. The latter is well suited for the carriage of real-time services such as television and supports the addition of FEC gracefully. Furthermore, MPEG packets are short enough to allow the interleaving of video and audio packets, so lip synch can be preserved. The left part of Figure 10.20 shows an ordinary MPEG packet as the right side packet supports the addition of a few Reed-Salomon Forward Error Correction bits, both are represented as a train wagon as they can just be serialized at the transmitter end and de-serialized and be routed to different destinations (e.g. video or audio decoders) at the receivers end.

10.2.9.3   LDM, a different multiplexing technique

Layer division multiplex (LDM) takes place at the modulated carrier level instead of at the bitstream level. Under this smart technique, two modulated signals are overlaid at the transmitter level and decoded at the receiver end in two phases:

1. In the first phase, only the more robust of the two signals is decoded; the second signal is just a noise to the demodulator that strives to decode the first.
2. In the second phase, the conjugate of the first decoded signal, error-corrected signal, is added to the composite signal making. Only the second service appears in a virtually noise-free environment.

Composite modulated signal $= (A + ME(A) + B + Me(B)) + $ Noise

The most robust signal is then decoded and error corrected, then the conjugate of this re-modulated signal gets added to the equation

2nd layer signal $= (A + ME(A) + B + Me(B)) + $ Noise $- A$

2nd layer signal $= B + ME(A) + Me(B) + $ Noise

in other words, a slightly noisier B.

where A is the first signal, B is the second signal, and $ME(x)$ is the modulation, the error noise like component of each modulated signal.

In practice, if the first signal *A* is a high-rate signal that uses a modulation with a high threshold of (in the order of 10–15 dB), and the second signal *B* is a low-rate robust signal with a threshold of something like 5 dB, the easiest signal to decode B will be decoded first.

Figures 10.21–10.23 show the constellation diagram of the core layer ((B)high bit rate low FEC), the enhanced layer (A) (low bit rate high FEC) and the sum of both respectively.

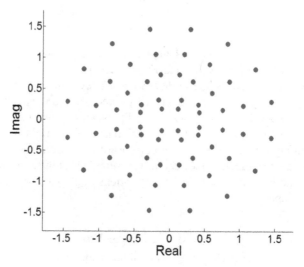

Figure 10.21   Core layer (ATSC)

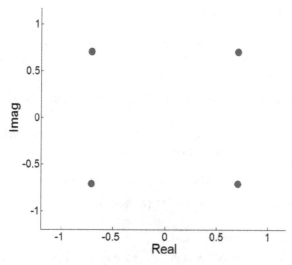

Figure 10.22   Enhanced layer (ATSC)

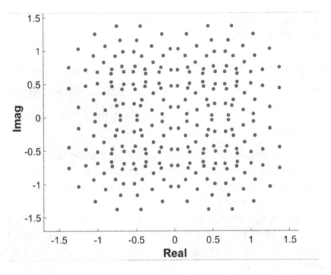

Figure 10.23   LDM overlaid constellation (ATSC)

10.3   Putting the blocks together

10.3.1   ATSC 1.0

10.3.1.1   Genesis

ATSC was created by a US industry consortium composed of broadcasters, Consumer Electronics Manufacturers, and regulators called the grand Alliance in the early 1990s. Their mandate create a digital television system that would replace the NTSC system that has been the base of the television for over 50 years.

The mandate was to replace NTSC. Their starting parameters were:

- Fixed reception
- NTSC compatible spectrum (which pretty much dictated the use of a 6 MHz channel) with minimized interference to NTSC
- Noise limited service contour (operation in a Gaussian channel)

The original mandate was not to:

- Provide an evolution path (retro-compatible or not)
- Support viewers mobility (pedestrian or vehicular)

Once the system adopted, it soon became the standard for United States, Canada, South Korea, and many other countries.

Putting the puzzle together: Once US regulators started to put together a transition plan, a huge engineering undertaking, they were hit by a few realities:

- The density of television stations was remarkably high in some areas

  o Noise limited contours soon became interference limited contour

- The tools used in analog television (dating from the 1940s) were inadequate for planning Digital television (DTV) as:

  o The F(a,b) prediction model was used for decades in broadcast service planning, it devines 2 factors, where a is the percentage of households and b the percentage of the time. So an F(50,10) is a prediction valid for 50% of homes, 10% of the time)

  o Analog services where based on an F(50,50) prediction. However, statistical distribution F(50,50)‡ was not applicable because of the cliff effect§ of the digital signal. This planning factor soon became F(50,90) a more reliable service was required.

  o The existing planning tools F(50,50) were not accounting for local topography. A new tools emerged to account for topographical isolation, the longly rice propagation model. The latter was found remarkably representative of the real terrain-based EM wave propagation.

10.3.1.2   ATSC 1.0 piece by piece

Allocated bandwidth: 6 MHz, to be implementable in NTSC countries such as the United States and other countries using a 6 MHz channelization.

Modulation type: The objective was to deliver a payload between 15 and 20 Mb/s to the entire NTSC audience in the old NTSC contour. The modulation chosen was a specific case of amplitude modulation called 8VSB (eight levels Vestigal Side Band). The latter was found to have the best Gaussian noise performance at the desired density function.

What is 8VSB? Spectrally, 8VSB is based on an amplitude modulation keyed on eight levels, remarkably similar to upper sideband SSB with a partially suppressed carrier. The modulation density is 3 bit/symbol; remember, in SSB each symbol, it occupies ½ Hz/symbol.

Symbol rate: 10.76 MS/S at eight amplitude levels, it carries 3 bit/symbol

FEC: The chosen FEC is TCM at a rate of 2/3.

---

‡F 50,50 means the signal delivered to 50% of receivers 50% of the times, the latter was usable in an analog environment as people that fells short of the 50% criterions the viewers were receiving a degraded signal.

§In an analog systems, the noise performance exhibits a graceful degradation of service quality; in other words, as the signal-to-noise ratio degrades, the picture just get a little snowy. As the signal-to-noise ratio degrades, digital systems are working flawlessly until their thresholds, fractions of dB's below threshold the signal disappear. We call this phenomenon as the cliff effect.

The ATSC signal synchronization structure is remarkably strong (Figures 10.24–10.26). The RF synch can be achieved close to 0 dB carrier-to-noise well below the receiver threshold. To make this stunt possible, the ATSC signal has to synch level a frame synch that takes place about 40 times per second and a segment synch that gets repeated every 77 µs:

Data transport structure: A segment is the base of the transport mechanism.

The segments are carried in a frame super structure:

The field synch is described below

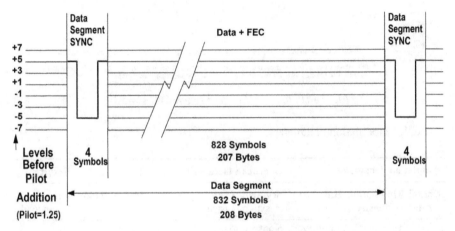

Figure 10.24   ATSC segment (ATSC A53 standard)

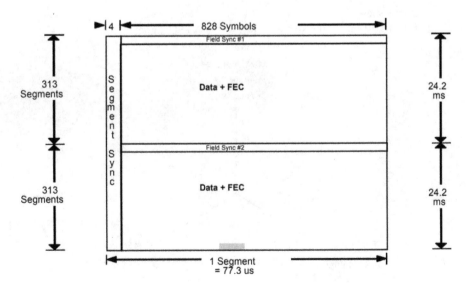

Figure 10.25   ATSC frame structure (ATSC A/53)

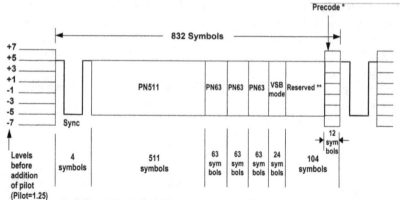

Figure 10.26   ATSC field synch (ATSC A53)

ATSC 1.0 Modulation parameters:

Modulation parameters			Overhead factors (Figure 10.27)				Result
Symbol rate	Modulation density	Raw bit rate	Segment synch	Frame synch	TCM FEC	Reed-Solomon FEC	Net payload
10.76 MS/s	3 Bits/symbol	32.28 E6 Bits/second	0.9952 828/832	0.9968 312/313	0.6666 2/3	0.9082 188/207	19.39 E6 Bits/second

Spectral View Figure 10.28.
Figure 10.29 shows the spectral view of the ATSC signal.

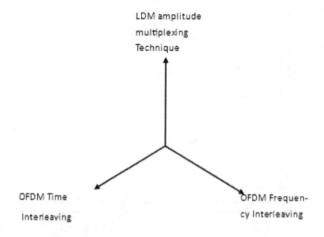

Figure 10.27   ATSC 3.0 diversity model

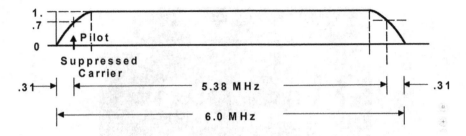

Figure 10.28 ATSC Spectrum

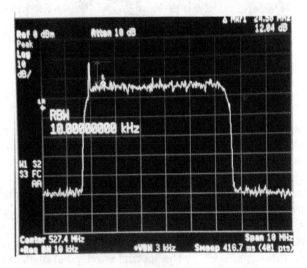

Figure 10.29 ATSC spectrum

Few things to bear in mind:

- Modulation density is 3 bits/symbols
- Spectral density factor is ½ GHz per symbol as only one sideband is present, so a symbol rate of 10.76 MS/s actually occupies 5.38 MHz, (a roll-off factor of 11% was allowed to allow filtration, so 5,38 MHz + 11% = 6MHz)
- The pilot is the carrier re-inserted at a level of 11% of the envelope power
- The peak-to-average ratio of this signal is 6 dB 99.9% of the time

A word about the pilot-to-carrier ratio:

Figures 10.28 and 10.30 show a typical ATSC modulated signal, note that the pilot-to-carrier ratio will vary when your spectrum analyzer resolution bandwidth changes. This is normal as the ATSC signal is made of energy that spreads evenly across the 5.38 MHz bandwidth; however, the pilot is CW (continuous Wave) energy

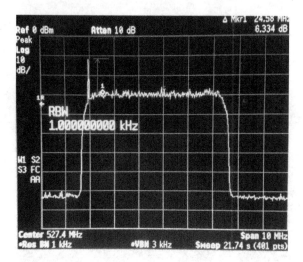

Figure 10.30   ATSC pilot variation

is unspread) When the RBW (resolution Bandwidth) of your analyzer is doubled[¶], all the noise-like components will move up 3 dB as more energy is detected. On the other hand, the amplitude of the CW of the carrier will remain the same. In this case, the pilot-to-carrier ratio will go down 3 dB.

A word for analog communication systems engineers (the rest of us):

I am always stunned to notice how flat the spectral envelope of digital modulation signals is. In analog communications, the same modulations were used, but the resulting spectrum varied in time as a function of the signal statistics. How come the digital modulations are so flat? The spectral envelope of digital modulation is flat because the statistics of the input signals are flat. It makes sure that the sounds illogical to the statistics of an unknown signals are known. The ATSC modulation engine includes the use of a randomizer; the latter flattens the signal statistics in a reversible manner as the receivers apply the conjugate transforms of the modulator and still deliver the same signals. In the meantime, the signal spectral envelope is pretty flat, which is an highly desirable signal characteristics specially in the cases of an urban interference dominated environment.

The vocabulary of this block diagram has been introduced earlier in this chapter:

- Randomizer
- Reed=Salomon (RS encoder is a block form of FEC
- Data interleaver introduce time interleaving
- Trellis coder is another form of FEC
- The service multiplex is a tight TDM

[¶]A spectrum analyzer resolution bandwidth dictates the amount of energy that each points on the spectrum analyzer display sees. If the RBW is 10 kHz, each points on the spectrum analyzer display sees 10 kHz of energy. The noise floor of the SA is becoming direct function of the RBW.

- Pilot insertion is a part of the modulation process
- 8VSB modulator

Input data: The data that get injected in the modulator input are a standard MPEG-2 transport stream with industry known components (Figure 10.31):

- A video service in MPEG-2 in a not so limited number of profiles
- An Dolby AC-3 audio service (from 2 to 5.1 channels)

The components are wrapped in an standard MPEG-2 packet-based transports stream. Figure 10.32 shows a typical Three diagram taken at the output of an in-service ATSC transmitter. The various signals are represented in PID (Packet IDentifiers), in this case the video is carried on PID 49, audio on 52 & 53, etc. Please note that the components identification as the station call sign are interpreted from the PSIP (Program Service & Information Protocol) metadata.

Metadata: In order to offer the viewers a smooth navigational experience, a chunk of navigational data had to be inserted to the signal to permit signal discovery and navigation. The latter is called as PSIP.

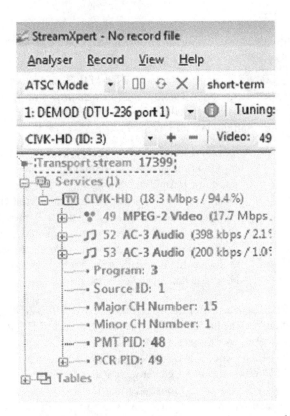

Figure 10.31   ATSC transport stream representation (courtesy of Télé-Québec)

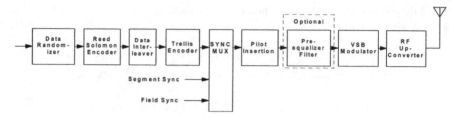

Figure 10.32   ATSC modulator block Diagram (ATSC A53)

There are two types of tables used in an ATSC A/78 compliant transport stream:

• Classic MPEG-2 TS tables:

   ○ Program association tables show how many programs are in the TS and points to their respective service tables (program map table (PMT))
   ○ PMTs lists all program components (audio, video and their respective Pids, and service descriptors)
   ○ Clock, the infamous PCR (Program Clock reference) is an absolute must to build the service timing

• ATSC specific tables:

   ○ Master guide table (MGT), the mother table of ATSC declares various other metadata service components:

      ▪ Television virtual channel tables (TVCTs), the latter builds the navigational structure and dictates how viewers will see the contents mapped on their TV set
      ▪ Program guide components:

         □ Event information tables (EITs), the EITs carry all the info required to build a program guide, program name duration, rating, and timing (e.g. "the lone ranger," English, 30 minutes and three frames, rated violent)
         □ Extended text table (ETT), in a nutshell, the ETT carries the information the user sees when the info key is pressed on the remote control (eg. "the lone ranger, this week the lone ranger is after a troop of bank robbers who take hostages in the process")

   System issues: There are information repeated between the ATSC specific tables (MGT and children) and the basic MPEG TS tables (PAT and PMT). It is very important that those two tables bare the exact same information; if they are not, some receivers will get confused and some other won't, a nightmare in customer support in view….
   The static PSIP component is a clever technical solution to a marketing issue. Study made across US viewers demonstrates that the average viewer identifies its station by its channel number. Not surprising since that in the last 4 decades, the station marketers have been hammering that their channel number is their branding Fox44, WVNY22, etc.

The issue is that, in the digital transition, most station actually change RF channels.

For most people switching to a program guide, environment is the natural thing to do, but some viewers would keep the old rotary tuner if they could; there was a way to accommodate both, built-in static PSIP.

Static PSIP permits to preserve the investment in station branding while accommodate RF channel changes.

The station could operate in any physical channel that the user would never see on the screen; the user would see a virtual channel. The virtual channel has been divided in two parts. The major channel, which is the old physical channel, had been using for decades, and the minor channel which represents the service number, e.g. a station has been operating in ch-44 for the last 20 years. It was assigned physical ch-27, their virtual channels would read as : 44-1 if the station elect to add more services they could be named 44-2, 44-3, etc. Detailed description of PSIP metadata can be found in ATSC A/65 document.

This information is hidden in one of the MGT children table called TVCT. You will notice that Figure 10.33 shows the use of virtual channel 15-1 for station call sign CIVK-HD.

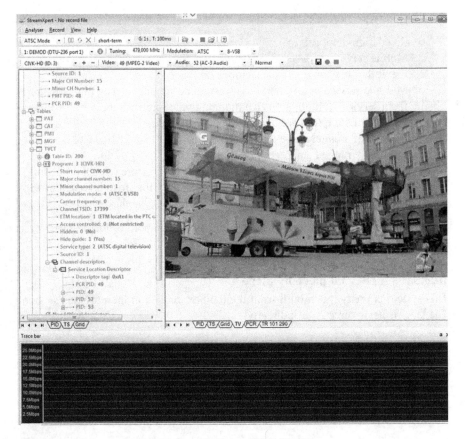

Figure 10.33   ATSC TVCT

10.3.2   ATSC 2.0

The mandate of ATSC 3.0 was to come up with improvements that would be retro compatible with ATSC 1.0. Although it was, at the beginning, a promising project, the advance of ATSC 3.0 soon made ATSC 2.0 little too late.

10.3.3   ATSC 3.0

Describing the key principle of ATSC3.0 in a few pages is a real challenge as this standard is very well engineered, very flexible, and so complex that making it looks simple is a stunt by itself.

10.3.3.1   Genesis

The mandate of the ATSC 3.0 group was simple, starting with a blank sheet and designing the best over-the-air media broadcast standard.

Don't worry about retro compatibility with other standards; however, it would be nice if you can be compatible with existing 6 MHz channelization plan. Adding the support of 7 and 8 MHz channel plan would be an asset.

However, the proposed standard must offer:

- Flexibility to address each broadcaster's business case

  - Fixed high-debit service
  - Mobile service, pedestrian, and vehicular
  - Flexibility to change broadcast mode on the fly
  - Multiple service in different broadcast mode simultaneously (e.g. keeping a fixed service +2 mobile service running in the same channel)
  - Support of store forward and non-real time services
  - Support of an interactive television service
  - Extremely rugged emergency alert system
  - Support and 6 MHz channel plan and allow smarter use (channel bonding)
  - Last but foremost, the system has to have all the hooks to evolve while keeping the legacy receiver operational in the service type they support

This ambitious project was taken by a fantastic team built of the best expertise they could find worldwide.

10.3.3.2   ATSC 3.0 piece by piece

This system is pretty much built from the toolbox defined in the previous section. The latter is using directly or indirectly:

- Variable time diversity
- Variable frequency diversity
- Pilot insertion
- Multiple Input Single Output (MISO) antenna diversity
- Single frequency Networks (SFN)

- Three types of FEC
- Time Division Multiplex (TDM)
- Layer Division Multiplex (LDM)

  o Packet-based multiplex
  o Multiple Physical Layer Pipes (PLP)

Figure 10.34 shows a vectorial representation of the diversity tools used in ATSC 3.0.

First and foremost, ATSC 3.0 service carriage mechanism has changed since ATSC 1.0. In ATSC, multiple services were carried in the unique system bitstream. In ATSC 3.0, the basic unit of data carriage has changed. Instead of a single transport layer, the system offers the possibility to carry multiple independents bitstreams that don't even talk to each other; they are called PLP (Physical Layer Pipes), each PLP has their own operation parameters set according to the specific application that the PLP has to served.

Signal detection issue:

In such flexible system, a mechanism called Discovery and signaling has been designed to ease signal acquisition.

The ATSC 3.0 signal relies on a signal section mechanism called a bootstrap (Figure 10.35). It is sent to a Spread Spectrum (SS) using a a Zadoff-Chu sequence modulated in very robust BPSK (Binary Phase Modulation keying) modulation at a code rate that has so much coding gain that it can be read few dB's below noise floor, quite a few dB before any signal can be decoded. The purpose of this signal is to declare step-by-step what is next.

What is the bootstrap signal? first and foremost, it is a signal known to all receivers.

- A fixed bandwidth (4.5 MHz) whatever the bandwidth of the rest of the signal will be

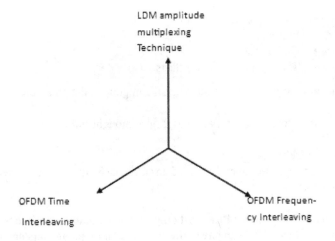

Figure 10.34   ATSC 3.0 diversity model

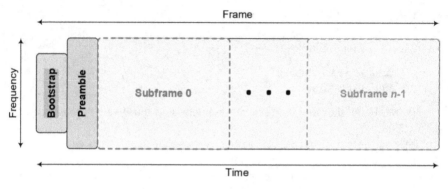

Figure 10.35   ATSC 3.0 signal discovery mechanism (ATSC A/322)

- Known parameters:
  - Symbol rate 6.144 MS/s
  - Zadoff-Chu sequence
  - N FFT = 2048 which results in a carrier spacing of 3kHz
  - Bootstrap duration 500 µs

- The information is carried in an orderly manner,
  - The first symbol is a synch
  - The second symbol defines the system bandwidth (6, 7, or 8 MHz)
  - The third symbol start the modulation parameter process:
    - Version of standard used
    - Frame length
    - Use of pilots symbols
    - FEC types in use
    - Use of parities
    - FFT size
    - Guard interval
    - Time, precision, and offsets

- All this info only defines the next signal component, the preamble

The preamble carries a similar dataset, just enought information to decode the desired PLP:
What is a PLP?
The PLP or Physical Layer Pipe is a data carriage channel that is defined in basic elements such as:

- Time (as TDM is used, a PLP exist only for a fixed amount of time per frame)
- Modulation (each PLP can have different modulation parameters and may vary in signal density and robustness)
- Layer, as LDM may be used, the PLP may be carried in the core or enhanced layer

The introduction of the PLP system creates an incredibly flexible tool. The choice of the PLP really depends on the broadcaster application:

- One very low rate and very robust PLP may be aimed at mobile services and even radio broadcasting
- One very robust fairly high-debit signal may be up all night to carry software updates
- One average protection mid-size channel may be up all night to carousel podcasts that the anchor will promote the next day
- One high debit low robustness may be used to carry the main 4k television program
- The possibility are infinite and can vary on the fly

Additional tools required:

Pilot's symbols: Pilot's symbols are symbols that the receivers already know. It may be used as a delimiter to derive channel impulse response, etc.

Group wise interleaving: This is a structure technique of interleaving (taking place at the transmit end) that uses a grouping technique that is LDPC FEC aware, so the critical info is interleaved in a smarter manner to permit best signal re-construction at the other end of the chain (receiver).

Building-up the room to evolve: Previous television standard were deposited to the local telecom authorities (FCC in US, CRTC in Canada, etc.) as a complete standard. The ATSC 3.0 standard was submitted based on the bootstrap only, leaving the option to add elements to it that may or may not be compatible with legacy receivers as long as the operability of legacy receivers was preserved on services they currently support. This approach facilitates a swift standards evolution, the closest thing from streaming standards evolution that could be made in the current regulatory framework.

Building ATSC 3.0 from the toolbox:

Although very complex, the ATSC 3.0 system uses most techniques defined, one-by-one in section 2

- Discovery and signaling defines the signals that the receiver will have to bear with
- Time and frequency diversity are achieved via the use of OFDM signals

  - Pilots signals (symbols) are used as delimiters and aimed to improve channel equalization
  - Modulation parameters (mod types, guard interval, and FFT) are varied depending on service types. All parameters are known in advance thanks to the preamble

- BCH and LDPC FEC are used to correct errors in the received bitstream
- TDM is achieved at the creation of the PLP
- Frequency division multiplex is achieved at the creation of the OFDM-based PLP
- LDM is also used as declared in the service preamble

Figure 10.36   ATSC Standards

- Both entropy and lossy compression techniques are used to reduce the data rate of video and audio signals
- A clever IP-based transport permits the carriage of both real-time and non-real-time services as well as interactive applications.

The entire standard is described in the following two documents, both available for free in the ATSC.org website Figure 10.36.

The information included is very detailed. any information not included in this chapter could probably be found in those documents.

10.3.3.3   ATSC signal generation
Figure 10.37 shows, in minimalist terms, what an ATSC 3.0 exciter should look like. Please note that tight time scheduling is kept at all stages in order to ensure time

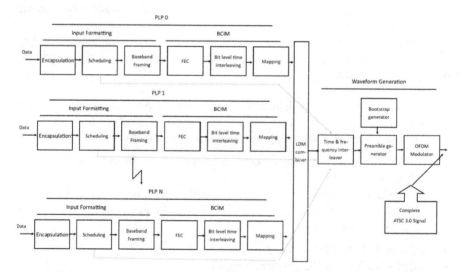

Figure 10.37   Simplified block diagram of an ATSC 3.0 signal generator

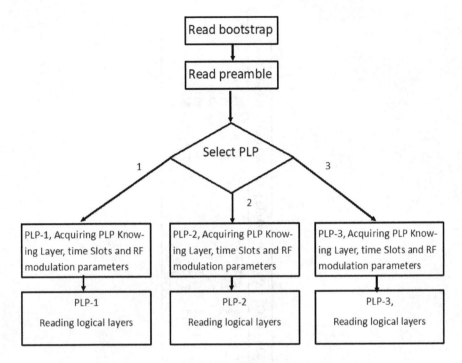

Figure 10.38   ATSC 3.0 signal acquisition flowchart

diversity in a recoverable manner across all. This modern-day signal blender pushes the time and diversity principles to the limit in such an ordered manner that the receiver can recover only the elements it is aimed to receive in the adverse channel conditions that the service is aimed to serve.

10.3.3.4   ATSC 3.0 signal acquisition

For the sake of this exercise, we wear the receiver shoe. Figure 10.38 describes, in simple terms, the steps that the receiver has to take to receive the service:

- Different receivers receiving different services from the same stations will just choose to decode a different PLP.

10.3.3.5   Typical use

ATSC 3.0 can deliver up to 47 Mb/s in a 6 MHz channel and services decodable at sub zero Eb/n0 levels. Naturally, there are a few rules to follow… for the sake of remaining representative of real-life conditions. Table 10.5 illustrates a few possible use cases I taught were realistic.

Table 10.5   (Courtesy Enensys ATSC 3.0 calculator)

Case	Numbers of PLP	PLP	Data Rate (Mb/s)	Mode	ModCod	FEC	AWGN threshold (dB)	GI (μs)	Application
1	1	1	20.3	8k	Nuq256	9/15	15.55	222 μs	Fixed service
2	2	1	14.58	16k	Mu256k	9/15	15.55	222	Fixed service
		2	3.225	8k	Nuq16	8/15	6.3	111	Mobile service
3	3	1	12.438	32k	Nu256	9/15	15.56	222	Fixed service
		2	3.941	16k	Nuq16	9/15	7.32	222	Multimedia broad
		3	0.4	8k	QPSK	4/15	-2.9	111	Radio/eas

To enlighten the flexibility built in ATSC 3.0, Table 10.5 demonstrates three different possible use case of an ASTC 3.0 multiplex.

- Case 1 is a basic replacement of an ATSC 1.0 service. You will notice a modest increase of data rate (ATSC 1.0 use to deliver 19.3 Mb/s), a comparable performance in a Gaussian channel; however, a significant gain is achieved in the mobility standpoints offering a guard interval of 222 μs, which pretty much guarantee the possibility of using a Single Frequency Network (SFN).
- Case 2 illustrates the case where a broadcaster wants to start a mobile multimedia service while preserving its basic service aimed at a fixed audience. Service 1 is the fixed service with service parameters slightly superior to ATSC 1.0, plus a fairly rugged mobile service delivering a 4 Mb/s service (just bare in mind that advance in video coding efficiency makes a 4 Mb/s service deliver a pristine picture quality).
- Case 3 delivers a full HD fixed service, one or two mobile multimedia services and a radio service with impressive characteristics decodable at -3 dB Eb/n0, enough to make digital radio operators jealous of ATSC 3.0 robustness.
- The best part in all of this is that the broadcaster can use all our three cases in a broadcast day as service changes can be scheduled and just happens seamlessly.
- Not to forget emergency alert applications that is very robust and smart enough to force the viewers TV set of the affected area to display the emergency information even if the receiver is in the off position.
- Bare in mind that the PAPR of the resulting signal is close to 9 dB.

10.3.3.6 Layered structure

Although this chapter focused its energy on physical layer, the next section (Figure 10.39) illustrates in very broad terms the unavoidable layered structure of ATSC 3.0.

Keeping the ultimate goal to be evolutive and flexible in mind, the ATSC 3.0 committee chose a structure under which they just design the physical and adaptation

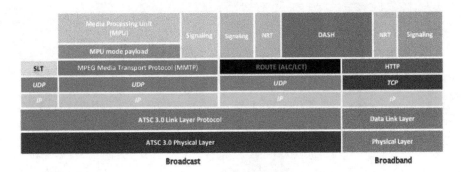

Figure 10.39   Simplified ATSC protocol Stack (ATSC A/331 Signaling, Delivery, Synchronization, and Error Protection

layers (PLP and ALP). The PLP has been covered earlier. The ALP is the ATSC 3.0 link layer protocol which delivers the IP layer just above it, a structure comparable to the ethernet data link layer. The rest of the magic is similar to whatever happens in the public Internet; however, the forward link is real-time with no possibility of asking the transmitter to repeat a lost packet. At the application level, the return path (usually the public Internet) is of a non-real time nature.

Non-real time services and interactive applications are catered for just like they would be on a smart phone or a computer; however, they benefit from a robust real-time forward link.

The world has changed since the inception of ATSC 1.0. There is just so many stones to build on...the ATSC 3.0 committee just designed their own stone-stepping protocol and let the walkers find their own way, thus building an extremely flexible ecosystem.

Index

Printed in the USA
CPSIA information can be obtained
at www.ICGtesting.com
JSHW010009290224
58256JS00003B/4